Ervin Győri
Gyula O. H. Katona
László Lovász (Eds.)

More Sets, Graphs and Numbers

A Salute to Vera Sós and András Hajnal

 Springer

JÁNOS BOLYAI MATHEMATICAL SOCIETY

Editors

Ervin Győri
Hungarian Academy of Sciences
Alfréd Rényi Inst. of Mathematics
Reáltanoda u. 13–15
1053 Budapest, Hungary

László Lovász
Eötvös University (Budapest)/
Microsoft (Seattle)
Pázmány Péter sétány 1/C
1117 Budapest, Hungary

Gyula O. H. Katona
Hungarian Academy of Sciences
Alfréd Rényi Inst. of Mathematics
Reáltanoda u. 13–15
1053 Budapest, Hungary

Managing Editor

Tamás Fleiner
Budapest University of Technology
and Economics
Pázmány Péter sétány 1/D
1117 Budapest, Hungary

Mathematics Subject Classification (2000): 05, 11, 03Exx

Library of Congress Control Number: 2006921044

.

ISSN 1217-4696
ISBN 3-540-32377-5 Springer Berlin Heidelberg New York
ISBN 963 9453 05 6 János Bolyai Mathematical Society, Budapest

Springer is a part of Springer Science+Business Media
springer.com

Cover design: Erich Kirchner, Heidelberg

Printed on acid-free paper   44/3142/db – 5 4 3 2 1 0

BOLYAI SOCIETY
MATHEMATICAL STUDIES

15

BOLYAI SOCIETY MATHEMATICAL STUDIES

CONTENTS

PREFACE

The present volume is slightly connected to the conference organized in Budapest, January 2001 to the honour of Vera Sós and András Hajnal on the occasion of their 70th birthdays. Namely, we mainly asked the invited speakers of that conference to write survey papers on their favorite subjects. Therefore the volume contains strong and well-written surveys in the areas of the celebrated colleagues: mostly in combinatorics, graph theory, less in number theory and set theory. The authors gave the up-to-date state of the art in their subjects, put the recent results into integral framework. Examples are listed below. The other papers contain original research results.

Matthias Beck, Xueqin Wang, and Thomas Zaslavsky find a nice, so-called unifying generalization of different versions of Sperner's theorem. They found a uniform handling of several different generalizations.

Béla Bollobás and Alexander Scott summarize different results on discrepancies of graphs and hypergraphs.

Éva Czabarka, Ondrej Sýkora, László A. Székely and Imrich Vrto survey some bounds on biplanar crossing numbers of graphs which is the sum of the crossing numbers over all partitions of a graph into two planar graphs.

András Frank studies the different notions of edge-connectivity of graphs, digraps and hypergraphs and uses properties of submodular functions to get different theorems on them. He gives an extensive survey of the results concerning orientations and connectivity augmentations in a general setting.

Kálmán Győry surveys when we can get (almost) complete powers as the product of consecutive terms of an arithmetic progression or binomial coefficients. The results are mostly negative as it turns out from the nice overview of classical papers of Erdős and Selfridge as well as the recent ones of the surveyer and others.

István Juhász and Andrzej Szymanski present a purely topological generalization of Fodor's theorem called "the pressing down lemma". By means of it, the authors prove a partial generalization of this framework of Solovay's celebrated stationary set decomposition theorem.

In his extensive survey paper, Alexandr Kostochka summarizes the results on the minimum number of edges in color-critical graphs and hypergraphs.

Michael Krivelevich and Benny Sudakov give an extensive survey on pseudo random graphs with emphasis on the results obtained by means of the investigation of the eigenvalues of the adjacency matrix.

Jaroslav Nešetřil deals with questions and results concerning order-theoretic properties of the homomorphism order of graphs, but the author surveys upper bounds, suprema and maximal elements of the homomorphism order lattice in other interesting finite structures too. The author also studies minor closed classes of graphs, shows how the order setting captures Hadwiger conjecture and suggests some new problems too.

András Recski and Dávid Szeszlér investigate VLSI routing algorithms, especially the influence of Gallai's Algorithm on them. They show the first forty years of the influence on VLSI design of the classic result on the perfectness of interval graphs.

András Sárközy's paper describes advance in a specific question, the possible behaviour of representation functions. We take a set A of positive integers, and consider $r_k(n)$, the number of representations of n as a sum of k elements of A, or variants where the order is neglected or where an element can be used only once. Typical questions are whether such a function can be monotonic, or can be very near to a given regular function. The author presents plenty of results and unsolved problems.

Andrew Thomason presents results and methods concerning the minimum number of edges guaranteeing a given graph minor. It turns out that the extremal graphs are pseudo-random. The survey describes what is known about the extremal function and discusses some related matters.

Robert Tijdeman's survey covers a broad area, with main emphasis on tilings and balanced words. We learn how words with small complexity (that is, with a small number of different subwords of length n for every n) are connected with balanced words, where the number of occurrences of any fixed letter in subwords of given length is almost constant, and with sequences given by the integer part of a linear function.

The organizers of the conference gratefully acknowledge the financial support of the High Level Scientific Conferences program of the European Union (contract No. HPCF-CT-2000-00419).

The editors

BOLYAI SOCIETY
MATHEMATICAL STUDIES, 15

Conference on Finite
and Infinite Sets
Budapest, pp. 9–24.

A Unifying Generalization of Sperner's Theorem

M. BECK, X. WANG and T. ZASLAVSKY*

Dedicated to the memories of Pál Erdős and Lev Meshalkin

Sperner's bound on the size of an antichain in the lattice $\mathcal{P}(S)$ of subsets of a finite set S has been generalized in three different directions: by Erdős to subsets of $\mathcal{P}(S)$ in which chains contain at most r elements; by Meshalkin to certain classes of compositions of S; by Griggs, Stahl, and Trotter through replacing the antichains by certain sets of pairs of disjoint elements of $\mathcal{P}(S)$. We unify these three bounds with a common generalization. We similarly unify their accompanying LYM inequalities. Our bounds do not in general appear to be the best possible.

1. Sperner-type theorems

Let S be a finite set with n elements. In the lattice $\mathcal{P}(S)$ of all subsets of S one tries to estimate the size of a subset with certain characteristics. The most famous such estimate concerns **antichains**, that is, subsets of $\mathcal{P}(S)$ in which any two elements are incomparable.

Theorem 1.1 (Sperner [11]). *Suppose $A_1, \ldots, A_m \subseteq S$ such that $A_k \not\subseteq A_j$ for $k \neq j$. Then $m \leq \binom{n}{\lfloor n/2 \rfloor}$. Furthermore, this bound can be attained for any n.*

We attain the bound by taking all $\lfloor \frac{n}{2} \rfloor$-element subsets of S, or all $\lceil \frac{n}{2} \rceil$-element subsets, but in no other way. There are many ways to prove

---

*Research supported by National Science Foundation grant DMS-0070729.

Sperner's bound and the near-uniqueness of the maximal example; several
of them will be found in the opening chapters of Anderson's lovely intro-
ductory book [1]. The most famous approach is perhaps that of the "LYM
inequality"; see Theorem 2.1 below.

Sperner's theorem has been generalized in many different directions.
Here are three: Erdős extended Sperner's inequality to subsets of $\mathcal{P}(S)$ in
which chains contain at most r elements. Meshalkin proved a Sperner-like
inequality for families of compositions of S into a fixed number of parts, in
which the sets in each part constitute an antichain. Finally, Griggs, Stahl,
and Trotter extended Sperner's theorem by replacing the antichains by sets
of pairs of disjoint elements of $\mathcal{P}(S)$ satisfying an intersection condition.
In this paper we unify the Erdős, Meshalkin, and Griggs–Stahl–Trotter
inequalities in a single generalization. However, except in special cases
(among which are generalizations of the known bounds), our bounds are
not the best possible.

For a precise statement of Erdős's generalization, call a subset of $\mathcal{P}(S)$
r-chain-free if its chains (i.e., linearly ordered subsets) contain no more
than r elements; that is, no chain has length r.[1] In particular, an antichain
is 1-chain-free. The generalization of Theorem 1.1 to r-chain-free families is

Theorem 1.2 (Erdős [4]). *Suppose $\{A_1, \ldots, A_m\} \subseteq \mathcal{P}(S)$ contains no
chains with $r + 1$ elements. Then m is bounded by the sum of the r largest
binomial coefficients $\binom{n}{k}$, $0 \le k \le n$. The bound is attainable for every n
and r.*

Sperner's theorem is the case $r = 1$. To attain the bound take all subsets
of sizes $\lfloor \frac{n-r+1}{2} \rfloor \le k \le \lfloor \frac{n+r-1}{2} \rfloor$ or all of sizes $\lceil \frac{n-r+1}{2} \rceil \le k \le \lceil \frac{n+r-1}{2} \rceil$; these
are the only ways.

Going in a different direction, Sperner's inequality can be generalized
to certain ordered weak partitions of S. We define a **weak partial composition of S into p parts** as an ordered p-tuple $(A_1, \ldots, A_p)$ of sets A_k,
possibly void (hence the word "weak"), such that $A_1, \ldots, A_p$ are pairwise
disjoint and $A_1 \cup \cdots \cup A_p \subseteq S$. If $A_1 \cup \cdots \cup A_p = S$, we have a **weak composition** of S. A Sperner-like inequality suitable for this setting was proposed
by Sevast'yanov and proved by Meshalkin (see [9]). By a **p-multinomial
coefficient for n** we mean a multinomial coefficient $\binom{n}{a_1, \ldots, a_p}$. where $a_i \ge 0$
and $a_1 + \cdots + a_p = n$. Let $[p] := \{1, 2, \ldots, p\}$.

---

[1] The term "r-family" or "k-family", depending on the name of the forbidden length,
has been used in the past, but we think it is time for a distinctive name.

Theorem 1.3 (Meshalkin). *Let $p \geq 2$. Suppose $(A_{j1}, \ldots, A_{jp})$ for $j = 1, \ldots, m$ are different weak compositions of S into p parts such that, for each $k \in [p]$, the set $\{A_{jk} : 1 \leq j \leq m\}$ (ignoring repetition) forms an antichain. Then m is bounded by the largest p-multinomial coefficient for n. Furthermore, the bound is attainable for every n and p.*

This largest multinomial coefficient can be written explicitly as

$$\frac{n!}{((\lfloor \frac{n}{p} \rfloor + 1)!)^{\rho}(\lfloor \frac{n}{p} \rfloor!)^{p-\rho}},$$

where $\rho = n - p\lfloor \frac{n}{p} \rfloor$. We attain the bound by choosing any set $K \subseteq [p]$ of size ρ and taking all weak compositions $(A_{j1}, \ldots, A_{jp})$ in which $|A_{jk}| = \lfloor \frac{n}{p} \rfloor$ if $k \in K$ and $|A_{jk}| = \lceil \frac{n}{p} \rceil$ if $k \notin K$. Hochberg and Hirsch [6] showed that no other family of weak compositions of S has maximum size. Meshalkin's theorem and the completion by Hochberg and Hirsch are curiously neglected: we have not seen them mentioned in any book except [7].

To see why Meshalkin's inequality generalizes Sperner's Theorem, suppose $A_1, \ldots, A_m \subseteq S$ form an antichain. Then $S - A_1, \ldots, S - A_m$ also form an antichain. Hence the m weak compositions $(A_j, S - A_j)$ of S into two parts satisfy Meshalkin's conditions and Sperner's inequality follows.

Yet another generalization of Sperner's Theorem is

Theorem 1.4′ (Griggs–Stahl–Trotter [5]). *Suppose $\{A_{j0}, \ldots, A_{jq}\}$ for $j = 1, \ldots, m$ are chains of size $q + 1$ in $\mathcal{P}(S)$ such that $A_{ji} \not\subseteq A_{kl}$ for all i and l and all $j \neq k$. Then $m \leq \binom{n-q}{\lfloor (n-q)/2 \rfloor}$. Furthermore, this bound can be attained for all n and q.*

An equivalent, simplified form of this result (in which $A_j = A_{j0}$, $B_j = S - A_{jq}$, and n replaces $n - q$) is

Theorem 1.4. *Let $n > 0$. Suppose (A_j, B_j) are m pairs of sets such that $A_j \cap B_j = \varnothing$ for all j, $A_j \cap B_k \neq \varnothing$ for all $j \neq k$, and all $|A_j| + |B_j| \leq n$. Then $m \leq \binom{n}{\lfloor n/2 \rfloor}$ and this bound can be attained for every n.*

Sperner's inequality follows as the special case in which $A_1, \ldots, A_m \subseteq S$ form an antichain and $B_j = S - A_j$. To attain the bound in Theorem 1.4′ take $\{A_{j0}\}$ to consist of all subsets of $[n - q]$ of size $\lfloor \frac{n-q}{2} \rfloor$, or all of size $\lceil \frac{n-q}{2} \rceil$. Then let $A_{jk} = A_{j0} \cup \{n - q + 1, \ldots, n - q + k\}$. In Theorem 1.4, take $A_j = A_{j0}$ and $B_j = [n] - A_j$.

Theorems 1.2, 1.3, and 1.4 are *incomparable* generalizations of Sperner's Theorem. We wish to combine (and hence further generalize) these generalizations. To state our main result, we define a **weak set composition** as a weak composition of any set S. Our generalization of Sperner's inequality is:

Theorem 1.5. *Fix integers $p \geq 2$ and $r \geq 1$. Suppose $(A_{j1}, \ldots, A_{jp})$ for $j = 1, \ldots, m$ are different weak set compositions into p parts with the condition that, for all $k \in [p]$ and all $I \subseteq [m]$ with $|I| = r + 1$, there exist distinct $i, j \in I$ such that either $A_{ik} = A_{jk}$ or*

$$(1) \qquad A_{ik} \cap \bigcup_{l \neq k} A_{jl} \neq \varnothing \neq A_{jk} \cap \bigcup_{l \neq k} A_{il},$$

and let $n := \max_{1 \leq j \leq m} \left(|A_{j1}| + \cdots + |A_{jp}| \right)$. Then m is bounded by the sum of the r^p largest p-multinomial coefficients for integers less than or equal to n.

Think of the p-multinomial coefficients as a sequence arranged in weakly descending order. Then if r^p is larger than $\binom{r+p}{p}$, the number of p-multinomial coefficients, we regard the sequence of coefficients as extended by 0's.

The reader may find the statement of this theorem somewhat difficult. We would first like to show that it does generalize Theorems 1.2, 1.3, and 1.4 simultaneously. The last follows easily as the case $r = 1$, $p = 2$. Theorem 1.3 can be deduced by choosing $r = 1$ and restricting the weak compositions to be compositions of a fixed set S with n elements. Finally, Theorem 1.2 follows by choosing $p = 2$ and the weak compositions to be compositions of a fixed n-set into 2 parts. What we find most interesting, however, is that specializing Theorem 1.5 yields three corollaries that generalize two at a time of Theorems 1.2, 1.3, and 1.4 yet are easy to state and understand. Section 4 collects these corollaries.

We came to Theorem 1.5 through seeking a common generalization of Erdős's and Meshalkin's theorems (see Corollary 4.1); our original motivation was, in part, surprise at the lack of general awareness of Meshalkin's result. When we learned of the Griggs–Stahl–Trotter theorem, we could not be satisfied until we succeeded in extending our result to include it as well. (Fortunately for us, we did not encounter a fourth kind of Sperner generalization.)

The condition of the theorem implies that each set $\mathcal{A}_k = \{ A_{jk} : j \in [m] \}$ (ignoring repetition) is r-chain-free. We suspect that the converse is not true in general. (It is true if all the weak set compositions are weak compositions of the same set of order n, as in Corollary 4.1.)

All the theorems we have stated have each a slightly stronger companion, an *LYM inequality*. In Section 2, we state these inequalities and show how Theorems 1.1–1.5 can be deduced from them. The proofs of Theorem 1.5 and the corresponding LYM inequality are in Section 3. After the corollaries of Section 4, in Section 5 we show that some, at least, of our upper bounds cannot be attained.

2. LYM INEQUALITIES

In attempting to estimate the order of the free distributive lattice with n generators, Yamamoto came up with the following result, which was rediscovered by Meshalkin in the course of proving his Sperner generalization (Theorem 1.3) and still later by Lubell with a classic short proof. In the meantime Bollobás had independently proved even a generalization (Theorem 2.4 below). The result is the famous LYM inequality, that has given its name to a whole class of similar relations.

Theorem 2.1 (Yamamoto [12, §6], Meshalkin [9, Lemma], Lubell [8]). *Suppose $A_1, \ldots, A_m \subseteq S$ such that $A_k \not\subseteq A_j$ for $k \neq j$. Then*

$$\sum_{k=1}^{m} \frac{1}{\binom{n}{|A_k|}} \leq 1.$$

Sperner's inequality follows immediately by noting that $\max_k \binom{n}{k} = \binom{n}{\lfloor n/2 \rfloor}$.

An LYM inequality corresponding to Theorem 1.2 appeared to our knowledge first in [10]:

Theorem 2.2 (Rota–Harper). *Suppose $\{A_1, \ldots, A_m\} \subseteq \mathcal{P}(S)$ contains no chains with $r + 1$ elements. Then*

$$\sum_{k=1}^{m} \frac{1}{\binom{n}{|A_k|}} \leq r.$$

Deducing Erdős's Theorem 1.2 from this inequality is not as straightforward as the connection between Theorems 2.1 and 1.1. It can be done through Lemma 3.1, which we also need in order to deduce Theorem 1.5.

The LYM companion of Theorem 1.3 first appeared in [6]; again, Meshalkin's Theorem 1.3 follows immediately.

Theorem 2.3 (Hochberg–Hirsch). *Suppose* $(A_{j1}, \ldots, A_{jp})$ *for* $j = 1, \ldots, m$ *are different weak compositions of* S *into* p *parts such that for each* $k \in [p]$ *the set* $\{A_{jk} : 1 \leq j \leq m\}$ *(ignoring repetitions) forms an antichain. Then*

$$\sum_{j=1}^{m} \frac{1}{\binom{n}{|A_{j1}|, \ldots, |A_{jp}|}} \leq 1.$$

The LYM inequality corresponding to Theorem 1.4 is due to Bollobás.

Theorem 2.4 (Bollobás [3]). *Suppose* (A_j, B_j) *are* m *pairs of sets such that* $A_j \cap B_j = \varnothing$ *for all* j *and* $A_j \cap B_k \neq \varnothing$ *for all* $j \neq k$. *Then*

$$\sum_{j=1}^{m} \frac{1}{\binom{|A_j|+|B_j|}{|A_j|}} \leq 1.$$

Once more, the corresponding upper bound, the Griggs–Stahl–Trotter Theorem 1.4, is an immediate consequence.

Naturally, there is an LYM inequality accompanying our main Theorem 1.5. Like its siblings, it constitutes a refinement.

Theorem 2.5. *Let* $p \geq 2$ *and* $r \geq 1$. *Suppose* $(A_{j1}, \ldots, A_{jp})$ *for* $j = 1, \ldots, m$ *are different weak compositions (of any sets) into* p *parts satisfying the same condition as in Theorem 1.5. Then*

$$\sum_{j=1}^{m} \frac{1}{\binom{|A_{j1}|+\cdots+|A_{jp}|}{|A_{j1}|, \ldots, |A_{jp}|}} \leq r^p.$$

Example 2.1. The complicated hypothesis of Theorem 2.5 cannot be replaced by the assumption that each $\mathcal{A}_k$ is r-chain-free, because then there is no LYM bound independent of n. Let $n \gg p \geq 2$, $S = [n]$, and $\mathcal{A} = \big\{ \big(A, \{n\}, \{n-1\}, \ldots, \{n-p+2\} \big) : A \in \mathcal{A}_1 \big\}$ where $\mathcal{A}_1$ is a largest r-chain-free family in $[n-p+1]$, specifically,

$$\mathcal{A}_1 = \bigcup_{j \in I} \mathcal{P}_j \big([n-p+1] \big)$$

where

$$I = \left\{ \left[\frac{n-p+1-r}{2} \right], \left[\frac{n-p+1-r}{2} \right] + 1, \left[\frac{n-p+1-r}{2} \right] + r - 1 \right\}.$$

The LYM sum is

$$\sum_{A \in \mathcal{A}_1} \frac{1}{\binom{|A|+p-1}{|A|,1,\dots,1}} = \sum_{A \in \mathcal{A}_1} \frac{|A|!}{(|A|+p-1)!}$$

$$= \sum_{j \in I} \binom{n-p+1}{j} \frac{j!}{(j+p-1)!}$$

$$= \sum_{j \in I} \frac{(n-p+1) \cdots (n-p-j+2)}{(p-1+j)!}$$

$$\to \infty \quad \text{as} \quad n \to \infty.$$

There is no possible upper bound in terms of n.

3. PROOF OF THE MAIN THEOREMS

Proof of Theorem 2.5. Let S be a finite set containing all A_{jk} for $j = 1, \dots, m$ and $k = 1, \dots, p$, and let $n = |S|$. We count maximal chains in $\mathcal{P}(S)$. Let us say a maximal chain **separates** the weak composition $(A_1, \dots, A_p)$ if there exist elements $\varnothing = X_0 \subseteq X_{l_1} \subseteq \cdots \subseteq X_{l_p} = S$ of the maximal chain such that $A_k \subseteq X_{l_k} - X_{l_{k-1}}$ for each k. There are

$$(2) \qquad \binom{n}{|A_1| + \cdots + |A_p|} |A_1|! \cdots |A_p|! \left(n - |A_1| - \cdots - |A_p|\right)!$$

maximal chains separating $(A_1, \dots, A_p)$. (To prove this, replace maximal chains $\varnothing \subset \{x_1\} \subset \{x_1, x_2\} \subset \cdots \subset S$ by permutations $(x_1, x_2, \dots, x_n)$ of S. Choose $|A_1| + \cdots + |A_p|$ places for $A_1 \cup \cdots \cup A_p$; then arrange A_1 in any order in the first $|A_1|$ of these places, A_2 in the next $|A_2|$, etc. Finally, arrange $S - (A_1 \cup \cdots \cup A_p)$ in the remaining places. This constructs all maximal chains that separate $(A_1, \dots, A_p)$.)

We claim that every maximal chain separates at most r^p weak partial compositions of $|S|$. To prove this, assume that there is a maximal chain that separates N weak partial compositions $(A_{j1}, \ldots, A_{jp})$. Consider all first components A_{j1} and suppose $r+1$ of them are different, say $A_{11}, A_{21}, \ldots, A_{r+1,1}$. By the hypotheses of the theorem, there are $i, i' \in [r+1]$ such that A_{i1} meets some $A_{i'l'}$ where $l' > 1$ and $A_{i'1}$ meets some A_{il} where $l > 1$. By separation, there are q_1 and q_1' such that $A_{i1} \subseteq X_{q_1} - X_0$ and $A_{i'1} \subseteq X_{q_1'} - X_0$, and there are $q_{l-1}, q_l, q_{l'-1}', q_{l'}'$ such that $q_1 \leq q_{l-1} \leq q_l$, $q_1' \leq q_{l'-1}' \leq q_{l'}'$, and

$$A_{il} \subseteq X_{q_l} - X_{q_{l-1}} \qquad \text{and} \qquad A_{i'l'} \subseteq X_{q_{l'}'} - X_{q_{l'-1}'}.$$

Since A_{i1} meets $A_{i'l'}$, there is an element $a_{i1} \in X_{q_{l'}'} - X_{q_{l'-1}'}$; it follows that $q_{l'-1}' < q_1$. Similarly, $q_{l-1} < q_1'$. But this is a contradiction. It follows that, amongst the N sets A_{j1}, there are at most r different sets. Hence (by the pigeonhole principle) there are $\lceil N/r \rceil$ among the N weak partial compositions that have the same first set A_{j1}.

Looking now at these $\lceil N/r \rceil$ weak partial compositions, we can repeat the argument to conclude that there are $\lceil \lceil N/r \rceil / r \rceil \geq \lceil N/r^2 \rceil$ weak partial compositions for which both the A_{j1}'s and the A_{j2}'s are identical. Repeating this process $p-1$ times yields $\lceil N/r^{p-1} \rceil$ weak partial compositions into p parts whose first $p-1$ parts are identical. But now the hypotheses imply that the last parts of all these weak partial compositions are at most r different sets; in other words, there are at most r distinct weak partial compositions. Hence $\lceil N/r^{p-1} \rceil \leq r$, whence $N \leq r^p$. (If we know that all the compositions are weak—but not partial—compositions of S, then the last parts of all these $\lceil N/r^{p-1} \rceil$ weak compositions are identical. Thus $N \leq r^{p-1}$.)

Since at most r^p weak partial compositions of S are separated by each of the $n!$ maximal chains, from (2) we deduce that

$$r^p n! \geq \sum_{j=1}^{m} \binom{n}{|A_{j1}| + \cdots + |A_{jp}|} |A_{j1}|! \cdots |A_{jp}|! \big(n - |A_{j1}| - \cdots - |A_{jp}|\big)!$$

$$= \sum_{j=1}^{m} \frac{n!}{\binom{|A_{j1}| + \cdots + |A_{jp}|}{|A_{j1}|, \ldots, |A_{jp}|}}.$$

The theorem follows. ∎

To deduce Theorem 1.5 from Theorem 2.5, we use the following lemma, which originally appeared in somewhat different and incomplete form in [10], used there to prove Erdős's Theorem 1.2 by means of Theorem 2.2, and appeared in complete form in [7, Lemma 3.1.3]. We give a very short proof, which seems to be new.

Lemma 3.1 (Harper–Klain–Rota). *Suppose $M_1, \ldots, M_N \in \mathbb{R}$ satisfy $M_1 \geq M_2 \geq \cdots \geq M_N \geq 0$, and let R be an integer with $1 \leq R \leq N$. If $q_1, \ldots, q_N \in [0, 1]$ have sum*

$$q_1 + \cdots + q_N \leq R,$$

then

$$q_1 M_1 + \cdots + q_N M_N \leq M_1 + \cdots + M_R.$$

Proof. By assumption,

$$\sum_{k=R+1}^{N} q_k \leq \sum_{k=1}^{R} (1 - q_k).$$

Hence, by the condition on the M_k,

$$\sum_{k=R+1}^{N} q_k M_k \leq M_R \sum_{k=R+1}^{N} q_k \leq M_R \sum_{k=1}^{R} (1 - q_k) \leq \sum_{k=1}^{R} (1 - q_k) M_k,$$

which is equivalent to the conclusion. ∎

Proof of Theorem 1.5. Let S be any finite set that contains all A_{jk}. Write down the LYM inequality from Theorem 2.5.

From the m weak partial compositions $(A_{j1}, \ldots, A_{jp})$ of S, collect those whose shape is $(a_1, \ldots, a_p)$ into the set $C(a_1, \ldots, a_p)$. Label the p-multinomial coefficients for integers $n' \leq n$ as $M_1', M_2', \ldots$ so that $M_1' \geq M_2' \geq \cdots$. If M_k' is $\binom{n'}{a_1, \ldots, a_p}$, let $q_k' := |C(a_1, \ldots, a_p)| / M_k'$. By Theorem 2.5, the q_k''s and M_k''s satisfy all the conditions of Lemma 3.1 with N replaced by the number of p-tuples $(a_1, \ldots, a_p)$ whose sum is at most n, that is $\binom{n+p}{p}$, and R replaced by $\min(N, r^p)$. Hence

$$\sum_{a_1 + \cdots + a_p \leq n} |C(a_1, \ldots, a_p)| \leq M_1' + \cdots + M_R'.$$

The conclusion of the theorem now follows, since

$$m = \sum_{a_1 + \cdots + a_p \leq n} |C(a_1, \ldots, a_p)|. \qquad \blacksquare$$

4. CONSEQUENCES

As promised in Section 1, we now state special cases of Theorems 1.5/2.5 that unify pairs of Theorems 1.2, 1.3, and 1.4 as well as their LYM companions.

The first special case unifies Theorems 1.2/2.2 and 1.3/2.3. (It is a corollary of the proof of the main theorems, not of the theorems themselves. See [2] for a very short, direct proof.)

Corollary 4.1. *Suppose $(A_{j1}, \ldots, A_{jp})$ are m different weak compositions of S into p parts such that for each $k \in [p-1]$, the set $\{A_{jk} : 1 \leq j \leq m\}$ is r-chain-free. Then*

$$\sum_{j=1}^{m} \frac{1}{\binom{n}{|A_{j1}|, \ldots, |A_{jp}|}} \leq r^{p-1}.$$

Consequently, m is bounded by the sum of the r^{p-1} largest p-multinomial coefficients for n.

Proof. We note that, for a family of m weak compositions of S, the condition of Theorem 2.5 for a particular $k \in [p-1]$ is equivalent to $\{A_{jk}\}_j$ being r-chain-free. Thus by the hypothesis of the corollary, the hypothesis of the theorem is met for $k = 1, \ldots, p-1$. Then the proof of Theorem 2.5 goes through perfectly with the only difference, explained in the proof, that (even without a condition on $k = p$) we obtain $N \leq r^{p-1}$. In the proof of Theorem 1.5, under our hypotheses the sets $C(a_1, \ldots, a_p)$ with $a_1 + \cdots + a_p < n$ are empty. Therefore we take only the p-multinomial coefficients for n, labelled $M_1 \geq M_2 \geq \cdots$. In applying Lemma 3.1 we take $R = \min(N, r^{p-1})$ and summations over $a_1 + \cdots + a_p = n$. With these alterations the proof fits Corollary 4.1. ∎

A good way to think of Corollary 4.1 is as a theorem about partial weak compositions, obtained by dropping the last part from each of the weak compositions in the corollary.

Corollary 4.2. *Fix $p \geq 2$ and $r \geq 1$. Suppose $(A_{j1}, \ldots, A_{jp})$ are m different weak partial compositions of an n-set S into p parts such that for each $k \in [p]$, the set $\{A_{jk} : 1 \leq j \leq m\}$ is r-chain-free. Then m is bounded by the sum of the r^p largest $(p+1)$-multinomial coefficients for n.* ∎

A difference between this and Theorem 1.5 is that Corollary 4.2 has a weaker and simpler hypothesis but a much weaker bound. But the biggest difference is the omission of an accompanying LYM inequality. Corollary 4.1 obviously implies one, but it is weaker than that in Theorem 2.5 because, since the top number in the latter can be less than n, the denominators are much smaller. We do not present in Corollary 4.2 an LYM inequality of the kind in Theorem 2.5 for the very good reason that none is possible; that is the meaning of Example 2.1.

The second specialization constitutes a weak common refinement of Theorems 1.2/2.2 and 1.4/2.4. We call it weak because its specialization to the case $B_j = S - A_j$, which is the situation of Theorems 1.2/2.2, is weaker than those theorems.

Corollary 4.3. *Let r be a positive integer. Suppose (A_j, B_j) are m pairs of sets such that $A_j \cap B_j = \varnothing$ and, for all $I \subseteq [m]$ with $|I| = r + 1$, there exist distinct $i, j \in I$ for which $A_j \cap B_k \neq \varnothing \neq A_k \cap B_j$. Let $n = \max_j \left(|A_j| + |B_j| \right)$. Then*

$$\sum_{j=1}^{m} \frac{1}{\binom{|A_j|+|B_j|}{|A_j|}} \leq r.$$

Consequently, m is bounded by the sum of the r largest binomial coefficients $\binom{n'}{k}$ for $0 \leq k \leq n' \leq n$. This bound can be attained for all n and r.

Proof. Set $p = 2$ in Theorems 1.5/2.5. To attain the bound, let A_j range over all k-subsets of $[n]$ and let $B_j = [n] - A_j$. ∎

The last special case of Theorems 1.5/2.5 we would like to mention is that in which $r = 1$; it unifies Theorems 1.3/2.3 and 1.4/2.4.

Corollary 4.4. *Suppose $(A_{j1}, \ldots, A_{jp})$ are m different weak set compositions into p parts with the condition that, for all $k \in [p]$ and all distinct $i, j \in [m]$, either $A_{ik} = A_{jk}$ or*

$$A_{ik} \cap \bigcup_{l \neq k} A_{jl} \neq \varnothing \neq A_{jk} \cap \bigcup_{l \neq k} A_{il}.$$

and let $n \geq \max_j \left(|A_{j1}| + \cdots + |A_{jp}| \right)$. Then

$$\sum_{j=1}^{m} \frac{1}{\binom{|A_{j1}|+\cdots+|A_{jp}|}{|A_{j1}|,\ldots,|A_{jp}|}} \leq 1.$$

Consequently, m is bounded by the largest p-multinomial coefficient for n. The bound can be attained for every n and p.

Proof. Everything follows from Theorems 1.5/2.5 except the attainability of the upper bound, which is a consequence of Theorem 1.3. ∎

5. THE MAXIMUM NUMBER OF COMPOSITIONS

Although the bounds in all the previously known Sperner generalizations of Section 1 can be attained, for the most part that seems not to be the case in Theorem 1.5. The key difficulty appears in the combination of r-families with compositions as in Corollary 4.1. (We think it makes no difference if we allow partial compositions but we have not proved it.) We begin with a refinement of Lemma 3.1. A weak set composition has **shape** $(a_1, \ldots, a_p)$ if $|A_k| = a_k$ for all k.

Lemma 5.1. *Given values of n, r, and p such that $r^{p-1} \leq \binom{n+p-1}{p-1}$, the bound in Corollary 4.1 can be attained only by taking all weak compositions of shape $(a_1, \ldots, a_p)$ that give p-multinomial coefficient larger than the $(r^{p-1}+1)$-st largest such coefficient $M_{r^{p-1}+1}$, and none whose shape gives a smaller coefficient than the (r^{p-1})-st largest such coefficient $M_{r^{p-1}}$.*

Proof. First we need to characterized sharpness in Lemma 3.1. Our lemma is a slight improvement on [7, Lemma 3.1.3].

Lemma 5.2. *In Lemma 3.1, suppose that $M_R > 0$. Then there is equality in the conclusion if and only if*

$$q_k = 1 \text{ if } M_k > M_R \qquad \text{and} \qquad q_k = 0 \text{ if } M_k < M_R$$

and also, letting $M_{R'+1}$ and $M_{R''}$ be the first and last M_k's equal to M_R,

$$q_{R'+1} + \cdots + q_{R''} = R - R'. \qquad ∎$$

In Lemma 5.1, all $M_k > 0$ for $k \leq \binom{n+p-1}{p-1}$. (We assume N is no larger than $\binom{n+p-1}{p-1}$. The contrary case is easily derived from that one.) It is clear that, when applying Lemma 3.1, we have to have in our set of weak compositions all those of the shapes $(a_1, \ldots, a_p)$ for which $\binom{n}{a_1, \ldots, a_p} > M_{r^{p-1}}$ and none for which $\binom{n}{a_1, \ldots, a_p} < M_{r^{p-1}}$. The rest of the m weak compositions can have any shapes for which $\binom{n}{a_1, \ldots, a_p} = M_{r^{p-1}}$. If $M_{r^{p-1}} > M_{r^{p-1}+1}$ this means we must have all weak compositions with shapes for which $\binom{n}{a_1, \ldots, a_p} > M_{r^{p-1}+1}$. ∎

To explain why the bound cannot usually be attained, we need to define the "first appearance" of a size a_i in the descending order of p-multinomial coefficients for n.

Fix $p \geq 3$ and n and let $n = \nu p + \rho$ where $0 \leq \rho < p$. In $\left(\begin{smallmatrix} n \\ a_1,\ldots,a_p \end{smallmatrix}\right)$, the a_i are the **sizes**. The multiset of sizes is the **form** of the coefficient. Arrange the multinomial coefficients in weakly decreasing order: $M_1 \geq M_2 \geq M_3 \geq \cdots$. (There are many such orderings; choose one arbitrarily, fix it, and call it **the descending order** of coefficients.) Thus, for example,

$$M_1 = \binom{n}{\nu,\ldots,\nu} > M_2 = \binom{n}{\nu+1,\nu,\ldots,\nu,\nu-1}$$

$$= M_3 = \cdots = M_{p(p-1)+1} \quad \text{if} \quad p \mid n$$

since $M_3,\ldots,M_{p(p-1)+1}$ have the same form as M_2, and

$$M_1 = \binom{n}{\nu+1,\ldots,\nu} = \cdots = M_{\binom{p}{\rho}} > M_{\binom{p}{\rho}+1} \quad \text{if} \quad p \nmid n,$$

where the form of M_1 has ρ sizes equal to $\nu+1$, so $M_1,\ldots,M_{\binom{p}{\rho}}$ all have the same form.

As we scan the descending order of multinomial coefficients, each possible size κ, $0 \leq \kappa \leq n$, appears first in a certain M_i. We call M_i the **first appearance** of κ and label it L_κ. For example, if $p \mid n$, $L_\nu = M_1 > L_{\nu+1} = L_{\nu-1} = M_2$, while if $p \nmid n$ then $L_\nu = L_{\nu+1} = M_1$. It is clear that $L_\nu > L_{\nu-1} > \ldots$ and $L_{\nu+1} > L_{\nu+2} > \ldots$, but the way in which the lower L_κ's, where $\kappa \leq \nu$, interleave the upper ones is not obvious. We write L_k^* for the k-th L_κ in the descending order of multinomial coefficients. Thus $L_1^* = L_\nu$; $L_2^* = L_{\nu+1}$ and $L_3^* = L_{\nu-1}$ (or vice versa) if $p \mid n$, and $L_2^* = L_{\nu+1}$ if $p \nmid n$ while $L_3^* = L_{\nu+2}$ or $L_{\nu-1}$.

Theorem 5.1. *Given $r \geq 2$, $p \geq 3$, and $n \geq p$, the bound in Corollary 4.1 cannot be attained if $L_r^* > M_{rp-1+1}$.*

The proof depends on the following lemma.

Lemma 5.3. *Let $r \geq 2$ and $p \geq 3$, and let $\kappa_1,\ldots,\kappa_r$ be the first r sizes that appear in the descending order of p-multinomial coefficients for n. The number of all coefficients with sizes drawn from $\kappa_1,\ldots,\kappa_r$ is less than r^{p-1} and their sum is less than $M_1 + \cdots + M_{r^{p-1}}$.*

Proof. Clearly, $\kappa_1, \ldots, \kappa_r$ form a consecutive set that includes ν. Let κ be the smallest and κ' the largest. One can verify that, in $\binom{n}{\kappa,\ldots,\kappa,x}$ and $\binom{n}{\kappa',\ldots,\kappa',y}$, it is impossible for both x and y to lie in the interval $[\kappa, \kappa']$ as long as $(r-1)(p-2) > 0$. ∎

Proof of Theorem 5.1. Suppose the upper bound of Corollary 4.1 is attained by a certain set of weak compositions of S, an n-element set. For each of the first r sizes $\kappa_1, \ldots, \kappa_r$ that appear in the descending order of p-multinomial coefficients, L_{κ_i} has sizes drawn from $\kappa_1, \ldots, \kappa_r$ and at least one size κ_i. Taking all coefficients M_k that have the same forms as the L_{κ_i}, κ_i will appear in each position j in some M_k. By hypothesis and Lemma 5.1, among our set of weak compositions, every κ_i-subset of S appears in every position in the weak compositions. If any subset of S of a different size from $\kappa_1, \ldots, \kappa_r$ appeared in any position, there would be a chain of length r in that position. Therefore we can only have weak compositions whose sizes are among the first r sizes. By Lemma 5.3, there are not enough of these to attain the upper bound. ∎

Theorem 5.1 can be hard to apply because we do not know $M_{r^{p-1}+1}$. On the other hand, we do know L_κ since it equals $\binom{n}{\kappa, a_2, \ldots, a_p}$ where $a_2, \ldots, a_p$ are as nearly equal as possible. A more practical criterion for nonattainment of the upper bound is therefore

Corollary 5.1. *Given $r \geq 2$, $p \geq 3$, and $n \geq p$, the bound in Corollary 4.1 cannot be attained if $L_r^* > L_{r+1}^*$.*

Proof. It follows from Lemma 5.3 that L_{r+1}^* is one of the first r^{p-1} coefficients. Thus $L_r^* > L_{r+1}^* \geq M_{r^{p-1}+1}$ and Theorem 5.1 applies. ∎

It seems clear that L_r^* will almost always be larger than L_{r+1}^* (if $r \geq 3$ or $p \nmid n$) so our bound will not be attained. However, cases of equality do exist. For instance, take $p = 3$, $r = 3$, and $n = 10$; then $L_5^* = L_1 = \binom{10}{5,4,1} = 1260$ and $L_6^* = L_6 = \binom{10}{6,2,2} = 1260$. Thus if $r = 5$, Corollary 5.1 does not apply here. (We think the bound is still not attained but we cannot prove it.) We can isolate the instances of equality for each r, but as r grows larger the calculations quickly become extensive. Thus we state the results only for small values of r.

Proposition 5.1. *The bound in Corollary 4.1 cannot be attained if $2 \leq r \leq 5$ and $p \geq 3$ and $n \geq r-1$, except possibly when $r = 2$, $p \mid n$, and $p = 3, 4, 5$, or when $r = 4$, $p \geq 4$, and $n = 2p-1$, or when $r = 5$, $p = 3$, and $n = 10$.*

Proof sketch. Suppose $p \nmid n$. We have verified (by long but routine calculations which we omit) that $L_1^* = L_2^* > L_3^* > L_4^* > L_5^* > L_6^*$ except that $L_4^* = L_5^*$ if $\rho = p - 1$ and $p \geq 4$ and $\nu = 1$ and $L_5^* = L_6^*$ when $p = \nu = 3$ and $\rho = 1$.

If $p \mid n$ then $L_1^* > L_2^* = L_3^* > L_4^* > L_5^* > L_6^*$. This implies the proposition for $r = 3$, 4, or 5. We approach $r = 2$ differently. The largest coefficients are

$$M_1 = \binom{n}{\nu, \ldots, \nu} > M_2 = \binom{n}{\nu + 1, \nu, \ldots, \nu, \nu - 1} = \cdots$$

$$= M_{p(p-1)+1} > M_{p(p-1)+2}.$$

If $p(p-1) + 1 \leq r^{p-1}$, the bound is unattainable by Theorem 5.1. That is the case when $p \geq 6$.

REFERENCES

[1] I. Anderson, *Combinatorics of Finite Sets,* Clarendon Press, Oxford, 1987.

[2] M. Beck and T. Zaslavsky, A shorter, simpler, stronger proof of the Meshalkin–Hochberg–Hirsch bounds on componentwise antichains, *J. Combin. Theory Ser. A,* **100** (2002), 196–199.

[3] B. Bollobás, On generalized graphs. *Acta Math. Acad. Sci. Hung.,* **16** (1965), 447–452.

[4] P. Erdős, On a lemma of Littlewood and Offord, *Bull. Amer. Math. Soc.,* **51** (1945), 898–902.

[5] J. R. Griggs, J. Stahl and W. T. Trotter, A Sperner theorem on unrelated chains of subsets, *J. Combinatorial Theory Ser. A,* **36** (1984), 124–127.

[6] M. Hochberg and W. M. Hirsch, Sperner families, s-systems, and a theorem of Meshalkin, *Ann. New York Acad. Sci.,* **175** (1970), 224–237.

[7] D. A. Klain and G.-C. Rota, *Introduction to Geometric Probability,* Cambridge University Press, Cambridge, Eng., 1997.

[8] D. A. Lubell, A short proof of Sperner's theorem, *J. Combinatorial Theory,* **1** (1966), 209–214.

[9] L. D. Meshalkin, Generalization of Sperner's theorem on the number of subsets of a finite set (in Russian), *Teor. Verojatnost. i Primenen,* **8** (1963), 219–220. English trans.: *Theor. Probability Appl.,* **8** (1963), 203–204.

[10]   G.-C. Rota and L. H. Harper, Matching theory, an introduction, in: P. Ney, ed., *Advances in Probability and Related Topics,* Vol. 1, pp. 169–215, Marcel Dekker, New York, 1971.

[11]   E. Sperner, Ein Satz über Untermengen einer endlichen Menge, *Math. Z.,* **27** (1928), 544–548.

[12]   K. Yamamoto, Logarithmic order of free distributive lattice, *J. Math. Soc. Japan,* **6** (1954), 343–353.

Matthias Beck

Department of Mathematical Sciences
Binghamton University (SUNY)
Binghamton, NY 13902-6000
U.S.A.

Present address:
Mathematical Sciences Research
Institute
17 Gauss Way, Berkeley
CA 94720-5070
U.S.A.

matthias@msri.org

Xueqin Wang

Department of Mathematical Sciences
Binghamton University (SUNY)
Binghamton, NY 13902-6000
U.S.A.

Present address:
Department of Mathematics
University of Mississippi
P.O. Box 1848
University, MS 38677-1848
U.S.A.

xueqin@olemiss.edu

Thomas Zaslavsky

Department of Mathematical Sciences
Binghamton University (SUNY)
Binghamton, NY 13902-6000
U.S.A.

zaslav@math.binghamton.edu

BOLYAI SOCIETY
MATHEMATICAL STUDIES, 15

Conference on Finite
and Infinite Sets
Budapest, pp. 25–32.

A QUICK PROOF OF SPRINDZHUK'S DECOMPOSITION THEOREM

Y. F. BILU and D. MASSER

Dedicated to the memory of V. G. Sprindzhuk

In [11] Sprindzhuk proved the following striking theorem.

Theorem 1 (Sprindzhuk [11]). *Let $F(x,y) \in \mathbb{Q}[x,y]$ be a $\mathbb{Q}$-irreducible polynomial satisfying*

$$(1) \qquad F(0,0) = 0, \qquad \frac{\partial F}{\partial y}(0,0) \neq 0.$$

Then for all but finitely many prime numbers p, the polynomial $F(p,y)$ is $\mathbb{Q}$-irreducible.

Actually, prime numbers can be replaced by prime powers, as well as by numbers of the form $1/t$, where $t \in \mathbb{Z}$, $t \neq 0$: see Corollary 3.

In the subsequent paper [12] (see also [13, 14] for a more detailed exposition) Sprindzhuk obtained an even more amusing result. To formulate it, recall that the *height* of a rational number $\alpha = a/b$ (where a and b are coprime integers) is defined by

$$(2) \qquad H(\alpha) = \max\left\{ |a|, |b| \right\}.$$

One immediately verifies that

$$(3) \qquad H(\alpha) = \prod_{v \in M_{\mathbb{Q}}} \max\left\{ 1, |\alpha|_v \right\} = \prod_{v \in M_{\mathbb{Q}}} \max\left\{ 1, |\alpha|_v^{-1} \right\},$$

where $M_{\mathbb{Q}}$ is the set of all places of the field $\mathbb{Q}$ (that is, $M_{\mathbb{Q}} = \{\text{primes}\} \cup \{\infty\}$).

For $\alpha \in \mathbb{Q}$ put $V(\alpha) = \left\{ v \in M_{\mathbb{Q}} : |\alpha|_v < 1 \right\}$.

Theorem 2 (Sprindzhuk [12]). *Let $F(x, y)$ be as in Theorem 1 and ε a positive number. For every $\alpha \in \mathbb{Q}$ let $d_1(\alpha), \ldots, d_k(\alpha)$ be the degrees of the $\mathbb{Q}$-irreducible factors of $F(\alpha, y)$ (so that $d_1(\alpha) + \cdots + d_k(\alpha) = \deg_y F$). Then for all but finitely many $\alpha \in \mathbb{Q}$ there is a partition $V(\alpha) = V_1 \cup \ldots \cup V_k$ such that*

$$(4) \qquad \left| \frac{-\sum_{v \in V_i} \log |\alpha|_v}{\log H(\alpha)} - \frac{d_i(\alpha)}{\deg_y F} \right| < \varepsilon \qquad (i = 1, \ldots, k).$$

We do not formally assert that the partition sets $V_1, \ldots, V_k$ are non-empty. However, (4) implies that they are indeed non-empty when ε is sufficiently small (in fact, when $\varepsilon < 1/ \deg_y F$).

Theorem 1 easily follows from Theorem 2. Put $\Omega = \{\text{prime powers}\} \cup \{1/t : t \in \mathbb{Z}, |t| > 1\}$.

Corollary 3. *Let $F(x, y)$ be as in Theorem 1. Then $F(\omega, y)$ is $\mathbb{Q}$-irreducible for all but finitely many $\omega \in \Omega$.*

Proof. As we observed above, the partition sets $V_1, \ldots, V_k$ are non-empty when ε is sufficiently small. But for every $\omega \in \Omega$ the set $V(\omega)$ consists of a single element, and cannot be partitioned into more than one non-empty part. ∎

Here is another amazing consequence of Theorem 2 (the proof is immediate).

Corollary 4. *Let $F(x, y)$ be as in Theorem 1 and let $\{q_\nu\}$, $\{r_\nu\}$ be two sequences of prime powers such that $\lim_{\nu \to \infty} \log q_\nu / \log r_\nu$ exists and is irrational. Then $F(q_\nu r_\nu, y)$ is $\mathbb{Q}$-irreducible for all but finitely many ν.* ∎

We invite the reader to invent many other corollaries of this wonderful theorem.

Actually Sprindzhuk in [12] obtained a yet sharper version of Theorem 2 with ε replaced by an error term of order $\left(\log H(\alpha) \right)^{-1/2}$. To prove this he used Siegel's Lemma and some sophisticated machinery from the theory of Diophantine approximation and transcendence such as the cancellation of factorials and a zero estimate (Lemma 6 of [11]). He also used Eisenstein's theorem, which is easy when (1) is assumed.

In the final paragraph of the Russian edition of his book [13], Sprindzhuk wrote that, while methods of Diophantine approximation are used in the proof of Theorem 2, its formulation

"...involves no concepts related to the theory of Diophantine approximation. This gives hope that a different proof exists, which is independent of the theory of Diophantine approximation."

Indeed, such a proof was soon after found by Bombieri [1], who used the machinery of Weil functions and Néron-Tate height. Weil functions were also employed by Fried [9] in the prime-power case. It was Bombieri who pointed out the connection with G-functions and Fuchsian differential operators of arithmetic type. This connection was further developed by Dèbes (and Zannier) [3, 4, 5, 6].

The object of the present note is to point out that Theorem 2 itself can be established rather quickly, also along the lines of Sprindzhuk's original articles, but without most of the sophisticated machinery. Our proof relies only on the simplest properties of heights (see Proposition 5 below) and Eisenstein's theorem.

Recall the definition of the height of an algebraic number. This is

$$(5) \qquad H(\alpha) = \left(\prod_{v \in M_K} \max \left\{ 1, |\alpha|_v^{[K_v:\mathbb{Q}_v]} \right\} \right)^{1/[K:\mathbb{Q}]},$$

where K is a number field containing α and M_K is the set of valuations on K, which are normalized to extend the standard valuations of $\mathbb{Q}$. As usual, K_v and $\mathbb{Q}_v$ stand for the topological completions with respect to $v \in M_K$.

It is straightforward to verify that the right-hand side of (5) does not depend on the choice of the field K. Also (3) implies that this definition is compatible with the definition of the height of a rational number from (2).

The product formula

$$\prod_{v \in M_K} |\alpha|_v^{[K_v:\mathbb{Q}_v]} = 1 \qquad (\alpha \in K^*)$$

implies that for any $V \subset M_K$ and $\alpha \in K^*$ one has the following "Liouville inequality":

$$(6) \qquad \prod_{v \in V} |\alpha|_v^{[K_v:\mathbb{Q}_v]} \geq H(\alpha)^{-[K:\mathbb{Q}]}.$$

The following two well-known properties of the height function are (almost) immediate consequences of its definition (5).

Proposition 5. *Let α, β be algebraic numbers and $F(x,y)$ a polynomial with algebraic coefficients. Put $m = \deg_x F$ and $n = \deg_y F$.*

1. *For $\gamma = F(\alpha, \beta)$ one has $H(\gamma) \ll H(\alpha)^m H(\beta)^n$.*

2. *Assume that F is not divisible by $x - \alpha$. Then $F(\alpha, \beta) = 0$ implies that $H(\beta) \ll H(\alpha)^m$.*

Constants implied by "$\ll$" depend only on the polynomial F.

Proof. Part "1" is straightforward. To prove "2", write $F(x,y) = f_n(x)y^n + \cdots + f_0(y)$. By the assumption, not all of the numbers $f_0(\alpha), \ldots, f_n(\alpha)$ vanish. Put $\nu = \max \{ j : f_j(\alpha) \neq 0 \}$.

Let K be a number field containing α, β and the coefficients of F. The equality $f_\nu(\alpha)\beta^\nu + f_{\nu-1}(\alpha)\beta^{\nu-1} + \cdots + f_0(\alpha) = 0$ implies that

$$\max \left\{ 1, |\beta|_v \right\} \leq \max \left\{ 1, |\nu|_v \right\}$$

$$\max \left\{ 1, \left| f_{\nu-1}(\alpha)/f_\nu(\alpha) \right|_v, \ldots, \left| f_0(\alpha)/f_\nu(\alpha) \right|_v \right\} \qquad (v \in M_K).$$

Using the product formula, we obtain

$$H(\beta) \leq H(\nu) \Bigg(\prod_{v \in M_K} \max \left\{ 1, \left| f_{\nu-1}(\alpha)/f_\nu(\alpha) \right|_v, \ldots, \right.$$

$$\left. \left| f_0(\alpha)/f_\nu(\alpha) \right|_v \right\}^{[K_v : \mathbb{Q}_v]} \Bigg)^{1/[K : \mathbb{Q}]}$$

$$= \nu \Bigg(\prod_{v \in M_K} \max \left\{ \left| f_\nu(\alpha) \right|_v, \left| f_{\nu-1}(\alpha) \right|_v, \ldots, \right.$$

$$\left. \left| f_0(\alpha) \right|_v \right\}^{[K_v : \mathbb{Q}_v]} \Bigg)^{1/[K : \mathbb{Q}]} \ll H(\alpha)^m,$$

as wanted. ∎

Recall also Eisenstein's theorem.

Theorem 6. *Let $Y(x) = a_0 + a_1 x + a_2 x^2 + \cdots$ be a power series with co-efficients in a number field K, algebraic over the field $K(x)$. Then for* every $v \in M_K$ there exists $c_v \geq 1$ such that all but finitely many c_v are equal to 1, and

(7) $$|a_j|_v \leq c_v^j \qquad (v \in M_K, \quad j = 1, 2, \ldots).$$

Classically, Eisenstein's theorem reads as follows: there exists a positive integer T such that $T^j a_j$ are algebraic integers for $j = 1, 2, \ldots$. This immediately implies Theorem 6. Indeed, for non-archimedean v one may put $c_v = |T|_v^{-1}$. For archimedean v, the existence of c_v follows from the fact that the convergence radius of a complex algebraic power series is positive.

Eisenstein's theorem goes back to Eisenstein's paper [8]. See [10, page 151] for an old-fashioned proof and [7] for a modern quantitative argument. See also [2, page 28] for an especially quick proof when $K = \mathbb{Q}$, which suffices for the present note. In addition, if $a_0 = 0$ and $F(x, Y(x)) = 0$, where $F(x, y) \in \mathbb{Z}[x, y]$ satisfies (1), then a very easy induction gives the value $T = (\partial F / \partial y(0, 0))^2$, and in fact this case suffices as well.

Proof of Theorem 2. Put $m = \deg_x F$ and $n = \deg_y F$. To prove the theorem, it is sufficient to find a partition $V(\alpha) = V_1 \cup \ldots \cup V_k$ satisfying

(8) $$\frac{-\sum_{v \in V_i} \log |\alpha|_v}{\log H(\alpha)} \leq \frac{d_i(\alpha)}{n} + \varepsilon \qquad (i = 1, \ldots, k).$$

Indeed, by the second equality in (3),

$$\sum_{i=1}^{k} \frac{-\sum_{v \in V_i} \log |\alpha|_v}{\log H(\alpha)} = 1 = \sum_{i=1}^{k} \frac{d_i(\alpha)}{n}.$$

Hence (8) implies that

$$\frac{-\sum_{v \in V_i} \log |\alpha|_v}{\log H(\alpha)} \geq \frac{d_i(\alpha)}{n} - (k - 1)\varepsilon \qquad (i = 1, \ldots, k),$$

and (4) follows after redefining ε.

It follows from (1) that there exists a power series $Y(x) = a_1 x + a_2 x^2 + \cdots$ with rational coefficients satisfying $F(x, Y(x)) = 0$. Put

(9) $$N = \lceil 4m(n - 1)/\varepsilon \rceil.$$

There is a non-zero polynomial $G(x, y) \in \mathbb{Q}[x, y]$ satisfying

(10) $$\deg_y G \leq n - 1, \quad \deg_x G \leq N,$$

(11) $$\mathrm{ord}_{x=0} G(x, Y(x)) \geq nN.$$

(Indeed, the vector space of polynomials satisfying (10) is of dimension $n(N + 1)$, while (11) is equivalent to nN linear relations.) In the sequel, constants implied by "$O(\cdot)$", "$\ll$" and "$\gg$" may depend only on F, G and ε.

Put $U(x) = G\big(x, Y(x)\big)$. By Eisenstein's theorem, for every $v \in M_{\mathbb{Q}}$ there exists $c_v \geq 1$ such that all but finitely many c_v are equal to 1, and the coefficients of the power series $Y(x) = \sum_{j=1}^{\infty} a_j x^j$ and $U(x) = \sum_{j=nN}^{\infty} b_j x^j$ satisfy

(12) $$|a_j|_v, |b_j|_v \leq c_v^j \qquad (v \in M_{\mathbb{Q}}).$$

For $\alpha \in \mathbb{Q}$ put

$$V'(\alpha) = \left\{ v \in V(\alpha) : |\alpha|_v \begin{array}{l} \leq 1/(2c_v) \text{ if } v = \infty, \\ < 1/c_v \quad \text{ if } v < \infty . \end{array} \right\}, \qquad V''(\alpha) = V(\alpha) \backslash V'(\alpha).$$

Since $-\sum_{v \in V''(\alpha)} \log |\alpha|_v \ll 1$, for all but finitely many α we have

$$\frac{-\sum_{v \in V''(\alpha)} \log |\alpha|_v}{\log H(\alpha)} \leq \frac{\varepsilon}{2}.$$

Hence it is sufficient to find a partition $V'(\alpha) = V_1' \cup \ldots \cup V_k'$ such that

(13) $$\frac{-\sum_{v \in V_i'} \log |\alpha|_v}{\log H(\alpha)} \leq \frac{d_i(\alpha)}{n} + \frac{\varepsilon}{2} \qquad (i = 1, \ldots, k).$$

for then putting, say, $V_1 = V_1' \cup V''(\alpha)$ and $V_i = V_i'$ for $i \geq 2$, we obtain (8).

Thus, fix $\alpha \in \mathbb{Q}$ and let $F(\alpha, y) = f_1(y) \cdots f_k(y)$ be the decomposition of $F(\alpha, y)$ into $\mathbb{Q}$-irreducible factors. We may assume (discarding finitely many α at which the y-discriminant of $F(x, y)$ vanishes) that the polynomials f_i are pairwise coprime. We put $d_i = \deg f_i$.

For any $v \in V'(\alpha)$ the series $Y(x)$ converges v-adically at α. Its sum in $\mathbb{Q}_v$, denoted by $Y_v(\alpha)$, is a zero of $F(\alpha, y)$. Define the partition $V'(\alpha) = V_1' \cup \ldots \cup V_k'$ as follows:

$$V_i' = \big\{ v \in V'(\alpha) : Y_v(\alpha) \text{ is a zero of } f_i(y) \big\} \qquad (i = 1, \ldots, k).$$

Now fix i and let $\beta = \beta_i$ be a zero of $f_i(y)$. Again discarding finitely many α, we may assume that $\eta := G(\alpha, \beta) \neq 0$. Indeed, since $F(x, y)$ is irreducible, and $\deg_y G < \deg_y F$, the system of algebraic equations $F(\alpha, \beta) = G(\alpha, \beta) = 0$ has only finitely many solutions.

Extend every $v \in V_i$ to the number field $K = \mathbb{Q}(\beta)$ to have $Y_v(\alpha) = \beta$. Notice that

$$(14) \qquad K_v = \mathbb{Q}_v$$

for this extension of v. Then for $v \in V_i'$ we have $\eta = U_v(\alpha)$, the v-adic sum of $U(x)$ at α. Using (11) and (12), we obtain

$$(15) \qquad |\eta|_v \le \max \left\{ 1, |2|_v \right\} \left(c_v |\alpha|_v \right)^{nN} \qquad (v \in V_i').$$

The equality $F(\alpha, \beta) = 0$ together with Proposition 5: 2 implies that $H(\beta) \ll H(\alpha)^m$ (since the polynomial F is irreducible, it is not divisible by $x - \alpha$). Now Proposition 5: 1 implies that $H(\eta) \ll H(\alpha)^{N+\nu}$, where $\nu = (n-1)m$. Using (14) and "Liouville inequality" (6), we obtain

$$(16) \qquad \prod_{v \in V_i'} |\eta|_v = \prod_{v \in V_i'} |\eta|_v^{[K_v : \mathbb{Q}_v]} \ge H(\eta)^{-d_i} \gg H(\alpha)^{-(N+\nu)d_i}.$$

Combining this with (15) and (9), we conclude that

$$(17)$$
$$-\sum_{v \in V_i'} \log |\alpha|_v \le \left(1 + \frac{\nu}{N}\right) \frac{d_i}{n} \log H(\alpha) + O(1) \le \left(1 + \frac{\varepsilon}{4}\right) \frac{d_i}{n} \log H(\alpha) + O(1).$$

When $H(\alpha)$ is sufficiently large, we obtain

$$(18) \qquad \frac{-\sum_{v \in V_i'} \log |\alpha|_v}{\log H(\alpha)} \le \frac{d_i}{n} + \frac{\varepsilon}{2},$$

which is (13). ∎

REFERENCES

[1] E. Bombieri, On Weil's "Théorème de Décomposition", *Amer. J. Math.,* **105** (1983), 295–308.

[2] J. W. S. Cassels, *Local fields,* Cambridge, 1986.

[3] P. Dèbes, Quelques remarques sur un article de Bombieri concernant le Théorème de Décomposition de Weil, *Amer. J. Math.,* **107** (1985), 39–44.

[4] P. Dèbes, *G*-fonctions et théorème d'irréductibilité de Hilbert, *Acta Arith.*, **47** (1986), 371–402.

[5] P. Dèbes, Hilbert subsets and *s*-integral points, *Manuscripta Math.*, **89** (1996), 107–137.

[6] P. Dèbes and U. Zannier, Hilbert's irreducibility theorem and *G*-functions, *Math. Ann.*, **309** (1997), 491–503.

[7] B. M. Dwork and A. J. van der Poorten, The Eisenstein constant. *Duke Math. J.*, **65**(1992), 23–43; Corrections: **76** (1994), 669–672.

[8] G. Eisenstein, Über eine allgemeine Eigenschaft der Reihen-Entwicklungen aller algebraischen Funktionen, *Bericht Königl. Preuß. Akad. Wiss. Berlin* 1852, 411–413; *Mathematische Werke,* Band **II**, Chelsea, New York, 1975, 765–767.

[9] M. Fried, On the Sprindzhuk-Weissauer approach to universal Hilbert subsets, *Israel J. Math.*, **51** (1985), 347–363.

[10] E. Heine, *Theorie der Kugelfunktionen,* zweite Auflage, Reimer, Berlin, 1878.

[11] V. G. Sprindzhuk, Hilbert's irreducibility theorem and rational points on algebraic curves (Russian), *Dokl. Akad. Nauk SSSR,* **247** (1979), 285–289; (English transl.: *Soviet Math. Dokl.,* **20** (1979), 701–705).

[12] V. G. Sprindzhuk, Reducibility of polynomials and rational points on algebraic curves (Russian), *Dokl. Akad. Nauk SSSR,* **250** (1980), 1327–1330: English transl.: *Soviet Math. Dokl.,* **21** (1980), 331–334.

[13] V. G. Sprindzhuk, *Classical Diophantine equations in two unknowns* (Russian), "Nauka", Moscow, 1982; English transl.: *Lecture Notes in Mathematics,* **1559**, Springer-Verlag, Berlin, 1993.

[14] V. G. Sprindzhuk, Arithmetic specializations in polynomials, *J. Reine Angew. Math.,* **340** (1983), 26–52.

Yuri F. Bilu

Institute of Mathematics
University of Bordeaux 1
33405 Talence
France

yuri@math.u-bordeaux1.fr

David Masser

Department of Mathematics
University of Basel
Petersplatz 1
4003 Basel
Switzerland

David.Masser@unibas.ch

BOLYAI SOCIETY
MATHEMATICAL STUDIES, 15

Conference on Finite
and Infinite Sets
Budapest, pp. 33–56.

DISCREPANCY IN GRAPHS AND HYPERGRAPHS

B. BOLLOBÁS and A. D. SCOTT[*]

Let G be a graph with n vertices and $p\binom{n}{2}$ edges, and define the discrepancies $\mathrm{disc}_p^+(G) = \max_{Y \subset V(G)} \{e(Y) - p\binom{|Y|}{2}\}$ and $\mathrm{disc}_p^-(G) = \max_{Y \subset V(G)} \{p\binom{|Y|}{2} - e(Y)\}$. We prove that if $p(1-p) \geq 1/n$ then $\mathrm{disc}_p^+(G)\,\mathrm{disc}_p^-(G) \geq p(1-p)n^3/6400$. We also prove a similar inequality for k-uniform hypergraphs, and give related results concerning 2-colourings of k-uniform hypergraphs. Our results extend those of Erdős, Goldberg, Pach and Spencer [6] and Erdős and Spencer [7].

1. INTRODUCTION

The *discrepancy* of a graph G is $\mathrm{disc}\,(G) = \max_{Y \subset V(G)} |e(Y) - \frac{1}{2}\binom{|Y|}{2}|$, where we write $e(Y) = e(G[Y])$ for the number of edges of G spanned by Y. If G has edge density $1/2$ then the discrepancy can be seen as a measure of how uniformly the edges are distributed among the vertices; see Sós [11] and Beck and Sós [1] for more discusssion and a general account of discrepancy. Erdős and Spencer [7] showed that for some constant $c > 0$ every graph G of order n satisfies $\mathrm{disc}\,(G) \geq cn^{3/2}$. More generally, they showed that for every $k \geq 3$ there is a constant $c_k > 0$ such that if H is a k-uniform hypergraph of order n then $\mathrm{disc}\,(H) \geq c_k n^{(k+1)/2}$, where $\mathrm{disc}\,(H) = \max_{Y \subset V(H)} |e(Y) - \frac{1}{2}\binom{|Y|}{k}|$. By considering random graphs they showed that this bound is sharp up to the value of the constant.

Now suppose that G is a graph with $e(G) = m = p\binom{|G|}{2}$, where $p < 1/2$, so that we expect a random subset $Y \subset V(G)$ to span a subgraph with

---

[*]Research supported in part by NSF grant DSM 9971788 and DARPA grant F33615-01-C-1900.

$p\binom{|Y|}{2}$ edges. Then a more appropriate measure of edge distribution is given
by the quantity $\mathrm{disc}_p(G) = \max_{Y \subset V(G)} \left| e(Y) - p\binom{Y}{2} \right|$. Erdős, Goldberg,
Pach and Spencer [6] showed that in this case $\mathrm{disc}_p(G) \geq c\sqrt{mn}$, where c is
an absolute constant.

A subset of vertices with large discrepancy can clearly be either more
or less dense than the whole graph. Let us define the *positive discrep-
ancy* by $\mathrm{disc}^+(G) = \max_{Y \subset V(G)} \left\{ e(Y) - \frac{1}{2}\binom{|Y|}{2} \right\}$ and the *negative discrep-
ancy* by $\mathrm{disc}^-(G) = \max_{Y \subset V(G)} \left\{ \frac{1}{2}\binom{|Y|}{2} - e(Y) \right\}$. Then a random graph
$G \in \mathcal{G}(n, 1/2)$ shows that it is possible to have $\max \left\{ \mathrm{disc}^+(G), \mathrm{disc}^-(G) \right\} \leq$
$cn^{3/2}$. The one-sided discrepancy can be smaller: for instance. the complete
bipartite graph $K_{n/2,n/2}$ has positive discrepancy $O(n)$, although its negative discrepancy is cn^2. Similarly, the graph $2K_{n/2}$ has positive discrepancy
$O(n)$ but negative discrepancy cn^2. These examples show that we can guarantee small discrepancy on one side provided we allow large discrepancy on
the other. In this paper we shall prove that positive discrepancy substantially smaller than $n^{3/2}$ guarantees negative discrepancy substantially larger
than $n^{3/2}$; indeed, we shall quantify the trade-off between positive and negative discrepancies. Surprisingly, the correct measure turns out to be the
product $\mathrm{disc}^+(G)\,\mathrm{disc}^-(G)$.

We remark that a different type of negative discrepancy was considered
by Erdős, Faudree, Rousseau and Schelp [5] with the idea of showing that
graphs with small negative discrepancy contain complete subgraphs of fixed
size. For further recent results in this direction see Krivelevich [9] and
Keevash and Sudakov [8].

We begin with some definitions. For a k-uniform hypergraph G, a real
$p \in [0, 1]$ and $X \subset V(G)$ let

$$d_p(X) = e(X) - p\binom{|X|}{k}.$$

For disjoint sets of vertices X and Y, let

$$d_p(X, Y) = e(X, Y) - p|X|\,|Y|.$$

Then we define

(1) $$\mathrm{disc}_p^+(G) = \max_{X \subset V(G)} d_p(X)$$

and

(2) $$\mathrm{disc}_p^-(G) = - \min_{X \subset V(G)} d_p(X),$$

and set

$$\mathrm{disc}_p(G) = \max_{X \subset V(G)} |d_p(X)| = \max \left\{ \mathrm{disc}_p^+(G), \mathrm{disc}_p^-(G) \right\}.$$

If p is not specified we assume $p = 1/2$, so for instance disc$(G) = \mathrm{disc}_{1/2}(G)$.
Note that the cases $p = 0$ and $p = 1$ are trivial, and that if $e(G) = p_0 \binom{|G|}{2}$
we have $\mathrm{disc}_p(G) \geq |d_p(V(G))| = |p - p_0| \binom{|G|}{2}$. We will therefore usually
take p with $e(G) = p \binom{|G|}{2}$. Note that, for any p, $\mathrm{disc}_p^+(G) = \mathrm{disc}_{1-p}^-(\overline{G})$ and
$\mathrm{disc}_p^-(G) = \mathrm{disc}_{1-p}^+(\overline{G})$. We shall usually assume $p \leq 1/2$, since if $p > 1/2$
we may replace G by $\overline{G}$ and p by $1 - p$.

We remark that it does not make much difference if we restrict the
definitions in (1) and (2) to sets X of size $n/2$ (or some other size cn): as
noted by Erdős, Goldberg, Pach and Spencer [6], this would change the
resulting discrepancy by at most a constant factor.

We shall frequently refer to a *random bipartition* $V = X \cup Y$. Unless
otherwise stated, this means a random bipartition in which each vertex is
assigned independently to X or Y with equal probability. Throughout the
paper we shall use ε_i and ρ_j for sequences of independent Bernoulli random
variables, with $\varepsilon_i \in \{+1, -1\}$ and $\rho_j \in \{0, 1\}$, each taking either value with
probability $1/2$.

The rest of the paper is organized as follows. In section 2 we give
lower bounds on $\mathrm{disc}_p(G)$ for graphs; in section 3 we turn our attention
to hypergraphs. Finally, in section 4, we consider some related results
concerning subgraphs of a fixed graph or hypergraph.

2. Discrepancy of graphs

In this section we prove our results on graph discrepancy. Let G be a graph
of order n and size $p \binom{n}{2}$. If G is very sparse, say $0 < p \leq 1/(n-1)$, then
taking the union of $p \binom{n}{2}/2$ edges from G gives a subgraph with at most $p \binom{n}{2}$
vertices, so $\mathrm{disc}_p^+(G) \geq p \binom{n}{2}/2 - p \left(p \binom{n}{2} \right)^2 / 2 \geq pn^2/5 - p^3n^4/8 > pn^2/20$ for
sufficiently large n, while since G has average degree at most 1 it contains
an independent set of size at least $n/2$, and so $\mathrm{disc}^-(G) \geq p \binom{n/2}{2} > pn^2/9$
for sufficiently large n. On the other hand, $\max \left\{ \mathrm{disc}^+(G), \mathrm{disc}^-(G) \right\} \leq
e(G) < pn^2/2$. Thus $\mathrm{disc}_p^+(G)$ and $\mathrm{disc}_p^-(G)$ are both $\Theta(pn^2)$. A similar

argument applies if G is very dense, with $p \geq 1 - 1/(n-1)$. (More precise bounds are given by Erdős, Goldberg, Pach and Spencer [6].)

We therefore restrict our attention to graphs with $p(1-p) \geq 1/n$. Our main result is the following.

Theorem 1. *Let G be a graph of order n and size $p\binom{n}{2}$, where $p(1-p) \geq 1/n$. Then*

(3) $$\mathrm{disc}_p^+(G)\,\mathrm{disc}_p^-(G) \geq p(1-p)n^3/6400.$$

As an immediate corollary we get the following result of Erdős, Goldberg, Pach and Spencer [6].

Corollary 2. *Let G be a graph of order n and size $p\binom{n}{2}$, where $p(1-p) \geq 1/n$. Then*
$$\mathrm{disc}_p(G) \geq \sqrt{p(1-p)}\, n^{3/2}/80.$$

We remark that the result of Erdős and Spencer for graphs can easily be deduced from Theorem 1: if $1/3 \leq p \leq 2/3$ then $\mathrm{disc}\,(G) \geq \frac{1}{2}\big(\mathrm{disc}_p^+(G) + \mathrm{disc}_p^-(G)\big) \geq n^{3/2}/160$, while otherwise $\mathrm{disc}\,(G) \geq \big|e(G) - \frac{1}{2}\binom{n}{2}\big| \geq \binom{n}{2}/6 \geq n^{3/2}/12$.

We also remark that, for $r \geq 2$, the Turán graph $T_r(n)$ gives a bound on the optimal constant in (3). Defining p by $t_r(n) = e\big(T_r(n)\big) = p\binom{n}{2}$, we have $p \sim 1 - \frac{1}{r}$. A little calculation shows that

(4) $$\mathrm{disc}_p^+\big(T_r(n)\big) = \frac{pn}{8} + O(r),$$

and, for r even,

$$\mathrm{disc}_p^-\big(T_r(n)\big) = \big(1 + o(1)\big)\frac{(1-p)n^2}{8},$$

which implies

$$\mathrm{disc}_p^+\big(T_r(n)\big)\,\mathrm{disc}_p^-\big(T_r(n)\big) \leq \big(1 + o(1)\big)\frac{p(1-p)n^3}{64}.$$

Before turning to the proof of Theorem 1, we make some comments about one-sided discrepancies. Since every graph with n vertices and $t_r(n)$ edges contains a subgraph of order u and size at least $t_r(u)$ for every $1 \leq u \leq n$, the Turán graphs $T_r(n)$ have minimal positive p-discrepancy among graphs of order n and size $t_r(n)$. Thus (4) gives an optimal bound in these cases, which have density $p \sim 1 - \frac{1}{r}$. To obtain a similar bound for arbitrary densities, we define an extension of the Turán numbers for non-integral r. Given an integer $n \geq 1$ and a real number $r \geq 1$, we can write $n = qr + s$, where q is an integer and $0 \leq s < r$. We define the fractional Turán number $t_r(n)$ by

$$t_r(n) = \binom{n}{2} - \bar{t}_r(n),$$

where

$$\bar{t}_r(n) = s\binom{q+1}{2} + (r-s)\binom{q}{2}.$$

Note that this is consistent with the definition of Turán numbers when r is integral; it is convenient to work with the quantity $\bar{t}_r(n)$ instead of $t_r(n)$.

A bound matching (4) will follow from the following result.

Lemma 3. *Suppose that $n \geq 1$ is an integer and $1 \leq r \leq n$. Let G be a graph with n vertices and at least $t_r(n)$ edges. Then, for $2 \leq u \leq n$, G contains a subgraph with u vertices and at least $t_r(u)$ edges.*

Proof. It is enough to prove the theorem when $u = n - 1$. Taking complements, this is equivalent to showing that if $e(G) \leq \bar{t}_r(n)$ then there is a vertex v such that $e(G \setminus v) \leq \bar{t}_r(n-1)$. We may also assume $r > 1$, or else G is empty.

Adding edges if necessary, we may assume that

(5) $$e(G) = \lfloor \bar{t}_r(n) \rfloor = \bar{t}_r(n) - \eta,$$

where $0 \leq \eta < 1$. Thus if $n = qr + s$,

$$e(G) = s\binom{q+1}{2} + (r-s)\binom{q}{2} - \eta$$

$$= \frac{1}{2}(rq^2 + (2s - r)q) - \eta.$$

A short calculation shows that

(6) $$\Delta(G) \geq \left\lceil \frac{2e(G)}{n} \right\rceil = q - 1 + \left\lceil \frac{qs + s - 2\eta}{qr + s} \right\rceil.$$

By (5) and (6), it is sufficient to show that

(7) $$q - 1 + \left\lceil \frac{qs + s - 2\eta}{qr + s} \right\rceil + \eta \geq \bar{t}_r(n) - \bar{t}_r(n - 1).$$

If $q = 0$ then we have a complete graph and are done immediately. Thus we may assume that $q \geq 1$. Now if $s \geq 1$, then it is easily seen that

$$\bar{t}_r(n) - \bar{t}_r(n - 1) = q,$$

while if $0 \leq s < 1$, then a simple calculation shows that

(8) $$\bar{t}_r(n) - \bar{t}_r(n - 1) = q - 1 + s.$$

Now if $s > \eta$ then $qs + s > 2\eta$, and so the left side of (7) is at least $q + \eta$, and thus (7) is satisfied. If $s \leq \eta$, however, then $0 \leq s < 1$, so (8) holds. It is then sufficient by (7) to show that

$$\left\lceil \frac{qs + s - 2\eta}{qr + s} \right\rceil + \eta \geq s,$$

which holds provided

$$\frac{qs + s - 2\eta}{qr + s} > -1.$$

But $qr + s = n$ and $qs + s - 2\eta \geq -2\eta > -2$, so this holds for $n \geq 2$. ∎

Calculating as in (4), we obtain the following result.

Corollary 4. For $0 \leq p \leq 1$, every graph G with n vertices and $p\binom{n}{2}$ edges satisfies

$$\mathrm{disc}_p^+(G) \geq \frac{pn}{8} + O\left(\frac{1}{1-p}\right).$$

We now turn to the proof of Theorem 1. We shall need two simple inequalities (these follow easily from the Littlewood–Khinchin inequality, see [10], [12], [13]; however, we give short proofs at the end of the section). Recall that ε_i and ρ_i are i.i.d. Bernoulli with $\varepsilon_i \in \{+1, -1\}$ and $\rho_i \in \{0, 1\}$.

Lemma 5. For $n \geq 1$,

$$\mathbb{E}\left|\sum_{i=1}^{n} \varepsilon_i\right| \geq \sqrt{n/2}.$$

Lemma 6. *Let* $\mathbf{a} = (a_i)_{i=1}^n$ *be a sequence of real numbers, and* A *a real number. Then*

$$\mathbb{E}\left|\sum_{i=1}^n \varepsilon_i a_i - A\right| \geq \|\mathbf{a}\|_1 / \sqrt{2n}$$

and

$$\mathbb{E}\left|\sum_{i=1}^n \rho_i a_i - A\right| \geq \|\mathbf{a}\|_1 / \sqrt{8n}.$$

Our main tool in the proof of Theorem 1 is the following lemma, which shows that in a random bipartition of a graph G, we do not expect the vertex neighbourhoods to split too evenly.

Lemma 7. *Let* G *be a graph of order* n *and size* $p\binom{n}{2}$, *where* $p(1-p) \geq 1/n$. *Let* $V(G) = X \cup Y$ *be a random bipartition. Then*

$$\mathbb{E}\sum_{x \in X} \left||\Gamma(x) \cap Y| - p|Y|\right| \geq \sqrt{p(1-p)}n^{3/2}/20.$$

Proof. We may assume $p \leq 1/2$ since we may take complements and replace p by $1 - p$. Suppose $x \in V(G)$ has degree $d = d(x) = p(n-1) + r(x)$. For $v \neq x$, define $e_v = 1$ if $xv \in E(G)$ and $e_v = 0$ otherwise. Then

$$\mathbb{E}\left||\Gamma(x) \cap Y| - p|Y \setminus \{x\}|\right| = \mathbb{E}\left|\sum_{v \neq x} \rho_v(e_v - p)\right|$$

$$= \mathbb{E}\left|\frac{1}{2}\sum_{v \neq x}(e_v - p) + \frac{1}{2}\sum_{v \neq x}\varepsilon_v(e_v - p)\right|$$

$$\geq \max\left\{\frac{1}{2}\left|d - (n-1)p\right|, \frac{1}{2}\left|\mathbb{E}\sum_{v \neq x}\varepsilon_v(e_v - p)\right|\right\},$$

since $\sum_{v \neq x}(e_v - p) = d - (n-1)p$ and the distribution of $\sum_{v \neq x}\varepsilon_v(e_v - p)$ is symmetric about 0. Now, by Lemma 5,

$$\mathbb{E}\left|\sum_{v \neq x}\varepsilon_v(e_v - p)\right| = \mathbb{E}\left|\sum_{i=1}^d \varepsilon_i(1-p) + \sum_{i=d+1}^{n-1}\varepsilon_i(-p)\right|$$

$$\geq \mathbb{E}\left|\sum_{i=1}^d \varepsilon_i(1-p)\right|$$

$$\geq (1-p)\sqrt{d/2}$$

and so

$$\mathbb{E}\left|\left|\Gamma(x)\cap Y\right|-p\left|Y\setminus\{x\}\right|\right|\geq\frac{1}{2}\max\left\{\left|r(x)\right|,(1-p)\sqrt{d(x)/2}\,\right\}.$$

Now for $x\in V=V(G)$ let $I(x)=1$ if $x\in X$ and $I(x)=0$ otherwise. Then, since $I(x)$ and $\left|\Gamma(x)\cap Y\right|$ are independent random variables,

$$\mathbb{E}\sum_{x\in X}\left|\left|\Gamma(x)\cap Y\right|-p\left|Y\right|\right|=\mathbb{E}\sum_{x\in V}I(x)\left|\left|\Gamma(x)\cap Y\right|-p\left|Y\setminus\{x\}\right|\right|$$

$$=\frac{1}{2}\sum_{x\in V}\mathbb{E}\left|\left|\Gamma(x)\cap Y\right|-p\left|Y\setminus\{x\}\right|\right|$$

$$\geq\frac{1}{4}\max\left\{\sum_{x\in V}\left|r(x)\right|,\sum_{x\in V}(1-p)\sqrt{d(x)/2}\right\}$$

$$\geq\frac{1}{8}\sum_{x\in V}\left(\left|r(x)\right|+(1-p)\sqrt{d(x)/2}\,\right).$$

Note that the first equality holds as $Y=Y\setminus\{x\}$ if $I(x)=1$. Furthermore, $\left|r(x)\right|+(1-p)\sqrt{d(x)/2}$ is minimized when $r(x)=0$ and so $d(x)=p(n-1)$. Thus

$$\frac{1}{8}\sum_{x\in V}\left(\left|r(x)\right|+(1-p)\sqrt{d(x)/2}\,\right)\geq\frac{1}{8}n(1-p)\sqrt{p(n-1)/2}$$

$$\geq\sqrt{p(1-p)}n^{3/2}/20,$$

since $p\leq1/2$ and we may assume $n\geq3$. $\blacksquare$

After this preparation, we are ready to prove Theorem 1.

Proof of Theorem 1. Since (3) is symmetric in p and $1-p$, we may replace G by its complement $\overline{G}$, and so we may assume that $\mathrm{disc}_p^+(G)\leq\mathrm{disc}_p^-(G)$. If $\mathrm{disc}_p^+(G)\geq\sqrt{p(1-p)}n^{3/2}/80$ we are done. Otherwise, suppose $\mathrm{disc}_p^+(G)=\sqrt{p(1-p)}n^{3/2}/80\alpha$, where $\alpha\geq1$. We shall show that

$$(9)\qquad\qquad\mathrm{disc}_p^-(G)\geq\alpha\sqrt{p(1-p)}n^{3/2}/80,$$

so $\mathrm{disc}_p^+(G)\,\mathrm{disc}_p^-(G)\geq p(1-p)n^3/6400.$

Let $V(G) = X \cup Y$ be a random bipartition. Then since $p(1-p) \geq 1/n$, it follows from Lemma 7 that

$$(10) \qquad \mathbb{E} \sum_{x \in X} \left| \left| \Gamma(x) \cap Y \right| - p|Y| \right| \geq \sqrt{p(1-p)} n^{3/2}/20.$$

Now let $X^+ = \left\{ x \in X : \left| \Gamma(x) \cap Y \right| \geq p|Y| \right\}$ and $X^- = X \setminus X^+$; so

$$d_p(X, Y) = \sum_{x \in X^+} \left(\left| \Gamma(x) \cap Y \right| - p|Y| \right) + \sum_{x \in X^-} \left(\left| \Gamma(x) \cap Y \right| - p|Y| \right).$$

Since $\mathbb{E} d_p(X, Y) = 0$, we have

$$\mathbb{E} \sum_{x \in X^+} \left| \left| \Gamma(x) \cap Y \right| - p|Y| \right| = \mathbb{E} \sum_{x \in X^-} \left| \left| \Gamma(x) \cap Y \right| - p|Y| \right|$$

and so by (10)

$$(11) \quad \mathbb{E} d_p(X^+, Y) = \mathbb{E} \sum_{x \in X^+} \left(\left| \Gamma(x) \cap Y \right| - p|Y| \right) \geq \sqrt{p(1-p)} n^{3/2}/40.$$

Now $\mathbb{E} d_p(Y) = 0$, so (11) implies

$$(12) \qquad \mathbb{E} \left(d_p(X^+, Y) + \alpha d_p(Y) \right) \geq \sqrt{p(1-p)} n^{3/2}/40.$$

Let X^+, Y be a pair of sets achieving at least the expectation in (12) and let Z be a random subset of X^+, where each vertex of X^+ is chosen independently with probability $1/\alpha$. Then it follows from (12) that

$$\mathbb{E} d_p(Z \cup Y) = \mathbb{E} \left(d_p(Z) + d_p(Z, Y) + d_p(Y) \right)$$

$$= \frac{1}{\alpha^2} d_p(X) + \frac{1}{\alpha} d_p(X, Y) + d_p(Y)$$

$$\geq \frac{1}{\alpha^2} d_p(X) + \frac{1}{\alpha} \sqrt{p(1-p)} n^{3/2}/40.$$

Since $\mathrm{disc}_p^+(G) = \sqrt{p(1-p)} n^{3/2}/80\alpha$, this implies

$$d_p(X)/\alpha^2 \leq -(1/\alpha) \sqrt{p(1-p)} n^{3/2}/80$$

and so $d_p(X) \leq -\alpha \sqrt{p(1-p)} n^{3/2}/80$, which gives the desired lower bound on $\mathrm{disc}_p^-(G)$. $\blacksquare$

Finally in this section we give the proofs of Lemmas 5 and 6, postponed from earlier.

Proof of Lemma 5. A simple calculation shows that for $n = 2k$ we have $\mathbb{E}\left|\sum_{i=1}^{n} \varepsilon_i\right| = 2^{1-2k} k \binom{2k}{k}$ and for $n = 2k+1$ we have $\mathbb{E}\left|\sum_{i=1}^{n} \varepsilon_i\right| = 2^{-2k}(2k+1)\binom{2k}{k} = \mathbb{E}\left|\sum_{i=1}^{n+1} \varepsilon_i\right|$. Let $s_n = \mathbb{E}\left|\sum_{i=1}^{n} \varepsilon_i\right|/\sqrt{n}$. Then, for $k \geq 1$, $s_{2k+2}/s_{2k} = (k+\frac{1}{2})/\sqrt{k(k+1)} > 1$ and, for $k \geq 0$, $s_{2k+3}/s_{2k+1} = \sqrt{\left(k+\frac{3}{2}\right)\left(k+\frac{1}{2}\right)}/(k+1) < 1$. Thus $(s_{2k})_{k=1}^{\infty}$ is increasing and $(s_{2k+1})_{k=0}^{\infty}$ is decreasing; both converge to $\mathbb{E}\left|N(0,1)\right| = \sqrt{2/\pi}$. Therefore $s_n \geq s_2 = 1/\sqrt{2}$ for all n. ∎

Proof of Lemma 6. We may clearly assume that all a_i are nonnegative. Since $\sum_{i=1}^{n} \varepsilon_i a_i$ is symmetric about 0, the expectation is minimized for a given $\mathbf{a}$ when $A = 0$. Now if $a_i \neq a_j$ then let $a_i' = a_j' = (a_i + a_j)/2$; it is easily checked that $\mathbb{E}|B + \varepsilon_i a_i + \varepsilon_j a_j| \geq \mathbb{E}|B + \varepsilon_i a_i' + \varepsilon_j a_j'|$ for every real B. It follows that $\mathbb{E}\left|\sum_{i=1}^{n} \varepsilon_i a_i\right| \geq \mathbb{E}\left|\sum_{i=1}^{n} \varepsilon_i a\right|$, where $a = \sum_{i=1}^{n} a_i/n$. Thus, by Lemma 5,

$$\mathbb{E}\left|\sum_{i=1}^{n} \varepsilon_i a_i - A\right| \geq a\mathbb{E}\left|\sum_{i=1}^{n} \varepsilon_i\right| \geq a\sqrt{n/2} = \|\mathbf{a}\|_1/\sqrt{2n}.$$

The second inequality follows directly from the first. ∎

Note that in fact proof of Lemma 5 implies the inequalities $\mathbb{E}\left|\sum_{i=1}^{n} a_i\right| \geq \sqrt{2/\pi n}\|\mathbf{a}\|_1$ if n is odd and $\mathbb{E}\left|\sum_{i=1}^{n} a_i\right| \geq \left(1 + o(1)\right)\sqrt{2/\pi n}\|\mathbf{a}\|_1$ for general n.

3. HYPERGRAPH DISCREPANCY

In this section we turn our attention to hypergraphs. After defining a little notation, we begin with a result for weighted hypergraphs; we then turn to the consideration of unweighted hypergraphs.

If G is the complete k-uniform hypergraph with edge-weighting w and $X \subseteq V(G)$, we define

$$d(X) = \sum_{K \in X^{(k)}} w(K).$$

As in definitions (1) and (2) we define $\mathrm{disc}^+(G) = \max_{X \subset V(G)} d(X)$ and $\mathrm{disc}^-(G) = -\min_{X \subset V(G)} d(X)$; we also define

$$\mathrm{disc}\,(G) = \max\left\{\,\mathrm{disc}^+(G), \mathrm{disc}^-(G)\right\}.$$

Note that this is consistent with the definitions for an unweighted hypergraph G by taking $w(e) = 1$ if $e \in E(G)$ and $w(e) = -1$ otherwise.

For disjoint sets $X_1, \ldots, X_t$ and integers $k_1, \ldots, k_t$ such that $\sum_{i=1}^{t} k_i = k$, we define

$$d_{k_1,\ldots,k_t}(X_1, \ldots, X_t) = \sideset{}{'}\sum w(e),$$

where the sum is over edges e with $|e \cap X_i| = k_i$ for every i.

We can now state the first result of the section.

Theorem 8. *Let G be the complete k-uniform hypergraph of order n with edge-weighting w such that $\sum w(e) = 0$ and $\sum |w(e)| = \binom{n}{k}$. Then*

$$\mathrm{disc}^+(G)\,\mathrm{disc}^-(G) \geq 2^{-14k^2} n^{k+1}.$$

We shall need three lemmas. In the first lemma we use the fact that if $P(x)$ is a polynomial of degree k with $\sup_{x \in [0,1]} |P(x)| \leq 1$ then every coefficient of $P(x)$ has absolute value at most $2^k k^{2k}/k!$. (Tamás Erdélyi [4] pointed out to us that this is an elementary consequence of Markov's Inequality; see [3].)

Lemma 9. *If G is a complete k-uniform hypergraph with edge-weighting w and $\mathrm{disc}\,(G) \leq M$ then for disjoint subsets X, Y of $V(G)$ and $0 \leq i \leq k$,*

$$\left|d_{i,k-i}(X,Y)\right| \leq 2^{2k^2} M.$$

Proof. Let Z be a random subset of X, where each vertex is chosen independently with probability p. Then

$$\mathbb{E}\big(d(Z \cup Y)\big) = \sum_{i=0}^{k} p^i d_{i,k-i}(X,Y).$$

Since $\mathrm{disc}\,(G) \leq M$, it follows that $\max_{0 \leq p \leq 1} \left|\sum_{i=0}^{k} p^i d_{i,k-i}(X,Y)\right| \leq M$ and so $\max_{0 \leq i \leq k} \left|d_{i,k-i}(X,Y)\right| \leq 2^k k^{2k} M/k! \leq 2^{2k^2} M.$ ∎

We also need an analogue of Lemma 7.

Lemma 10. *Let G be a complete k-uniform hypergraph of order n with edge-weighting w. Let $V(G) = U \cup W$ be a random bipartition. Then*

$$\mathbb{E} \sum_{K \in U^{(k-1)}} \left| d_{k-1,1}(K, W) \right| \geq k 2^{-k} \sum_{L \in V(G)^{(k)}} \left| w(L) \right| / \sqrt{2n}.$$

Proof. Let $V = V(G) = U \cup W$ be a random bipartition. Given $K \in V^{(k-1)}$, it follows from Lemma 6 that

$$\mathbb{E} \left| d_{k-1,1}(K, W \setminus K) \right| \geq \sum_{v \in V \setminus K} \left| w(K \cup \{v\}) \right| / \sqrt{8n}.$$

Since the event $\{K \subset U\}$ and the random variable $d_{k-1,1}(K, W \setminus K)$ are independent, and each edge $L \in V^{(k)}$ occurs k times as $K \cup \{v\}$, we have

$$\mathbb{E} \sum_{K \in U^{(k-1)}} \left| d_{k-1,1}(K, W) \right| = \sum_{K \in V^{(k-1)}} \mathbb{P}(K \subset U) \mathbb{E} \left| d_{k-1,1}(K, W \setminus K) \right|$$

$$\geq \sum_{K \in V^{(k-1)}} 2^{-k+1} \sum_{v \in V \setminus K} \left| w(K \cup \{v\}) \right| / \sqrt{8n}$$

$$= k 2^{-k} \sum_{L \in V^{(k)}} \left| w(L) \right| / \sqrt{2n}. \qquad \blacksquare$$

The following lemma will be useful several times.

Lemma 11. *Let G be a k-uniform hypergraph of order n with edge-weighting w. Suppose that $\alpha \geq 1$ and X, Y are disjoint subsets of $V(G)$ with*

$$(13) \qquad\qquad d_{1,k-1}(X, Y) + \alpha d(Y) = M \geq 0.$$

Then either

$$\mathrm{disc}^+(G) \geq 2^{-3k^2} M / \alpha$$

or

$$\mathrm{disc}^-(G) \geq 2^{-3k^2} M \alpha.$$

Proof. If $\left| d_{i,k-i}(X,Y) \right| \geq 2^{-k^2} \alpha M$ for some $0 \leq i \leq k$ then we are done by Lemma 9. Otherwise, let Z be a random subset of X, obtained by choosing each vertex of X independently with probability $1/\alpha$. Then

$$\mathbb{E}\, d(Z \cup Y) = \mathbb{E} \sum_{i=0}^{k} d_{i,k-i}(Z,Y)$$

$$= \sum_{i=0}^{k} d_{i,k-i}(X,Y)/\alpha^i$$

$$\geq d(Y) + d_{1,k-1}(X,Y)/\alpha - \sum_{i=2}^{k} 2^{-k^2} \alpha M/\alpha^i$$

$$\geq M/\alpha - (k-1)2^{-k^2} M/\alpha$$

$$\geq 2^{-3k^2} M/\alpha.$$

Since some set Z must achieve this bound, we obtain the desired bound on $\mathrm{disc}^+(G)$. ∎

We can now prove the main theorem of this section.

Proof of Theorem 8. As in the proof of Theorem 1, we may assume that $\mathrm{disc}^+(G) \leq \mathrm{disc}^-(G)$. If $\mathrm{disc}^+(G) \geq 2^{-7k^2} n^{(k+1)/2}$ we are done. Otherwise, suppose $\mathrm{disc}^+(G) = 2^{-7k^2} n^{(k+1)/2}/\alpha$ for some $\alpha > 1$: we shall show $\mathrm{disc}^-(G) \geq 2^{-7k^2} \alpha n^{(k+1)/2}$.

Note first that for disjoint sets $X, Y \subset V(G)$, if

(14) $$d_{1,k-1}(X,Y) + \alpha d(Y) \geq 2^{-4k^2} n^{(k+1)/2}$$

then we are done by Lemma 11. It is therefore enough to find disjoint X, Y satisfying (14).

Let $V(G) = X_k \cup W_{k-1}$ be a random bipartition and let $W_{k-1} = X_{k-1} \cup W_{k-2}, \ldots, W_2 = X_2 \cup W_1$ be random bipartitions where, as usual, in each bipartition each vertex is assigned independently to either vertex class with probability $1/2$. We define weightings w_i on the i-sets in W_i for each i by

(15) $$w_i(K) = d_{i,1,\ldots,1}(K, X_{i+1}, \ldots, X_k).$$

Let $W_k = V(G)$ and define $w_k = w$. Then for $1 \le i < k$ and $K \in W_i^{(i)}$,

$$w_i(K) = d_{i,1,\ldots,1}(K, X_{i+1}, \ldots, X_k)$$

$$= \sum_{x \in X_{i+1}} d_{i+1,1,\ldots,1}\big(K \cup \{x\}, X_{i+2}, \ldots, X_k\big)$$

$$= \sum_{x \in X_{i+1}} w_{i+1}\big(K \cup \{x\}\big).$$

It therefore follows from Lemma 10 that given W_{i+1} and w_{i+1},

$$(16) \qquad \mathbb{E} \sum_{K \in W_i^{(i)}} \big| w_i(K) \big| \ge (i+1) 2^{-(i+1)} \sum_{L \in W_{i+1}^{(i+1)}} \big| w_{i+1}(L) \big| / \sqrt{2n}.$$

It follows that

$$\mathbb{E} \sum_{x \in W_1} \big| d_{1,\ldots,1}\big(\{x\}, X_2, \ldots, X_k\big) \big| = \mathbb{E} \sum_{x \in W_1} \big| w_1(x) \big|$$

$$\ge k! 2^{-\binom{k+1}{2}} \sum_{K \in W_k^{(k)}} \big| w_k(K) \big| / \big(\sqrt{2n}\,\big)^{k-1}$$

$$= k! 2^{-\binom{k+1}{2}} \binom{n}{k} / (2n)^{(k-1)/2}.$$

Let $X_1^+ = \big\{ x \in W_1 : d_{1,\ldots,1}(x, X_2, \ldots, X_k) > 0 \big\}$. Then, as in (11),

(17)
$$\mathbb{E}\, d_{1,\ldots,1}(X_1^+, X_2, \ldots, X_k) \ge \frac{1}{2} k! 2^{-\binom{k+1}{2}} \binom{n}{k} / (2n)^{(k-1)/2} \ge 2^{-2k^2} n^{(k+1)/2}.$$

We partition the edges in $V_0 = X_1^+ \cup \bigcup_{i=2}^k X_i$ that meet X_1^+ in exactly one vertex as follows. For a nonempty $S \subset \{2, \ldots, k\}$, let $V_S = \bigcup_{i \in S} X_i$ and $E_S = \big\{ K \cup \{x\} : x \in X_1^+, K \in V_S^{(k-1)}, |K \cap X_i| > 0 \ \forall i \in S \big\}$. Let $d_S = \sum_{K \in E_S} w(K)$ and note that $d_{1,k-1}(X^+, V_S) = \sum_{\emptyset \ne T \subset S} d_T$ and $d_{\{2,\ldots,k\}} =$

$d_{1,\ldots,1}(X_1^+, X_2, \ldots, X_k)$. Let S_0 be minimal with $|d_{S_0}| \geq (2k)^{-k+|S|} d_{\{2,\ldots,k\}}$. Then

$$\max_{S \subset \{2,\ldots,k\}} |d_{1,k-1}(X_1^+, S)| \geq |d_{1,k-1}(X_1^+, V_{S_0})|$$

$$\geq |d_{S_0}| - \sum_{\emptyset \neq T \subsetneq S_0} |d_T|$$

$$\geq \left((2k)^{-k+|S_0|} - \sum_{i=1}^{|S_0|-1} k^{|S_0|-i} (2k)^{-k+i} \right) d_{\{2,\ldots,k\}}$$

$$\geq d_{\{2,\ldots,k\}} / 2(2k)^{k-1}$$

$$\geq 2^{-k^2} d_{\{2,\ldots,k\}}.$$

Thus it follows from (17) that

$$\mathbb{E} \max_{S \subset \{2,\ldots,k\}} |d_{1,k-1}(X_1^+, V_S)| \geq 2^{-3k^2} n^{(k+1)/2}$$

and so there is some $S \subset \{2, \ldots, k\}$ with

$$\mathbb{E} |d_{1,k-1}(X_1^+, V_S)| \geq 2^{-3k^2} n^{(k+1)/2} / 2^k.$$

Now let $Y = V_S$ and $X_S^+ = \{ x \in W_1 : d_{1,k-1}(\{x\}, V_S) > 0 \}$. Then, since $\mathbb{E}\, d_{1,k-1}(W_1, V_S) = 0$, we have

$$\mathbb{E}\, d_{1,k-1}(X_S^+, V_S) \geq 2^{-3k^2} n^{(k+1)/2} / 2^{k+1} \geq 2^{-4k^2} n^{(k+1)/2}.$$

Finally, since $\mathbb{E}\, d(V_S) = 0$, we have

$$\mathbb{E}\, d_{1,k-1}(X_S^+, V_S) + \alpha d(V_S) \geq 2^{-4k^2} n^{(k+1)/2}.$$

It follows that there are sets X, Y satisfying (14). ∎

We note that Theorem 8 implies the following bound on

$$\mathrm{disc}_p^+(G)\, \mathrm{disc}_p^-(G)$$

for unweighted hypergraphs G.

Corollary 12. *Let G be a k-uniform hypergraph with n vertices and $p\binom{n}{k}$ edges. Then*

$$\operatorname{disc}_p^+(G)\operatorname{disc}_p^-(G) \geq 2^{-14k^2+2}p^2(1-p)^2 n^{k+1}.$$

Proof. The result is trivial if $p = 0$ or $p = 1$. Otherwise, let H be the complete k-uniform hypergraph on the same vertex set as G with edge-weighting w defined by $w(e) = 1/2p$ if $e \in E(G)$ and $w(e) = -1/2(1-p)$ otherwise. Then $\sum w(e) = 0$ and $\sum |w(e)| = \binom{n}{k}$, and so, by Theorem 8,

$$\operatorname{disc}^+(H)\operatorname{disc}^-(H) \geq 2^{-14k^2}n^{k+1}.$$

Now for $Y \subset V(G)$,

$$d_p^{(G)}(Y) = e(Y) - p\binom{|Y|}{2}$$

$$= \sum_{K \in Y^{(k)}} (\mathbf{1}_{K \in E(G)} - p)$$

$$= \sum_{K \in Y^{(k)}} 2p(1-p)w(K)$$

$$= 2p(1-p)d^{(H)}(Y).$$

Thus

(18) $$\operatorname{disc}_p^+(G)\operatorname{disc}_p^-(G) = 4p^2(1-p)^2 \operatorname{disc}^+(H)\operatorname{disc}^-(H),$$

which implies the required bound. ∎

We can, however, improve upon the $p^2(1-p)^2$ term in Corollary 12 (at the cost of a slightly worse constant) to obtain a bound similar to that in Theorem 1. First, however, we need a version of Lemma 7 for unweighted hypergraphs.

Lemma 13. *Let G be a k-uniform hypergraph of order n with $p\binom{n}{k}$ edges, where $p(1-p) \geq 1/n$ and $n \geq 2k$. Let $V(G) = X \cup Y$ be a random bipartition. Then*

$$\mathbb{E} \sum_{K \in X^{(k-1)}} \left| d_{k-1,1}(K,Y) - p|Y| \right| \geq 2^{-2k^2}\sqrt{p(1-p)}n^{k-\frac{1}{2}}.$$

Proof. We follow the argument of Lemma 7. As before, we may assume $p \leq 1/2$. Let $V = V(G) = X \cup Y$ be a random bipartition. For $K \in V^{(k-1)}$, let $d(K)$ be the number of edges of G containing K and define $r(K)$ by $d(K) = p(n - k + 1) + r(K)$. Let $d = p(n - k + 1)$. Then, as in Lemma 7,

$$\mathbb{E}\big|d_{k-1,1}(K, Y \setminus K) - p|Y \setminus K|\big| \geq \frac{1}{2}\max\left\{|r(K)|, (1 - p)\sqrt{d(K)/2}\right\}.$$

For $K \in V^{(k-1)}$, we define $I(K) = 1$ if $K \subset X$ and $I(K) = 0$ otherwise. Then $I(K)$ and $d_{k-1,1}(K, Y \setminus K)$ are independent random variables, so

$$\mathbb{E}\sum_{K \in X^{(k-1)}}\big|d_{k-1,1}(K, Y) - p|Y|\big|$$

$$= \mathbb{E}\sum_{K \in V^{(k-1)}} I(K)\big|d_{k-1,1}(K, Y \setminus K) - p|Y \setminus K|\big|$$

$$= 2^{-k+1}\sum_{K \in V^{(k-1)}} \mathbb{E}\big|d_{k-1,1}(K, Y \setminus K) - p|Y \setminus K|\big|$$

$$\geq 2^{-k}\max\left\{\sum_{K \in V^{(k-1)}}|r(K)|, \sum_{K \in V^{(k-1)}}(1 - p)\sqrt{d(K)/2}\right\}$$

$$\geq 2^{-(k+1)}\sum_{K \in V^{(k-1)}}|r(K)| + (1 - p)\sqrt{d(K)/2}.$$

Since $|r(K)| + (1 - p)\sqrt{d(x)/2}$ is minimized when $r(K) = 0$ and $d(K) = p(n - k + 1)$,

$$\mathbb{E}\sum_{K \in X^{(k-1)}}\big|d_{k-1,1}(K, Y) - p|Y|\big| \geq \binom{n}{k-1}2^{-(k+1)}(1 - p)\sqrt{p(n - k + 1)/2}$$

$$> 2^{-2k^2}\sqrt{p(1 - p)}n^{k-\frac{1}{2}}. \qquad \blacksquare$$

Theorem 14. *Let G be a k-uniform hypergraph of order n with $p\binom{n}{k}$ edges, where $p(1 - p) \geq 1/n$. Then*

$$\mathrm{disc}_p^+(G)\,\mathrm{disc}_p^-(G) \geq 2^{-18k^2}p(1 - p)n^{k+1}.$$

Proof. Let H be the complete k-uniform hypergraph on $V(G)$ with weighting $w(e) = 1 - p$ if $e \in E(G)$ and $w(e) = -p$ otherwise. Then $\mathrm{disc}^+(H) = \mathrm{disc}_p^+(G)$ and $\mathrm{disc}^-(H) = \mathrm{disc}_p^-(G)$. Note that $w(H) = 0$. As usual we may assume $p \leq 1/2$ and $\mathrm{disc}^-(H) \geq \mathrm{disc}^+(H) = 2^{-9k^2}\sqrt{p(1-p)}n^{(k+1)/2}/\alpha$. If $\alpha \leq 1$ we are done, so we may assume $\alpha \geq 1$. We will show that $\mathrm{disc}^-(H) \geq 2^{-9k^2}\sqrt{p(1-p)}\alpha n^{(k+1)/2}$. If there are disjoint $X, Y \subset V(H)$ with

(19)
$$d_{1,k-1}(X,Y) + \alpha d(Y) \geq 2^{-6k^2}\sqrt{p(1-p)}n^{(k+1)/2}$$

then we are done by Lemma 11. Thus it is enough to find disjoint X, Y satisfying (19).

As in the proof of Theorem 8, we define random sets $W_k = X \supset W_{k-1} \supset \cdots \supset W_1$, where the i-sets in W_i are weighted as in equation (15). Then by Lemma 13,

(20)
$$\mathbb{E} \sum_{K \in W_{k-1}^{(k-1)}} \left| w_{k-1}(K) \right| \geq 2^{-2k^2}\sqrt{p(1-p)}n^{k-\frac{1}{2}},$$

while $W_1, \ldots, W_{k-2}$ satisfy (16). We have

$$\mathbb{E} \sum_{x \in W_1} \left| w_1(x) \right| \geq (k-1)! \, 2^{-\binom{k}{2}} \sum_{K \in W_{k-1}^{(k-1)}} \left| w_{k-1}(K) \right| / \left(\sqrt{2n}\right)^{k-2},$$

and so, defining X_1^+ as before, we can replace (17) by

(21)
$$\mathbb{E}\, d_{1,\ldots,1}(X_1^+, X_2, \ldots, X_k) \geq 2^{-4k^2}\sqrt{p(1-p)}n^{(k+1)/2}.$$

The argument is completed as before (with all bounds changed by a factor $2^{-2k^2}\sqrt{p(1-p)}$). ∎

The following corollary is immediate.

Corollary 15. *Let G be a k-uniform hypergraph of order n with $p\binom{n}{k}$ edges, where $p(1-p) \geq 1/n$. Then*

$$\mathrm{disc}_p(G) \geq 2^{-9k^2}\sqrt{p(1-p)}n^{(k+1)/2}.$$

We note that Corollaries 2 and 15 are best possible up to the value of the constant 2^{-9k^2}. To see this, let $G \in \mathcal{G}^{(k)}(n,p)$ be a random k-uniform hypergraph, where each possible edge is present independently with probability p, and let $S \subset V(G)$. Let $N = \binom{n}{k}$ and

$$h = (1+\varepsilon)k!^{-1/2}\sqrt{2p(1-p)\ln 2}\, n^{(k+1)/2}.$$

Then by standard bounds on the tail of the binomial distribution (see [2], Theorem 1.3), provided $p(1-p) \geq c_k n^{1-k}$, for any subset S of $V(G)$ we have

$$\mathbb{P}\big(\big|d_p(S)\big| \geq h\big) \leq \mathbb{P}\big(\big|B(N,p) - Np\big| \geq h\big) < 2^{-n}$$

for sufficiently large n. Thus there is some k-uniform hypergraph G of order n with $\operatorname{disc}_p(G) \leq h$.

Let us also note that the gain from $p^2(1-p)^2$ to $p(1-p)$ between Corollary 12 and Theorem 14 comes because a "typical " vertex in G has degree $p\binom{n-1}{k-1}$: so if p is small, then the weight around a typical vertex is concentrated in fairly few edges. We remark that no similar bound is possible for the larger class of k-uniform hypergraphs with $\sum|w(e)| = \binom{n}{k}$ such that $\sum \max\{w(e), 0\} = p\binom{n}{k}$: consider a random k-uniform hypergraph $H \in \mathcal{G}^{(k)}(n, 1/2)$, and let G be the weighted hypergraph obtained by giving each edge weight $2p$ and each non-edge weight $-2(1-p)$. Then if $e(H) = \frac{1}{2}\binom{n}{k}$ (which happens with probability at least $c'_k n^{-k/2}$ if $\binom{n}{k}$ is even) we have $\sum|w(e)| = \binom{n}{k}$ and $\sum \max\{w(e), 0\} = p\binom{n}{k}$. On the other hand, it follows from (18) that $\operatorname{disc}_p^+(G)\operatorname{disc}_p^-(G) = 4p^2(1-p)^2 \operatorname{disc}^+(H)\operatorname{disc}^-(H)$, while $\operatorname{disc}^+(H)$ and $\operatorname{disc}^-(H)$ are both $O(n^{(k+1)/2})$ with exponentially small failure probability.

It is interesting to ask about the range in which Theorem 1 and Theorem 14 are sharp (up to the constant). For instance, in the case of graphs the remarks above show that $\operatorname{disc}_p^+(G)$ and $\operatorname{disc}_p^-(G)$ can both be around $c\sqrt{p(1-p)}n^{3/2}$. When p is (about) $1/2$, the complete bipartite graph and its complement show that we can have discrepancy $O(n)$ on one side (and cn^2 on the other). Thus Theorem 1 is sharp in in middle of the the scale from cn to $c'n^2$, and (for $p = 1/2$) is sharp at the ends. How sharp is it at other parts of the scale, or at the ends when $p \neq 1/2$?

The constant in Theorem 14 is clearly not best possible. A more careful version of the argument should improve it to $2^{-ck\ln k}$; it would be of interest to know the correct order of magnitude. It would also be interesting to know what happens in the range $n^{1-k} \leq p \leq 1/n$.

4. SUBGRAPH DISCREPANCY

In previous sections we have been concerned with the discrepancy of subgraphs or, equivalently, 2-colourings of the complete graph. We begin this section by considering 2-colourings of an arbitrary graph: questions of this form were raised by Sós in [11].

For a k-uniform hypergraph G, a subgraph H of G and a real number $p \in [0, 1]$, we define

$$\mathrm{disc}_p^+(H, G) = \max_{S \subset V(G)} e\big(H[S]\big) - pe\big(G[S]\big)$$

and

$$\mathrm{disc}_p^-(H, G) = \max_{S \subset V(G)} pe\big(G[S]\big) - e\big(H[S]\big).$$

Note that if G is the complete k-uniform hypergraph then these two definitions agree with (1) and (2). We set

$$\mathrm{disc}_p(H, G) = \max\big\{ \mathrm{disc}_p^+(H, G), \mathrm{disc}_p^-(H, G)\big\}.$$

We begin with a fairly straightforward analogue to Theorem 8. Note that arguing as in Corollary 12 gives a bound with $p^2(1 - p)^2$ in place of $p(1 - p)$.

Theorem 16. *Let G be a k-uniform hypergraph with n vertices and m edges, and let H be a subgraph of G with pm edges, where $p(1 - p) \geq 1/n$. Then*

$$\mathrm{disc}_p^+(H, G)\, \mathrm{disc}_p^-(H, G) \geq 2^{-18k^2} p(1 - p)m^2/n^{k-1}.$$

We first need a version of Lemma 13.

Lemma 17. *Let G be a k-uniform hypergraph with n vertices and m edges, and let $H \subset G$ be a subhypergraph of G with pm edges, where $p(1 - p) \geq 1/n$. Let $V(G) = X \cup Y$ be a random bipartition. Then*

$$\mathbb{E} \sum_{K \in X^{(k-1)}} \big|d_{k-1,1}^{(H)}(K, Y) - pd_{k-1,1}^{(G)}(K, Y)\big| \geq 2^{-(k+1)}\sqrt{p(1 - p)}m/\sqrt{n}.$$

Proof. For a partition $V(G) = X \cup Y$, let us write

$$f(X, Y) = \sum_{K \in X^{(k-1)}} \big|d_{k-1,1}^{(H)}(K, Y) - pd_{k-1,1}^{(G)}(K, Y)\big|.$$

As in Lemma 13, we may assume that $p \leq 1/2$ or else replace H by its complement in G. For $K \in V^{(k-1)}$, let $d_H(K)$ be the number of edges of

H containing K and let $d_G(K)$ be the number of edges of G containing K. Define $r(K)$ by $d_H(K) = pd_G(K) + r(K)$. Then, as in Lemma 13,

$$\mathbb{E}\left|d^{(H)}_{k-1,1}(K, Y \setminus K) - pd^{(G)}_{k-1,1}(K, Y)\right| \geq \frac{1}{2} \max\left\{\left|r(K)\right|, (1-p)\sqrt{d_H(K)/2}\,\right\}.$$

Thus

$$\mathbb{E}f(X, Y) \geq 2^{-(k+1)} \sum_{K \in V^{(k-1)}} \left|r(K)\right| + (1-p)\sqrt{d_H(K)/2}.$$

Now $\left|r(K)\right| + (1-p)\sqrt{d_H(K)/2}$ is minimized when $r(K) = 0$ and so $d_H(K) = pd_G(K)$. Thus

$$\mathbb{E}f(X, Y) \geq 2^{-(k+1)} \sum_{K \in V^{(k-1)}} (1-p)\sqrt{pd_G(K)/2}$$

$$\geq 2^{-(k+1)} \sum_{K \in V^{(k-1)}} (1-p)d_G(K)\sqrt{p/2n}$$

since $d_G(K) < n$. Now $\sum_{K \in V^{(k-1)}} d_G(K) = km$, so

$$\mathbb{E}f(X, Y) \geq 2^{-(k+1)}km(1-p)\sqrt{p/2n}$$

$$\geq 2^{-(k+1)}\sqrt{p(1-p)}m/\sqrt{n}. \qquad \blacksquare$$

Theorem 16 now follows by a modification of the proof of Theorem 14.

Proof of Theorem 16. Let $V = V(G)$. We may assume $p \leq 1/2$ or replace H by its complement in G. We define, as in Theorem 14, an edge-weighting w on $V^{(k)}$ by $w(K) = 1-p$ if $K \in E(H)$, $w(K) = -p$ if $K \in E(G) \setminus E(H)$ and $w(K) = 0$ otherwise. Note that then $w(V) = 0$. We may assume $\mathrm{disc}_p^-(H, G) \geq \mathrm{disc}_p^+(H, G) = 2^{-9k^2}\sqrt{p(1-p)}e(G)/\sqrt{n}^{k-1}\alpha$. If $\alpha \leq 1$ we are done, so we may assume $\alpha \geq 1$. If there are disjoint X, Y with

$$d_{1,k-1}(X, Y) + \alpha d(Y) \geq 2^{-6k^2}\sqrt{p(1-p)}m/n^{(k-1)/2},$$

then we are done as before by Lemma 11. Once again, we define random subsets $W_k = X \supset W_{k-1} \supset \cdots \supset W_1$. Applying Lemma 17 instead of Lemma 13 to W_{k-1}, we can replace (20) by

$$(22) \qquad \mathbb{E} \sum_{K \in W^{(k-1)}_{k-1}} \left|w_{k-1}(K)\right| \geq 2^{-2k^2}\sqrt{p(1-p)}m/\sqrt{n}.$$

As before, $W_1, \ldots, W_{k-2}$ satisfy (16); applying this $k - 2$ times to (22), we see that (instead of (21)) we obtain

$$\mathbb{E}d(X_1^+, X_2, \ldots, X_n) \geq 2^{-4k^2} \sqrt{p(1-p)}m/n^{(k-1)/2},$$

and the argument is completed as before. ■

Corollary 18. *Let G be a k-uniform hypergraph with n vertices and m edges, and H a subgraph of G with pm edges, where $p(1 - p) \geq 1/n$. Then*

$$\mathrm{disc}_p(H, G) \geq 2^{-9k^2} \sqrt{p(1-p)}m/n^{(k-1)/2}.$$

We obtain stronger results when there is a restriction on the maximum overlap between edges of positive and negative weights.

Theorem 19. *Let G be a complete k-uniform hypergraph of order n with edge-weighting w. Suppose in addition that, for some $1 \leq s \leq r$, if $w(e) > 0$ and $w(e') < 0$ then $|e \cap e'| < s$. Let $M = \sum |w(e)|$ and $m = \sum w(e)$. If $m = (2p - 1)M$, where $p(1 - p) \geq 1/n$, then*

$$\mathrm{disc}_p^+(G)\, \mathrm{disc}_p^-(G) \geq 2^{-18k^2} p^2 (1 - p)^2 M^2/n^{s-1}.$$

Proof. Suppose first that $p = 1/2$, and let $E = \{e : w(e) \neq 0\}$. As in the proof of Theorem 8, we may assume $\mathrm{disc}^+(G) \leq \mathrm{disc}^-(G)$. Suppose $\mathrm{disc}^+(G) = 2^{-9k^2} e(H)/n^{(s-1)/2}\alpha$, where $\alpha \geq 1$. If there are disjoint X, Y with

$$d_{1,k-1}(X, Y) + \alpha d(Y) \geq 2^{-6k^2} e(H)/n^{(s-1)/2}$$

then we are done by Lemma 11. Otherwise, define W_i, X_i and w_i as before, and consider W_s and w_s. Since $w(e) > 0$ and $w(e') < 0$ implies $|e \cap e'| < s$, we have, for $K \in W_s^{(s)}$,

$$\left| w_s(K) \right| = \sum_{e \cap W_s = K,\ |e \cap X_i| = 1\, \forall i > s} \left| w(e) \right|$$

and so

$$\mathbb{E} \sum_{K \in W_s^{(s)}} \left| w_s(K) \right| = \mathbb{E} \sum_{|e \cap W_s| = s,\ |e \cap X_i| = 1\, \forall i > s} \left| w(e) \right|.$$

Let A_e be the event that $|e \cap W_s| = s$ and $|e \cap W_i| = 1$ for all $i > s$. Then $\mathbb{P} A_e > 2^{-k^2}$ and so $\mathbb{E} \sum_{K \in W_s^{(s)}} |w_s(K)| \geq 2^{-k^2} M$. Applying Lemma 10 as in (16), we obtain that

$$\mathbb{E} \sum_{x \in W_1} \left| d_{1,\ldots,1}\left(\{x\}, X_2, \ldots, X_k\right)\right| \geq \mathbb{E} \sum_{K \in W_s^{(s)}} |w_S(K)| / \left(\sqrt{2n}\right)^{s-1}$$

$$\geq 2^{-2k^2} M / n^{(s-1)/2}.$$

The rest of the argument follows as in the proof of Theorem 8.

Now suppose $p \neq 1/2$. As in the proof of Corollary 12, we multiply all positive edge-weights by $1/2p$ and all negative edge-weights by $-1/2(1-p)$ to obtain a new edge-weighting w'. The result follows immediately. ∎

As an application of Theorem 19, let us consider the complete subgraphs of a graph and its complement. For $t \geq 2$ and a graph G, we write $k_t(G)$ for the number of copies of K_t of G. We write

$$\operatorname{disc}_{K_k}(G) = \max_{S \subset V(G)} \left| k_k\left(G[S]\right) - k_k\left(\overline{G}[S]\right)\right|.$$

For instance, $\operatorname{disc}_{K_2}(G)$ is just $\operatorname{disc}(G)$. Clearly, complete subgraphs of G meet complete subgraphs of its complement in at most one vertex: applying Theorem 19 to the k-uniform hypergraph of complete or independent k-sets gives the following result.

Corollary 20. *For every graph G of order n,*

$$\operatorname{disc}_{K_k}(G) \geq c_k n^{k-\frac{1}{2}}.$$

For instance, in some subset S,

$$\left| k_3\left(G[S]\right) - k_3\left(\overline{G}[S]\right)\right| \geq cn^{5/2}.$$

Considering random graphs shows that this result is best possible up to the constant. A similar approach yields results in some cases for $\operatorname{disc}_H(G)$ where H is not a complete graph (and disc_H is defined in the obvious way). It would be interesting to determine the correct order of magnitude of disc_H for all graphs H. When H is fairly dense, so that copies of H and $\overline{H}$ cannot overlap very much, we obtain a lower bound on $\operatorname{disc}_H(G)$ using Theorem 19. However, when H is sparse this gives a much weaker bound; for instance, what can we say when H is a tree?

REFERENCES

[1] J. Beck and V. T. Sós, Discrepancy theory, in: *Handbook of Combinatorics,* Vol. 2, 1405–1446, Elsevier, Amsterdam, 1995.

[2] B. Bollobás, *Random Graphs,* Second Edition, Cambridge Studies in Advanced Mathematics, Cambridge University Press, 2001, xviii+498pp.

[3] P. Borwein and T. Erdélyi, *Polynomials and Polynomial Inequalities,* Graduate Texts in Mathematics, 161, Springer-Verlag, New York, 1995, x+480 pp.

[4] Tamás Erdélyi, Personal communication.

[5] P. Erdős, R. Faudree, C. Rousseau and R. Schelp, A local density condition for triangles, *Discrete Math.,* **127** (1994), 153–161.

[6] P. Erdős, M. Goldberg, J. Pach and J. Spencer, Cutting a graph into two dissimilar halves, *J. Graph Theory,* **12** (1988), 121–131.

[7] P. Erdős and J. Spencer, Imbalances in k-Colorations, *Networks,* **1** (1971/2), 379–385.

[8] P. Keevash and B. Sudakov, Local density in graphs with forbidden subgraphs, *to appear.*

[9] M. Krivelevich, On the edge distribution in triangle-free graphs, *J. Combinatorial Theory, Ser. B,* **63** (1995), 245–260.

[10] J. E. Littlewood, On bounded bilinear forms in an infinite number of variables, *Quart. J. Math. Oxford,* **1** (1930), 164–174.

[11] V. T. Sós, Irregularities of partitions: Ramsey theory, uniform distribution, in: *Surveys in Combinatorics* (Southampton, 1983), 201–246, London Math. Soc. Lecture Note Ser., 82, Cambridge Univ. Press, Cambridge–New York, 1983.

[12] S. J. Szarek, On the best constants in the Khinchin inequality, *Studia Math.,* **58** (1976), 197–208.

[13] B. Tomaszewski, A simple and elementary proof of the Kchintchine inequality with the best constant, *Bull. Sci. Math.,* **111** (1987), 103–109.

B. Bollobás

Trinity College
Cambridge CB2 1TQ
and
Department of Mathematical Sciences
University of Memphis
Memphis TN38152

bollobas@msci.memphis.edu

A. D. Scott

Department of Mathematics
University College London
Gower Street
London WC1E 6BT

scott@math.ucl.ac.uk

BOLYAI SOCIETY
MATHEMATICAL STUDIES, 15

Conference on Finite
and Infinite Sets
Budapest, pp. 57–77.

BIPLANAR CROSSING NUMBERS I:
A SURVEY OF RESULTS AND PROBLEMS

É. CZABARKA, O. SÝKORA*, L. A. SZÉKELY[†] and I. VRŤO[‡]

This paper is dedicated to the 70th birthdays of András Hajnal and Vera T. Sós

We survey known results and propose open problems on the biplanar crossing number. We study biplanar crossing numbers of specific families of graphs, in particular, of complete bipartite graphs. We find a few particular exact values and give general lower and upper bounds for the biplanar crossing number. We find the exact biplanar crossing number of $K_{5,q}$ for every q.

1. INTRODUCTION

During WWII in a forced work camp, Paul Turán [27] introduced the crossing number problem, in particular the Brick Factory Problem, which asks for the crossing number of complete bipartite graphs. The present paper surveys the few known results and proposes open problems on a variant of the crossing number, the *biplanar crossing number*, and solves the biplanar version of the Brick Factory Problem for $K_{5,q}$ exactly.

Recall that a graph G is *biplanar* [5], if one can write $G = G_1 \cup G_2$, where G_1 and G_2 are planar graphs. Let $\mathrm{cr}\,(G)$ denote the standard crossing number of the graph G, i.e. the minimum number of crossings of its edges over all possible drawings of G in the plane, under the usual rules for

_____

*This research was supported in part by the EPSRC grant GR/R37395/01.
[†]This research was supported in part by the NSF contract Nr. 007 2187 and 0302307.
[‡]This research was supported in part by the VEGA grant Nr. 2/3164/23.

drawings for crossing numbers [20, 26]. Motivated by printed circuit boards, Owens [15] introduced the *biplanar crossing number* of a graph G, that we denote by $cr_2(G)$. By definition $cr_2(G) = \min \left\{ cr(G_1) + cr(G_2) \right\}$, where the minimum is taken over all unions $G = G_1 \cup G_2$. A *biplanar drawing* of a graph G means drawings of two subgraphs, G_1 and G_2, of G, on two disjoint planes under the usual rules for drawings for crossing numbers, such that $G_1 \cup G_2 = G$. Owens described a biplanar drawing of the complete graph K_n with $cr_2(K_n) \leq 7n^4/1536 + O(n^3)$. One can define $cr_k(G)$ similarly for any $k \geq 2$, making G a union of k subgraphs. Determining $cr_k(G)$ would have application to the design of multilayer VLSI circuits [1]; but perhaps the case $k = 2$ is the most interesting, and even this simplest case is little explored so far. Note that one always can realize $cr_2(G)$ by drawing the edges of G_1 and G_2 on two different sides of the same plane, while identical vertices of G_1 and G_2 are placed to identical locations on the plane on the two sides.

The biplanar crossing number problem is related to the *thickness* and *book crossing number problems*. The *thickness* $\Theta(G)$ of G is the minimum number of planar graphs whose union is G. By definition, $cr_2(G) = 0$ if and only if $\Theta(G) \leq 2$, i.e. G is *biplanar*. The nature of the crossing number and the biplanar crossing number problems seems different, since testing whether $cr(G) = 0$ can be done in linear time, while testing biplanarity is an NP-complete problem [12]. Asano's result [3] implies that if a graph is toroidal, then $cr_2(G) = 0$. Surveys on biplanar graphs and the thickness problem can be found in [5, 13].

A k-book embedding of a graph G consists of placing vertices of G on the spine of a book and drawing each edge on one of the k pages. The *book crossing number* of G, denoted by $\nu_k(G)$, is the minimum total number of crossings on all pages among all k-page book embedding of G [21]. One can easily observe that $cr_2(G) \leq \nu_4(G)$.

We denote by n the order and by m the size of a graph, and we deviate from this rule only for complete bipartite graphs.

We are indebted to an anonymous referee for their comments.

2. GENERAL RESULTS

2.1. Variants of Euler's formula

Little is known about the biplanar crossing number in general. Some of the lower bounds for crossing numbers, *mutatis mutandis* apply to biplanar crossing numbers. For example, the lower bound resulting from Euler's formula, $\mathrm{cr}\,(G) \geq m - 3n + 6$ for $n \geq 3$, provides

$$(1) \qquad\qquad \mathrm{cr}_2(G) \geq m - 6n + 12.$$

There is a strengthening of the lower bound resulting from Euler's formula for graphs G with girth $\geq g$, $\mathrm{cr}\,(G) \geq m - g(n-2)/(g-2)$ for $n \geq g$; and we get

$$(2) \qquad\qquad \mathrm{cr}_2(G) \geq m - 2\frac{g}{g-2}(n-2)$$

for $n \geq g$ (it follows from combining Theorem 2.1 in [5] with the arguments in [20]). Pach and Tóth showed ([18] and personal communication from G. Tóth) that with $n \geq 3$

$$(3) \qquad\qquad \mathrm{cr}\,(G) \geq 6m - 33n + 66,$$

and for triangle-free graphs with $n \geq 4$

$$(4) \qquad\qquad \mathrm{cr}\,(G) \geq 6m - 27n + 54.$$

These results immediately imply their counterparts for the biplanar crossing number:

$$(5) \qquad\qquad \mathrm{cr}_2(G) \geq 6m - 66n + 132,$$

for $n \geq 3$; and for triangle-free graphs with $n \geq 4$

$$(6) \qquad\qquad \mathrm{cr}_2(G) \geq 6m - 54n + 108.$$

2.2. Other lower bounds

Using our (1) instead of formula (1) from [20] in the second proof of Theorem 3.2 in [20], one obtains the following biplanar counterpart of the Leighton [10] and Ajtai et al. [2] bound: for all $c > 6$, if $m \geq cn$, then

$$(7) \qquad\qquad \mathrm{cr}_2(G) \geq \frac{c-6}{c^3} \cdot \frac{m^3}{n^2}.$$

For somewhat denser graphs one can improve (7) using the Pach–Tóth's results cited above.

Pach, Spencer and Tóth [17] proved a conjecture of Simonovits, improving the bound of (7). If G has girth $> 2r$ and $m \geq 4n$, then

$$(8) \qquad\qquad \mathrm{cr}\,(G) = \Omega \left(\frac{m^{r+2}}{n^{r+1}} \right).$$

It is easy to see that (8) also hold for cr_2 instead of cr, if $m \geq 8n$.

Lower bounds for the crossing number based on the *counting method* [20] provide similar arguments setting lower bounds for the biplanar crossing number. Since we are going to use it, we review the counting method. Assume that we have a sample graph H. Take a graph G together with a biplanar drawing which realizes its biplanar crossing number. Without loss of generality we may assume that no adjacent edges cross and any two edges cross at most once in the drawing [26]. If we find A copies of H in G, and no crossing of the drawing belongs to more than B copies of H, then

$$\mathrm{cr}_2(G) \geq \mathrm{cr}_2(H)\frac{A}{B}.$$

However, important techniques as the *embedding method* [10] or the *bisection width method* [16], [24] (see also the survey [20]) do not seem to generalize to biplanar crossing numbers. Even worse, as Tutte noted [5], the biplanar crossing number is not an invariant for homeomorphic graphs; in fact, the edges of every graph can be subdivided such that the subdivided graph is biplanar! Furthermore, Beineke [5] shows that the minimum number of subdivisions needed to make a graph biplanar equals the minimum number of edges whose deletion leaves a biplanar graph.

Open Problem 1. *Find lower bound arguments for the biplanar crossing number based on structural properties of graphs, not merely on the density of graphs.*

J. Spencer [25] was the first to find such a lower bound. Say that a graph of order n and size m has property $(*)$, if for every vertex set A with $n/6 \leq |A| \leq 5n/6$, the number of edges between A and $\bar{A}$ is at least $m/10000$. Spencer showed that if $m > cn$ for a certain c, $\sum d_i^2 = o(m^2)$, and the graph has the $(*)$ property, then $\mathrm{cr}_2(G) = \Omega(m^2)$. Since random graphs have the $(*)$ property, the biplanar crossing number of the random graph is $\Omega(p^2 \binom{n}{2}^2)$ for $p \geq c'/n$. Bounded degree expander graphs also have property $(*)$.

2.3. Drawings, upper bounds

We showed [23] using a randomized algorithm, that for all graphs G,

(9) $$\mathrm{cr}_2(G) \leq \frac{3}{8} \, \mathrm{cr}\,(G).$$

However, one cannot give an upper bound for $\mathrm{cr}\,(G)$ in terms of $\mathrm{cr}_2(G)$, since there are graphs G of order n and size m, with crossing number $\mathrm{cr}\,(G) = \Theta(m^2)$ (i.e. as large as possible) and biplanar crossing number $\mathrm{cr}_2(G) = \Theta(m^3/n^2)$ (i.e. as small as possible), for any $m = m(n)$, where m/n exceeds a certain absolute constant. As [23] shows, such graphs G can be obtained from a certain graph H with $\mathrm{cr}\,(H) = \Theta(m^3/n^2)$, such that vertices of H are identified with identically named vertices of H^π, where H^π is obtained from H by permuting the vertices randomly.

Open Problem 2. *What is the smallest number c^* (in place of $3/8$), with which (9) is true?*

Owens [15] came up with a conjectured cr_2-optimal drawing of K_n which has about $7/24$ of the crossings of a conjectured cr-optimal drawing of K_n. This might give some basis to conjecture that $c^* \leq 7/24$. On the other hand, we will show in (19) that $\mathrm{cr}_2(K_n) \geq n^4/952$ for large n, and comparison with $\mathrm{cr}(K_n) \leq n^4/64$ [29] proves $c^* \geq 64/952$. We used (9) to prove that for any graph G, $\Theta(G) - 2 = O\big(\,\mathrm{cr}\,(G)^{.4057}\big)$ [23]. It is likely that $.4057$ can be replaced by smaller constants, perhaps with $.25$. The example of a complete graph shows that the constant cannot be smaller than $.25$.

We see a curious phenomenon. Call a biplanar drawing realizing the biplanar crossing number of a graph G *self-complementary*, if the subgraphs G_1 and G_2 are isomorphic in the graph theoretic sense. K_8 is biplanar, and a self-complementary drawing shows it [5], and the same can be told about

$K_{5,12}$. Self-complementary biplanar drawings are very convenient to draw. As G_1 and G_2 are isomorphic we only need to label the vertices by symbols like $(a : b)$, which means that the vertex in question is vertex a in the drawing on the first plane, and is vertex b in the drawing on the second plane. (See Figs. 1, 2, 3, 4.) Our drawing in Theorem 6 for the hypercube Q_k with even k—although clearly not optimal, but probably near-optimal— is also self-complementary.

Open Problem 3. *Show that if K_n or $K_{p,q}$ has an even number of edges, then it has an optimal biplanar drawing, which is self-complementary.*

Concerning upper bounds for $\mathrm{cr}_2(G)$, in terms of m, we proved in a joint paper with Shahrokhi [21] a general upper bound for the k-page book crossing numbers of graphs:

$$(10) \qquad \mu_k(G) \le \frac{1}{3k^2}\left(1 - \frac{1}{2k}\right)m^2 + O\left(\frac{m^2}{kn}\right),$$

which together with $\mathrm{cr}_2(G) \le \mu_4(G)$ gives a general upper bound on $\mathrm{cr}_2(G)$

$$(11) \qquad \mathrm{cr}_2(G) \le \frac{7}{384}m^2 + O\left(\frac{m^2}{n}\right).$$

3. Results and problems on complete bipartite graphs

The famous Zarankiewicz's Crossing Number Conjecture or Turán's Brick Factory Problem is as follows:

$$(12) \qquad \mathrm{cr}\,(K_{p,q}) = \left\lfloor\frac{p}{2}\right\rfloor\left\lfloor\frac{p-1}{2}\right\rfloor\left\lfloor\frac{q}{2}\right\rfloor\left\lfloor\frac{q-1}{2}\right\rfloor.$$

Kleitman showed that (12) holds for $q \le 6$ [9] and also proved that the smallest counterexample to the Zarankiewicz's conjecture must occur for odd p and q. Woodall used elaborate computer search to show that (12) holds for $K_{7,7}$ and $K_{7,9}$. Thus, the smallest unsettled instances of Zarankiewicz's conjecture are $K_{7,11}$ and $K_{9,9}$. The following remarkable construction suggests Zarankiewicz's conjecture: place $\lfloor p/2 \rfloor$ vertices to negative positions on the

x-axis, $\lceil p/2 \rceil$ vertices to positive positions on the x-axis, $\lfloor q/2 \rfloor$ vertices to negative positions on the y-axis, $\lceil q/2 \rceil$ vertices to positive positions on the y-axis, and draw pq edges by straight line segments to obtain a drawing of $K_{p,q}$.

In this section we work towards a biplanar analogue of the Zarankiewicz's Conjecture and make conjectures for the cases $q = 6$ and 8.

3.1. Lower bounds for complete bipartite Graphs

The girth formula (2) yields

$$(13) \qquad \mathrm{cr}_2(K_{p,q}) \geq pq - 4(p + q - 2).$$

One can use the counting argument with $H = K_{10,10}$, $G = K_{p,q}$, and the fact that $\mathrm{cr}_2(K_{10,10}) \geq 28$ from (13), to obtain:

Theorem 1. *For $10 \leq p \leq q$, we have*

$$(14) \qquad \mathrm{cr}_2(K_{p,q}) \geq \frac{p(p-1)q(q-1)}{290}.$$

For $p \leq 9$ we make a finer analysis of $\mathrm{cr}_2(K_{p,q})$.

3.2. Exact results for complete bipartite graphs

It is easy to see that $K_{4,q}$ is always biplanar. The result on the thickness of complete bipartite graphs of Harary et al. [4] implies that for $q \leq 12$, $\Theta(K_{5,q}) \leq 2$ and $\Theta(K_{5,13}) = 3$. Hence $\mathrm{cr}_2(K_{5,13}) \geq 1$. Paterson [19] observed that $\mathrm{cr}_2(K_{5,13}) = 1$. Determining the biplanar crossing number of $K_{5,q}$ for $q \geq 14$ is the main result of this paper.

Theorem 2. *For any $q \geq 1$, we have*

$$\mathrm{cr}_2(K_{5,q}) = \left\lfloor \frac{q}{12} \right\rfloor \left(q - 6 \left\lfloor \frac{q}{12} \right\rfloor - 6 \right),$$

and for even q there is an optimal drawing, which is self-complementary.

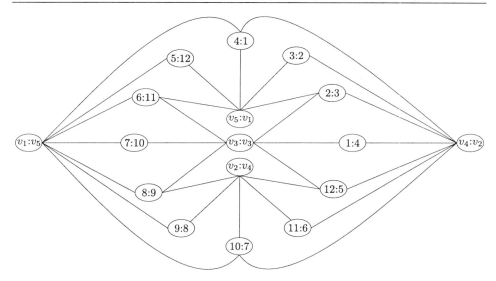

Fig. 1. Self-complementary drawing of $K_{5,12}$

Proof. We provide a drawing first. Assume that $q = 12a + b$, $0 \leq b < 12$. Partition the q vertices into 12 consecutive arcs, which are as equal as possible. Let these arcs be $S_1, S_2, \ldots, S_{12}$. Clearly b arcs contain $a + 1$ vertices and $12 - b$ arcs contain a vertices. Consider the regular 12-gon inscribed into the unit circle centered at $(0,0)$, with one vertex placed in $(1,0)$. Fig. 1 shows a self-complementary biplanar drawing of $K_{5,12}$, where the 12 vertices are placed into the vertices of the regular 12-gon. To draw $K_{5,q}$, we place the 5 vertices into the locations as they take in Fig. 1. We use small neighborhoods of the vertices of this regular 12-gon for the placement of the 12 arcs on the circumscribed circle of the 12-gon, starting with S_1 at $(1,0)$, and going counterclockwise, i.e. put S_i where the vertex is $(i : 5 - i)$ on the figure. Now we describe a drawing of $K_{5,q}$ on the first plane.
Place v_1 at $(-2,0)$ and join it to S_4, S_5, S_6, S_7, S_8, S_9, S_{10}.
Place v_2 at $\left(0, -\frac{1}{2}\right)$ and join it to S_8, S_9, S_{10}, S_{11}, S_{12}.
Place v_3 at $(0,0)$ and join it to S_{12}, S_1, S_2 and S_6, S_7, S_8.
Place v_4 at $(2,0)$ and join it to S_{10}, S_{11}, S_{12}, S_1, S_2, S_3, S_4.
Place v_5 at $\left(0, \frac{1}{2}\right)$ and join it to S_2, S_3, S_4, S_5, S_6.

On the second plane, place v_1 at $\left(0, \frac{1}{2}\right)$, v_2 at $(2,0)$, v_3 at $(0,0)$, v_4 at $\left(0, -\frac{1}{2}\right)$, and v_5 at $(-2,0)$. Put S_{5-i} (counting mod 12) where S_i was in the first plane and and draw the remaining edges exactly with the same curves that we used in the first plane.

In general, vertex $(i : 5 - i)$ represents an arc with S_i in the first plane and an arc with S_{5-i} in the second plane. Clearly the number of crossings— as we made the necessary crossings only—is exactly

$$\sum_{i=1}^{12} \binom{|S_i|}{2} = b \binom{a+1}{2} + (12-b) \binom{a}{2}.$$

Substituting $a = \lfloor q/12 \rfloor$ and $b = q - 12 \lfloor q/12 \rfloor$ into the previous formula we get the required upper bound.

We obtained above a self-complementary drawing of $K_{5,12q}$. To make this drawing self-complementary for every even q, the question is, where we put the extra $b = 2b'$ vertices. Whenever we have to add two new vertices, they must be added to arcs S_i and S_{5-i} for some i. Note that the twelve arcs make exactly 6 such pairs.

The lower bound is proved by induction on q. The claim is true for $12 \le q \le 24$, as formula (2) gives a lower bound of $q - 12$. Assume that it is true for some $q \ge 24$. Using the counting argument with $H = K_{5,q}$, $G = K_{5,q+1}$, we argue that

$$\mathrm{cr}_2(K_{5,q+1}) - \left\lfloor \frac{q+1}{12} \right\rfloor \left(q - 6 \left\lfloor \frac{q+1}{12} \right\rfloor - 5 \right)$$

$$\ge \left\lceil \frac{\binom{q+1}{q}}{\binom{q-1}{q-2}} \mathrm{cr}_2(K_{5,q}) \right\rceil - \left\lfloor \frac{q+1}{12} \right\rfloor \left(q - 6 \left\lfloor \frac{q+1}{12} \right\rfloor - 5 \right)$$

$$\ge \left\lceil \frac{q+1}{q-1} \left\lfloor \frac{q}{12} \right\rfloor \left(q - 6 \left\lfloor \frac{q}{12} \right\rfloor - 6 \right) - \left\lfloor \frac{q+1}{12} \right\rfloor \left(q - 6 \left\lfloor \frac{q+1}{12} \right\rfloor - 5 \right) \right\rceil.$$

To conclude the proof, one has to show that the expression inside the big brackets of the last line is greater than -1. This can be done by distinguishing two cases: whether $q = 11 \pmod{12}$, or not, and doing some algebra. ∎

Other exact results that we know about $\mathrm{cr}_2(K_{p,q})$ are summarized in the following table. In some interesting cases we also included lower and upper bounds.

p vs. q	7	8	9	10	11	12	13	14	15	16
6	0	0	2	4	6	8	10	12	14	16
7	1	4	7	10	13	16	19,21			
8	4	8	12	16	20	24	29,32			
9	7	12	17,19	22, 24						
10	10	16	22,24	28,32						

All the lower bounds in the table follow from the lower bound (13). Exactness for $p = 6$ follows from Theorem 3 in Subsection 3.3. Exactness for $p = 7$ follows from the drawing Fig. 2 of $K_{7,12}$ for $q = 12$; and optimal drawings for $K_{7,q}$ for $8 \leq q \leq 11$ can be obtained from Fig. 2 by successively erasing vertices 12, 11, 10, 9, in this order. Note that the drawings obtained for $K_{7,8}$, $K_{7,10}$, and $K_{7,12}$ are also self-complementary. Unfortunately, we do not have a biplanar drawing of $K_{7,q}$ that we would dare to think optimal.

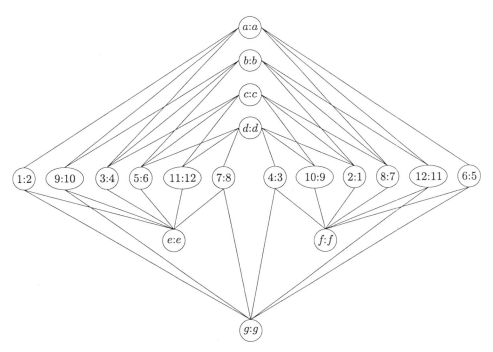

Fig. 2. Self-complementary drawing of $K_{7,12}$

Exactness for $p = 8$ follows from the self-complementary drawing Fig. 3 of $K_{8,12}$; optimal drawings for $K_{8,q}$ for $6 \leq q \leq 11$ can be obtained from that drawing by e.g. successively erasing vertices 12, 1, 7, 6, 10, 3, in this order.

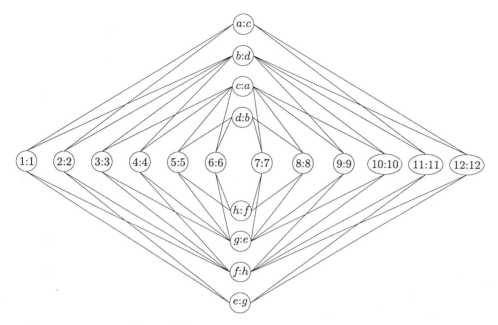

Fig. 3. Self-complementary drawing of $K_{8,12}$

One can get drawings for $K_{9,q}$ and $K_{10,q}$ from the general drawing described in Subsection 3.4. We know that as early as for $K_{11,11}$ or $K_{10,13}$, the estimation (13) is no longer the best lower bound. This follows from the arguments that lead to (6).

3.3. Conjectured exact results for complete bipartite graphs

Theorem 3. *For any $q \geq 1$, we have*

$$\mathrm{cr}_2(K_{6,q}) \leq 2 \left\lfloor \frac{q}{8} \right\rfloor \left(q - 4 \left\lfloor \frac{q}{8} \right\rfloor - 4 \right).$$

This bound is optimal for any $q \leq 16$.

Proof. We provide two different drawings. *First drawing.* On both planes we draw a "thinned out" copy of the drawing from the Zarankiewicz conjecture. Place the vertices v_1, v_2 and v_3 (resp. u_1, u_2 and u_3) on the positive (resp. negative) part of the x axis, in this order from the origin. Partition the q vertices into 8 almost equal sets, S_1, S_2, S_3, S_4 and T_1, T_2, T_3, T_4. Place $S_i(T_i)$, $i = 1, 2, 3, 4$ consecutively from the origin toward infinity (minus infinity) on the y axis. On both planes we connect any v_i, u_j to all or no vertices of any S_k or T_l, and all connections are straight line segments. For the drawing on the first plane join v_1 and u_1 with S_1, S_2, T_1, T_2; v_2 and u_2 with S_2, S_3, T_2, T_3; v_3 and u_3 with S_3, S_4, T_3, T_4. For the drawing on the second plane the locations of v_i's and u_i's are the same. But place the S_i's vertices in the order S_3, S_4, S_1, S_2, from the origin toward infinity; and place the T_i's vertices in the order T_3, T_4, T_1, T_2, from the origin toward minus infinity. Draw the remaining edges with straight line segments. The number of crossings is precisely

$$(15) \quad 2\left(\binom{|S_1|}{2} + \binom{|S_2|}{2} + \binom{|S_3|}{2} + \binom{|S_4|}{2} + \binom{|T_1|}{2}\right.$$

$$\left. + \binom{|T_2|}{2} + \binom{|T_3|}{2} + \binom{|T_4|}{2}\right).$$

Simple algebra shows that this is equal to the expresion in the statement of the Theorem.

Second drawing. Fig. 4 shows a crossing-free self-complementary drawing of $K_{6,8}$. We explain how to extend it into a self-complementary drawing with the same number of crossings as the first drawing. Assume first that $n = 8k$. Substitute every lettered vertex in Fig. 4 with k vertices on a very short straight line segment. We will join all three former neighbors of a lettered vertex to all k successors of the lettered vertex. Join one of the three from one side of the short straight line segment, and join the two others from the other side of the short straight line segment. Clearly the number of crossings is the same as in (15). If $q = 8k + r$ $(1 \le r \le 3)$, then use $k + 1$ successor vertices for r of the lettered vertices $(a : a)$ and $(c : c)$ and $(g : g)$. If $q = 8k + 4 + r$ $(1 \le r \le 3)$, then use $k + 1$ successor vertices for the lettered vertices $(e : b)$ and $(b : e)$ and $(d : f)$ and $(f : d)$; and also use $k + 1$ successor vertices for r of the lettered vertices $(a : a)$ and $(c : c)$ and $(g : g)$. The number of crossings is—in all cases—the same as in (15) again.

The optimality of the lower bound for $q \le 16$ follows from (2), which gives a lower bound of $2q - 16$. ∎

We would like to point out that if $cr_2(K_{6,q})$ *is even* for every q, then the counting argument from the proof of Theorem 2, *mutatis mutandis*, can be

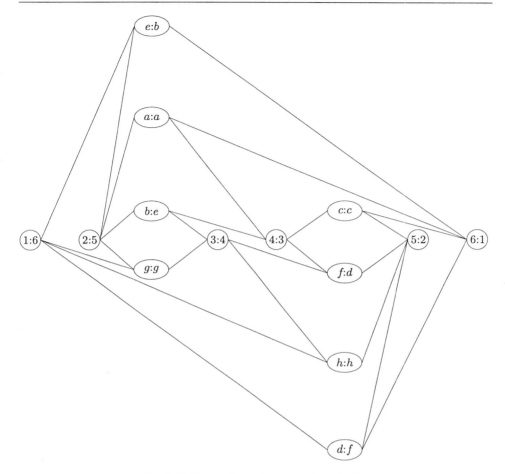

Fig. 4. Self-complementary drawing of $K_{6,8}$

repeated for Theorem 3. Note that if $K_{6,q}$ has an optimal biplanar drawing in which G_1 is isomorphic to G_2, as we conjecture, then $cr_2(K_{6,q})$ is even.

Theorem 4. *For any $q \geq 1$, we have*

$$cr_2(K_{8,q}) \leq 4 \left\lfloor \frac{q}{6} \right\rfloor \left(q - 3 \left\lfloor \frac{q}{6} \right\rfloor - 3 \right).$$

This bound is optimal for any $q \leq 12$.

Proof. Place the vertices v_1, v_2, v_3 and v_4 (resp. u_1, u_2, u_3 and u_4) on the positive (resp. negative) part of the y axis, in this order from the origin.

Partition the q vertices into 6 almost equal sets, S_1, S_2, S_3 and T_1, T_2, T_3. Place $S_i(T_i)$, $i = 1, 2, 3$, consecutively from the origin toward infinity (minus infinity) on the x axis. On both planes we connect any v_i, u_j to all or no vertices of any S_k or T_l, and all connections are straight line segments. For the drawing on the first plane join v_1 and u_1 with S_1, T_1; v_2 and u_2 with S_1, S_2, T_1, T_2; v_3 and u_3 with S_2, S_3, T_2, T_3 and v_4, u_4 to S_3, T_3. For the drawing on the second plane the locations of S_i's and T_i's are the same. But place the v_i's vertices in the order v_3, v_4, v_1, v_2, from the origin toward infinity; and place the u_i's vertices in the order u_3, u_4, u_1, u_2, from the origin toward minus infinity. Draw the remaining edges with straight line segments. The number of crossings is precisely

$$4\left(\binom{|S_1|}{2} + \binom{|S_2|}{2} + \binom{|S_3|}{2} + \binom{|T_1|}{2} + \binom{|T_2|}{2} + \binom{|T_3|}{2}\right).$$

The rest is similar as in the proof of Theorem 3. Optimality follows from (2), which gives a lower bound of $4q - 24$. ∎

Open Problem 4. *Prove that the upper bounds in Theorem 3 and in Theorem 4 are optimal. Make a first step in this direction by proving that* $\mathrm{cr}_2(K_{6,q}) = \left(\frac{1}{8} + o(1)\right)q^2$.

3.4. The best known drawings for other complete bipartite graphs

Theorem 5. *For any $p \geq 6$, $q \geq 8$, we have*

$$\mathrm{cr}_2(K_{p,q}) \leq \left\lceil \frac{p}{6} \right\rceil \left\lceil \frac{q}{8} \right\rceil \left(32 \left\lceil \frac{p}{6} \right\rceil \left\lceil \frac{q}{8} \right\rceil - 20 \left\lceil \frac{p}{6} \right\rceil - 24 \left\lceil \frac{q}{8} \right\rceil + 12\right)$$

$$\leq \frac{1}{144}(p + 5)(q + 7)(2pq + 4p + q - 7).$$

Proof. We generalize the drawings for $K_{6,q}$ and $K_{8,q}$. Partition the p vertices into almost equal sets $X_1, X_2, \ldots, X_6$. Place X_1, X_2, X_3 (resp. X_4, X_5, X_6) on the positive (negative) part of the x axis in this order from the origin towards infinity (minus infinity). Partition the q vertices into almost equal sets $Y_1, Y_2, \ldots, Y_8$. Place Y_1, Y_2, Y_3, Y_4 (resp. Y_5, Y_6, Y_7, Y_8) on the positive (negative) part of the y axis in this order from the origin towards infinity (minus infinity).

On both planes we connect all vertices of any X_i to all or no vertices of any Y_j, and all connections are straight line segments. For the drawing on the first plane join X_1 and X_4 with Y_1, Y_2, Y_5, Y_6; X_2 and X_5 with Y_2, Y_3, Y_6, Y_7; X_3 and X_6 with Y_3, Y_4, Y_7, Y_8. For the drawing on the second plane the locations of X_i's are the same. Place the Y_i's vertices in the order Y_3, Y_4, Y_1, Y_2, from the origin towards infinity; and Y_7, Y_8, Y_5, Y_6, from the origin towards minus infinity on the y axis. Draw the remaining edges with straight line segments. By counting up of all kinds of crossings in the drawing and by regrouping terms we get that the number of crossings is precisely

$$\sum_{i=1}^{6} \binom{|X_i|}{2} \sum_{j=1}^{8} \binom{|Y_j|}{2}$$

$$+ \left(\binom{|X_1|}{2} + \binom{|X_3|}{2} + \binom{|X_4|}{2} + \binom{|X_6|}{2} \right)$$
$$\times \left(|Y_1||Y_2| + |Y_3||Y_4| + |Y_5||Y_6| + |Y_7||Y_8| \right)$$

$$+ \left(\binom{|X_2|}{2} + \binom{|X_5|}{2} \right) \left(|Y_1||Y_4| + |Y_2||Y_3| + |Y_5||Y_8| + |Y_6||Y_7| \right)$$

$$+ \left(|X_1||X_2| + |X_4||X_5| \right) \left(\binom{|Y_2|}{2} + \binom{|Y_4|}{2} + \binom{|Y_6|}{2} + \binom{|Y_8|}{2} \right)$$

$$+ \left(|X_2||X_3| + |X_5||X_6| \right) \left(\binom{|Y_1|}{2} + \binom{|Y_3|}{2} + \binom{|Y_5|}{2} + \binom{|Y_7|}{2} \right).$$

First assume that p is divisible by 6 and q is divisible by 8. One can easily compute that the number of crossings is $pq(2pq - 10p - 9q + 36)/144$.

Now let p, q be arbitrary numbers. Let p' be the smallest number divisible by 6 such that $p' \geq p$ and q' be the smallest number divisible by 8 such that $q' \geq q$. Then the number of crossings is at most $p'q'(2p'q' - 10p' - 9q' + 36)/144$. Noting that $p' = 6\lceil \frac{p}{6} \rceil \leq p+5$ and $q' = 8\lceil \frac{q}{8} \rceil \leq q+7$ we get the claim. ∎

Open Problem 5. *Make a conjecture showing a pattern for optimal biplanar drawings of $K_{p,q}$, i.e. pose the biplanar version of the Zarankiewicz conjecture. A good conjecture for $K_{7,q}$ already seems to be hard to find.*

Open Problem 6. *Find an asymptotic formula for $cr_2(K_{p,q})$ for small fixed p.*

4. Results and problems on other specific families graphs

4.1. Complete graphs

Note that bounding $cr_2(K_n)$ is a Nordhaus-Gaddum type problem [14]. Owens gave an explicit biplanar drawing of K_n with

$$cr_2(K_n) \leq \frac{7}{1536}n^4 + O(n^3).$$

The same upper bound (up to the second order term), based on a different drawing follows immediately from our work with Shahrokhi [21] by setting $G = K_n$ in (11).

Harary et al. [4] and Tutte [28] showed that for $n \leq 8$, $\Theta(K_n) \leq 2$ and $\Theta(K_9) = 3$. Their construction actually also shows $cr_2(K_9) = 1$. Applying the counting argument for $H = K_{10,10}$, $G = K_n$, and using $cr_2(K_{10,10}) \geq 28$ from (13), we obtain

$$(16) \qquad cr_2(K_n) \geq \frac{1}{1158}n^4 + O(n^3).$$

We can do somewhat better than (16). Consider a biplanar drawing D of K_n. Then any subset of vertices induces a biplanar subdrawing, D', of the induced complete subgraph G'. Assume that G' has order n' and size $m' = \binom{n'}{2}$. According to (5),

$$(17) \qquad cr_2(G') \geq \begin{cases} 6m' - 66n' + 132 & \text{if } n' \geq 3 \\ 6m' - 66n' + 132 - 12 & \text{if } n' = 2 \\ 6m' - 66n' + 132 - 66 & \text{if } n' = 1 \\ 6m' - 66n' + 132 - 132 & \text{if } n' = 0. \end{cases}$$

Pick now independently with probability p vertices of K_n to obtain a random G'. Taking expectation of the inequality of two random variables, (17), we obtain:

$$(18) \qquad p^4 \, cr_2(K_n) \geq 6mp^2 - 66np + 132 - 12\binom{n}{2}p^2(1-p)^{n-2}$$

$$- 66np(1-p)^{n-1} - 132(1-p)^n.$$

Setting $p = 30.073871/n$ in (18) yields that for n sufficiently large,

$$\text{(19)} \qquad\qquad \text{cr}_2(K_n) \geq \frac{n^4}{952}.$$

It follows from the counting argument applied to $G = K_n$ and $H = K_{n-1}$, that $\text{cr}\,(K_n)/\binom{n}{4}$ is a non-decreasing function of n, and hence has finite limit. The same argument applies to $\text{cr}_2(K_n)$ as well

Open Problem 7. *Improve the lower bound in* (19). *Is*

$$\lim_{n\to\infty} \text{cr}_2(K_n) \Big/ \binom{n}{4} = \frac{7}{24} \cdot \frac{24}{64} = \frac{7}{64}?$$

Find exact values for the biplanar crossing numbers of complete graphs for small values $n = 10, 11, \ldots$.

4.2. Hypercubes

For the k-dimensional hypercube Q_k, it is known that $\Theta(Q_7) \leq 2$ and the estimation (2) gives $\text{cr}_2(Q_8) \geq 8$. We give a general upper bound for the biplanar crossing number of hypercubes.

Theorem 6. *For* $k \geq 8$

$$\text{cr}_2(Q_k) \leq \begin{cases} \dfrac{165}{512} 2^{\frac{3}{2}k} + O(k^2 2^k), & \text{if } k \text{ is even,} \\[2mm] \dfrac{176}{512} 2^{\frac{3}{2}k} + O(k^2 2^k), & \text{if } k \text{ is odd.} \end{cases}$$

Proof. Our biplanar drawing of Q_k is based on the best known planar drawing due to Faria and Figueiredo [6] satisfying

$$\text{(20)} \qquad\qquad \text{cr}\,(Q_k) \leq \frac{165}{1024} 4^k - (2k^2 - 11k + 34)2^{k-3}.$$

Let $0 \leq i \leq k$. Observe that all edges belonging to the first i dimensions in Q_k induce 2^i distinct hypercubes isomorphic to Q_{k-i}. Draw these hypercubes on the first plane and the 2^{k-i} hypercubes isomorphic to Q_i, induced by the last $k - i$ dimensions on the second plane, using (20). We get a biplanar drawing with

$$\text{cr}_2(Q_k) \leq \frac{165}{1024} 2^{2k-i} + \frac{165}{1024} 2^{k+i}.$$

Finally, by setting $i = \lceil k/2 \rceil$, we get the result. ∎

Unfortunately, the lower bound formula (7) gives only a weak estimation of order $\Omega(k^3 2^k)$, and even (8) improves it insignificantly to $\Omega(k^4 2^k)$. In order to use (8), we have to note that we can keep a positive percentage of edges of Q_k, while destroying all 4-cycles by throwing out edges, see [8]. We know that our drawing is not optimal: some edges between vertex disjoint copies of $Q_{\lfloor k/2 \rfloor}$ (resp. $Q_{\lceil k/2 \rceil}$) can be brought over from the other plane without making new crossings, and in this way their old crossings are eliminated.

Open Problem 8. *Is the upper bound in Theorem 6 still the best possible up to a constant multiplicative factor?*

4.3. Meshes

In the standard plane crossing number theory one of the most studied graph is the toroidal mesh, i.e. the Cartesian product of two cycles. See the recent paper [7] for the almost complete exact solution. We will concentrate on the biplanar crossing number of toroidal and ordinary meshes. It is an easy exercise to show that the graph $C_{n_1} \times C_{n_2} \times C_{n_3}$ is biplanar for any $3 \leq n_1, n_2, n_3$. On the other hand $C_{n_1} \times C_{n_2} \times C_{n_3} \times C_{n_4}$ has thickness at least 3. We do not know whether

Open Problem 9. *Is it true that* $\mathrm{cr}_2(P_n \times C_n \times C_n \times C_n) = 0$?

If it is nonzero, it is surprisingly small, since we have a biplanar drawing showing that $\mathrm{cr}_2(P_n \times C_n \times C_n \times C_n) = O(n^4)$, which is just linear in the number of edges. (Put edges from the first two dimensions on the first plane, and edges from the second two dimensions on the second plane.)

Theorem 7. *For even* k

$$
\mathrm{cr}_2 \left(\prod_{i=1}^{k} C_n \right) \leq 2^{\frac{k}{2}+5} n^{k-2}.
$$

Proof. Put the edges of the first $k/2$ dimensions on the first plane. They induce $2^{\frac{k}{2}}$ vertex disjoint subgraphs isomorphic to $\prod_{i=1}^{\frac{k}{2}} C_n$. Place the leftover edges on the second plane. Using the estimation

$$
\mathrm{cr} \left(\prod_{i=1}^{\frac{k}{2}} C_n \right) \leq 16 n^{k-2}
$$

from [22] we get the result. ∎

We leave it to the Reader to prove an analogue of Theorem 7 for odd k.

Open Problem 10. *Show that the upper bound in Theorem 7 is tight.*

5. CONCLUSION

Our knowledge on biplanar crossing numbers is as rudimentary as it was our knowledge on crossing numbers till Leighton's work [10] in the 70's. Bisection width and graph embedding methods cannot be used, only the counting method and density-based lower bounds are available. We hope that the development of structure-based lower bounds for the biplanar crossing numbers will shed light to some so far unknown properties of ordinary crossing numbers as well.

REFERENCES

[1] A. Aggarwal, M. Klawe and P. Shor, Multi-layer grid embeddings for VLSI, *Algorithmica,* **6** (1991), 129–151.

[2] M. Ajtai, V. Chvátal, M. Newborn and E. Szemerédi, Crossing-free subgraphs, *Annals of Discrete Mathematics,* **12** (1982), 9–12.

[3] K. Asano, On the genus and thickness of graphs, *J. Combinatorial Theory B,* **43** (1987), 187–192.

[4] J. Battle, F. Harary and Y. Kodama, Every planar graph with nine vertices has a nonplanar component, *Bulletin of the American Mathematical Society,* **68** (1962), 569–571.

[5] L. W. Beineke, Biplanar graphs: a survey, *Computers and Mathematics with Applications,* **34** (1997), 1–8.

[6] L. Faria and C. M. H. de Figueiredo, On the Eggleton and Guy conjectured upper bound for the crossing number of the n-cube, *Mathematica Slovaca,* **50** (2000), 271–287.

[7] L. Y. Glebski and G. Salazar, The conjecture cr $(C_m \times C_n) = (m - 2)n$ is true for all but finitely many n, for each m, *J. Graph Theory,* **47** (2004), 53–72.

[8] N. Graham, F. Harary, M. Livingston and Q. Stout, Subcube fault-tolerance in hypercubes, *Inform. and Comput.,* **102** (1993), no. 2, 280–314.

[9]   D. J. Kleitman, The crossing number of $K_{5,n}$, *J. Combinatorial Theory,* **9** 1970, 315–323.

[10]   F. T. Leighton, *Complexity Issues in VLSI,* MIT Press, Cambridge 1983.

[11]   T. Madej, Bounds for the crossing number of the n-cube, *J. Graph Theory,* **15** (1991), 81–97.

[12]   A. Mansfield, Determining the thickness of graphs is NP-hard, *Mathematical Proceedings of the Cambridge Philosophical Society,* **9** (1983), 9–23.

[13]   P. Mutzel, T. Odenthal and M. Scharbrodt, The thickness of graphs: a survey, *Graphs and Combinatorics,* **14** (1998), 59–73.

[14]   E. A. Nordhaus and J. W. Gaddum, On complementary graphs, *American Mathematical Monthly,* **63** (1956), 175–177.

[15]   A. Owens, On the biplanar crossing number, *IEEE Transactions on Circuit Theory,* **18** (1971), 277–280.

[16]   J. Pach, F. Shahrokhi and M. Szegedy, Applications of crossing numbers, *Algorithmica,* **16** (1996), 111–117.

[17]   J. Pach, J. Spencer and G. Tóth, New bounds on crossing numbers. *Discrete Comp. Geom.,* **24** (2000), 623–644.

[18]   J. Pach and G. Tóth, Graphs drawn with few crossings per edge, *Combinatorica,* **17** (1998), 427–439.

[19]   M. S. Paterson, Personal communication (2001).

[20]   F. Shahrokhi, O. Sýkora, L. A. Székely and I. Vrt'o, Crossing numbers: bounds and applications, in: *Intuitive Geometry,* Bolyai Society Mathematical Studies **6**, (I. Bárány and K. Böröczky, eds.), Akadémia Kiadó, Budapest, 1997, 179–206.

[21]   F. Shahrokhi, O. Sýkora, L. A. Székely and I. Vrt'o, The book crossing number of graphs, *J. Graph Theory,* **21** (1996), 413–424.

[22]   F. Shahrokhi, O. Sýkora, L. A. Székely and I. Vrt'o, Crossing numbers of meshes, in: *Proc. 4th Intl. Symposium on Graph Drawing,* Lecture Notes in Computer Science 1027, Springer Verlag, Berlin, 1996, 462–471.

[23]   O. Sýkora, L. A. Székely and I. Vrt'o, Crossing numbers and biplanar crossing numbers II: using the probabilistic method, submitted.

[24]   O. Sýkora and I. Vrt'o, On VLSI layouts of the star graph and related networks, *Integration, The VLSI Journal,* **17** (1994), 83–93.

[25]   J. Spencer, The biplanar crossing number of the random graph, in: *Towards a Theory of Geometric Graphs* (J. Pach, ed.), Contemporary Mathematics, **342** (2004), 269–271.

[26]   L. A. Székely, A successful concept for measuring non-planarity of graphs: the crossing number, *Discrete Math.,* **276** (2003), 1–3, 331–352.

[27]   P. Turán, A note of welcome, *J. Graph Theory,* **1** (1977), 7–9.

[28]   W. L. Tutte, On non-biplanar character of K_9, *Canadian Mathematical Bulletin,* **6** (1963), 319–330.

[29] A. T. White and L. W. Beineke, Topological graph theory, in: *Selected Topics in Graph Theory* (L. W. Beineke and R. J. Wilson, eds.) Academic Press, 1978, 15–50.

Éva Czabarka

Department of Mathematics
College of William & Mary
Williamsburg, VA 23187
U.S.A.

Ondrej Sýkora

Department of Computer Science
Loughborough University
Loughborough, Leicestershire LE11
3TU
The United Kingdom

László A. Székely

Department of Mathematics
University of South Carolina
Columbia, SC 29208
U.S.A.

Imrich Vrťo

Department of Informatics
Institute of Mathematics
Slovak Academy of Sciences
Dúbravská 9
842 35 Bratislava
Slovak Republic

BOLYAI SOCIETY
MATHEMATICAL STUDIES, 15

Conference on Finite
and Infinite Sets
Budapest, pp. 79–92.

An Exercise on the Average Number of Real Zeros of Random Real Polynomials

C. DOCHE and M. MENDÈS FRANCE

À Vera Sós et Andras Hajnal avec admiration et amitié

The average number of real zeros of random n degree real polynomials is well known since M. Kac's seminal article of 1943 [12] which states that it is $\log n + O(1)$. Some fifty years later, A. Edelman and E. Kostlan found a beautiful geometrical proof which allowed them to give many other related results [10]. Using their method we discuss the average number of real zeros of random real polynomials

$$\sum_{j=0}^{n} A_j X^j$$

where the A_j's are independent Gaussian variables with mean 0 and with variance

$$\sigma^2(A_j) = \binom{n}{j} n^{-\beta j}$$

where $\beta \in \mathbb{R}$ is a given parameter. The average number of real zeros in the interval (a, b) is shown to be

$$E(n; a, b) = \frac{1}{\pi} \sqrt{n} \left(\text{Arctan} \frac{b}{n^{\beta/2}} - \text{Arctan} \frac{a}{n^{\beta/2}} \right).$$

While discussing special polynomials we are led to show that under general conditions, polynomials of the type

$$\sum_{i=1}^{k} A_i(X)(a_i X + b_i)^n$$

have at most $O(1)$ real zeros as n increases to infinity.

1. THE GENERAL SETTING

Let Λ be a sequence of $n+1$ integers $\lambda_0, \lambda_1, \ldots, \lambda_n$; $\lambda_j \geq 0$, not necessarily distinct. Consider the real polynomial

$$P(X) = \sum_{j=0}^{n} a_j X^{\lambda_j}$$

where the coefficients a_j are real independent Gaussian variables with mean 0 and standard deviation $\sigma(a_j) = 1$. The object of the paper is to compute the expectation of the number of real zeros of P for a special sequence Λ which we shall describe shortly.

At this point it should be observed that the result is independent of the order of the λ_j's in Λ since the a_j's are independent identical random variables: any permutation on Λ leaves the expected number of real zeros invariant.

Define

$$\lambda^{-1}(k) = \{0 \leq j \leq n \mid \lambda_j = k\}$$

and

$$|\Lambda| = \max \lambda_j, \quad 0 \leq j \leq n.$$

Then

$$P(X) = \sum_{0 \leq k \leq |\Lambda|} X^k \sum_{j \in \lambda^{-1}(k)} a_j.$$

The random variables

$$A_k = \sum_{j \in \lambda^{-1}(k)} a_j$$

are independent Gaussian variables with mean 0 and with standard deviation

$$\sigma(A_k) = \left(\operatorname{card} \lambda^{-1}(k) \right)^{1/2}.$$

The most natural and interesting case is when $\Lambda = \{0, 1, \ldots, n\}$ and this is the one studied by M. Kac [12, 13] and later by A. Edleman and E. Kostlan [10]. For a short history we refer to the book of A. T. Bharuca–Reid and M. Sambandham [3] or to [8].

2. A SPECIAL CASE

The special case we wish to discuss is the following which even though it may be thought as artificial, it seems to have some relevance in Quantum Mechanics. This was already noticed in [10]. Let $s(j)$ be the sum of the binary digits of the integer $j \geqslant 0$. Choose

$$\Lambda = \big(s(0), s(1), s(2), \ldots, s(2^n - 1)\big)$$

so that $|\Lambda| = n$. Then

$$P(X) = \sum_{0 \leqslant j < 2^n} a_j X^{s(j)}$$

$$= \sum_{0 \leqslant k \leqslant n} X^k \sum_{s(j)=k} a_j = \sum_{0 \leqslant k \leqslant n} A_k X^k.$$

Quite obviously

$$\sum_{s(j)=k} 1 = \binom{n}{k}$$

so that the variance of A_k is $\binom{n}{k}$. This is precisely the case encountered by the three physicists E. Bogomolny, O. Bohigas and P. Lebœuf [2] even though the sum of the digits does not appear explicitly in their presentation.

Here we shall give a simple generalization of the above case. We are given a real parameter β and we assume that the coefficients A_k of the polynomial

$$\sum_{0 \leqslant k \leqslant n} A_k X^k$$

are independent random Gaussian variables with mean 0 and variance

$$\sigma^2(A_k) = \binom{n}{k} n^{-\beta k}.$$

The parameter β can be thought of as an "order parameter" which introduces some "noise" (inverse temperature) in the system. Negative noise or negative temperatures should not surprise the physicists; see for example [15, Chapter VI, § 71].

As β decreases from $+\infty$ to $-\infty$, the standard deviation $\sigma(A_k)$, $k \geqslant 1$, increases from 0 to $+\infty$. For large β, the A_k's have small standard deviation and are to some extent well determined. On the contrary, when β is in the vicinity of $-\infty$, the A_k's have large standard deviation and as such, are completely unpredictable. See also Figure 1.

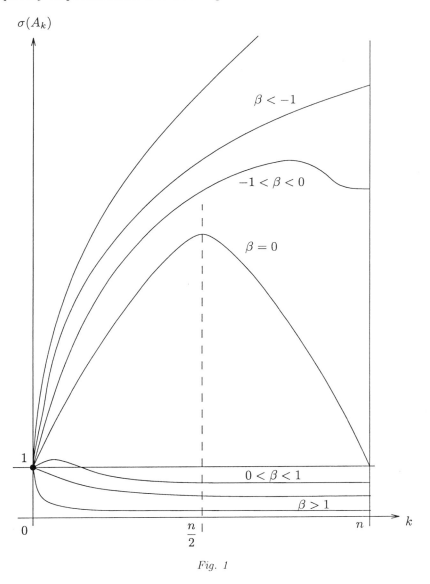

Fig. 1

If $|\beta| < 1$, the graph $k \mapsto \sigma(A_k)$ has a maximum in the open interval $]0, n[$; the values $\beta = \pm 1$ seem to play a special role.

3. A FIRST RESULT

Theorem 1. *Suppose $A_0, A_1, \ldots, A_n$ are $(n + 1)$ centered independent Gaussian variables with variance*

$$\sigma^2(A_k) = \binom{n}{k} n^{-\beta k}, \quad 0 \leqslant k \leqslant n.$$

Then the average number of real zeros of the polynomial

$$\sum_{k=0}^{n} A_k X^k$$

in the interval (a, b) is

$$E(n; a, b) = \frac{1}{\pi} \sqrt{n} \left(\text{Arctan} \frac{b}{n^{\beta/2}} - \text{Arctan} \frac{a}{n^{\beta/2}} \right).$$

In particular the average number of real zeros is independent of β: $E(n; \mathbb{R}) = \sqrt{n}$.

Proof. The proof is very simple since according to [10], the average number of zeros in (a, b) is

$$E(n; a, b) = \frac{1}{\pi} \int_a^b \left(\frac{\partial^2}{\partial x \partial y} \log \sum_{k=0}^{n} \sigma^2(A_k) x^k y^k \right)^{1/2}_{x=y=t} dt.$$

In our case

$$\sum_{k=0}^{n} \sigma^2(A_k) x^k y^k = \sum_{k=0}^{n} \binom{n}{k} \left(\frac{xy}{n^\beta} \right)^k$$

$$= \left(1 + \frac{xy}{n^\beta} \right)^n.$$

Therefore

$$E(n; a, b) = \frac{\sqrt{n}}{\pi} \int_a^b \frac{n^{\beta/2}}{t^2 + n^\beta} dt$$

and the result follows. ∎

4. Comments

The density of probability is

$$\rho_n(t) = \frac{\sqrt{n}}{\pi} \frac{n^{\beta/2}}{t^2 + n^\beta}$$

and the normalized density is

$$\tilde{\rho}_n(t) = \frac{1}{\sqrt{n}} \rho_n(t) = \frac{1}{\pi} \frac{n^{\beta/2}}{t^2 + n^\beta}.$$

For a fixed large n Figure 2 displays the aspects of the graphs of the functions $t \mapsto \tilde{\rho}_n(t)$. If $\beta < 0$ is fixed and if n tends to infinity, $\tilde{\rho}_n(t)$ converges to the Dirac measure at the origin. The zeros tend to concentrate on the

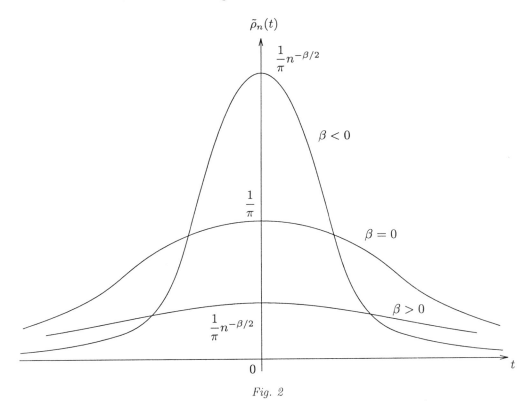

Fig. 2

neighbourhood of 0. A hand–waving argument could have predicted this behaviour. Indeed,

$$\frac{\sigma^2(A_{j+1})}{\sigma^2(A_j)} = \frac{n-j}{j+1} \frac{1}{n^\beta}$$

and therefore, since $\beta < 0$

$$\lim_{n \to \infty} \frac{\sigma^2(A_{j+1})}{\sigma^2(A_j)} = +\infty.$$

A_{j+1} is infinitely more dispersed than A_j. Divide the polynomial by A_n. For large n it behaves like

$$0 + 0X + 0X^2 + \cdots + 0X^{n-1} + X^n$$

and indeed, all the zeros are concentrated on $X = 0$.

In the same fashion, suppose $\beta > 0$. Then the graph $t \mapsto \tilde{\rho}_n(t)$ is close to the horizontal axis. The zeros are well dispersed on $\mathbb{R}$. This could have been foreseen. Indeed

$$\lim_{n \to +\infty} \sigma(A_j) = \begin{cases} 1 & \text{if } j = 0 \\ 0 & \text{if } 0 < j \end{cases}$$

so that for infinitely large n almost all polynomials coincide with

$$a_0 + 0X + 0X^2 + \cdots + 0X^n.$$

The zeros have infinite size, i.e. they are not confined in a bounded set in $\mathbb{R}$. In other terms they appear dispersed within the real line.

Let us analyze our results differently. Let $0 < a < b$. The average number of zeros in the interval (a, b) is, as we showed

$$E(n; a, b) = \frac{1}{\pi} \sqrt{n} \left(\text{Arctan} \frac{b}{n^{\beta/2}} - \text{Arctan} \frac{a}{n^{\beta/2}} \right).$$

When n increases to infinity, the limit of $E(n; a, b)$ depends on β:

$$\lim_{n \to \infty} E(n; a, b) = \begin{cases} 0 & \text{if } \beta < -1 \\ \dfrac{1}{\pi} \left(\dfrac{1}{a} - \dfrac{1}{b} \right) & \text{if } \beta = -1 \\ +\infty & \text{if } -1 < \beta < 1 \\ \dfrac{1}{\pi}(b - a) & \text{if } \beta = 1 \\ 0 & \text{if } \beta > 1. \end{cases}$$

Two critical values appear $\beta = \pm 1$. If $|\beta| > 1$ the limit vanishes for all interval (a, b), and then of course the same conclusion holds for all interval $(-b, -a)$, $0 < a < b$. Therefore, if the real zeros are sufficiently many, they must all be close to 0. Comparing this result with our previous ones we conclude that

$$\left.\begin{array}{c} \beta < -1 \\ n \text{ large} \end{array}\right\} \implies \left\{\begin{array}{l} \text{The zeros are mostly in} \\ \text{a small neighbourhood of } 0. \end{array}\right.$$

$$\left.\begin{array}{c} \beta > 1 \\ n \text{ large} \end{array}\right\} \implies \left\{\begin{array}{l} \text{The zeros are so dispersed and sparse} \\ \text{that for all } c > 0 \ E(n; -c, +c) \approx 0. \end{array}\right.$$

5. Random sequences and deterministic sequences

In relationship with our initial discussion it may be interesting to test the distribution of real zeros of the random polynomials

$$P_n(X) = \sum_{j=0}^{2^n-1} \pm X^{s(j)}$$

where the signs $(\pm)$ are chosen randomly according to the uniform probability $\left(\frac{1}{2}, \frac{1}{2}\right)$. We should expect $\sqrt{n}$ real zeros. However it is particularly difficult to compute these polynomials for large n since they involve a huge quantity of information. In practice it is hard to exceed the degree 30 for a given sequence. We have done some computations for the deterministic choice of signs $a_j = (-1)^{\lfloor \sqrt{2j} \rfloor}$ which seem consistant with our expectation.

For well chosen sequences we can sometimes compute exactly the number of real zeros. For example if $a_j = \lambda^j$ for some real λ, it is easily seen that

$$(1) \qquad \sum_{0 \leqslant j < 2^n} \lambda^j X^{s(j)} = \prod_{0 \leqslant k < n} \left(1 + X\lambda^{2^k}\right).$$

The zeros are all real: $X = -\lambda^{-2^k}$; $k = 0, 1, \ldots, n-1$. Another trivial example is $a_j = (-1)^{s(j)}$, the Thue–Morse sequence. Then clearly

$$\sum_{0 \leqslant j < 2^n} (-1)^{s(j)} X^{s(j)} = (1 - X)^n$$

and again the zeros are all real.

The Thue–Morse sequence is a special example of the so-called 2-automatic sequences, *i.e.* sequences generated by a finite basis 2 automaton (see [5, 6], [4] or [1] for the general theory of automatic sequences).

Another example is given by the Rudin–Shapiro sequence defined by

$$a_{2j} = a_j \quad \text{and} \quad a_{2j+1} = (-1)^j a_j.$$

It is possible to show that

$$P_0(X) = 1, \ P_1(X) = 1+X \quad \text{and} \quad P_n(X) = (1-X)P_{n-1}(X) + 2X P_{n-2}(X).$$

Indeed

$$P_n(X) = P_{n-1}(X) + \sum_{j=2^{n-1}}^{2^n-1} a_j X^{s(j)}.$$

This last term is trivially equal to

$$\sum_{k=0}^{2^{n-1}-1} a_{k+2^{n-1}} X^{s(k)+1}.$$

Now from the definition of $(a_j)_{j \geq 0}$ it is easy to ensure that

$$\begin{cases} a_{k+2^{n-1}} = a_k & \text{if } 0 \leqslant k \leqslant 2^{n-2} - 1 \\ a_{k+2^{n-1}} = -a_k & \text{if } 2^{n-2} \leqslant k \leqslant 2^{n-1} - 1 \end{cases}$$

so that

$$P_n(X) = P_{n-1}(X) + X \sum_{k=0}^{2^{n-2}-1} a_k X^{s(k)} - X \sum_{k=2^{n-2}}^{2^{n-1}-1} a_k X^{s(k)}$$

$$= P_{n-1}(X) + X P_{n-2}(X) - X P_{n-1}(X) + X P_{n-2}(X)$$

$$= (1 - X)P_{n-1}(X) + 2X P_{n-2}(X)$$

as claimed. Computations up to degree 200 with the `polsturm` command of PARI suggest that the number of real roots of $P_n(X)$ is about $n/2$ more precisely it seems to satisfy

$$2 \left\lceil \frac{n}{4} \right\rceil - \frac{(-1)^{n+1} + 1}{2}.$$

Unfortunately we are not able to establish this...

It is probable that 2–automatic sequences are too correlated to the sequence of exponents $s(j)$ to provide examples of polynomials which have $(1 + o(1)) \sqrt{n}$ real zeros as in the generic case.

The next example, namely the paperfolding sequence strengthens this guess. This is a $(\pm)$ sequence defined as [7, 9, 1]

$$a_{2n} = (-1)^n, \quad a_{2n+1} = a_n; \quad n \geqslant 0.$$

Let

$$P_n(X) = \sum_{0 \leqslant j < 2^n} a_j X^{s(j)}.$$

In the next paragraph we establish that

(2) $$P_n(X) = (1 - X)(1 + X)^{n-1} + 2X^n$$

and by use of a general theorem which we shall prove we manage to show that the number of real zeros of $P_n(X)$ is bounded independently of n.

6. THE PAPERFOLDING CASE AND A GENERAL THEOREM

We first establish Identity (2).

$$P_n(X) = \sum_{0 \leqslant j < 2^{n-1}} a_{2j} X^{s(2j)} + \sum_{0 \leqslant j < 2^{n-1}} a_{2j+1} X^{s(2j+1)}$$

$$= \sum_{0 \leqslant j < 2^{n-1}} (-1)^j X^{s(2j)} + X P_{n-1}(X).$$

The first sum on the right hand side is $(1 - X)(1 + X)^{n-2}$ as is clear from Identity (1) with $\lambda = -1$. Therefore

$$P_n(X) = X P_{n-1}(X) + (1 - X)(1 + X)^{n-2}$$

from where we conclude that indeed

$$\begin{cases} P_n(X) = (1 - X)(1 + X)^{n-1} + 2X^n, & n \geqslant 1 \\ P_0(X) = 1 \end{cases}$$

The fact that the number of distinct real zeros of $P_n(X)$ is bounded independently of n is a consequence of the following argument. Put $Y = 1 + \frac{1}{X}$. The zeros of $P_n(X)$ satisfy $Y^n - 2Y^{n-1} + 2 = 0$. Descartes' theorem asserts that this polynomial has at most 2 real zeros if n is even and 3 if n is odd. Actually for $n > 4$ these results are sharp. The zeros are distinct since P_n and P_n' have no common zero and therefore the number of real zeros is $O(1)$.

The above argument can be extended to establish the following result.

Theorem 2. *Let $k \geqslant 2$ be a given integer and let $A_{1n}(X), A_{2n}(X), \ldots,$ $A_{kn}(X)$ be an infinite family of nonzero real polynomials $(n = 0, 1, 2, \ldots)$ the degrees of which are bounded independently of n. Let $a_1, a_2, \ldots, a_k;$ $b_1, b_2, \ldots, b_k$ be $2k$ real numbers such that for all $i \neq j$ $a_i b_j - a_j b_i \neq 0$. Then the polynomials*

$$P_n(X) = \sum_{i=1}^{k} A_{in}(X)(a_i X + b_i)^n$$

each have a number of real zeros which is bounded independently of n.

Proof. We prove the theorem by induction on $k \geqslant 2$. Let

$$P_n(X) = A_n(X)(aX + b)^n + B_n(X)(cX + d)^n$$

with $ad - bc \neq 0$. Ignoring a finite number of real zeros, we can suppose with no loss of generality that $A_n(X)$ is coprime with $cX + d$. So $-d/c$ is not a zero of P_n.

$$P_n(X) = (cX + d)^n \left[A_n(X) \left(\frac{aX + b}{cX + d} \right)^n + B_n(X) \right].$$

Put

$$Y = \frac{aX + b}{cX + d}, \qquad X = \frac{-dY + b}{cY - a}.$$

$$(3) \qquad P_n(X) = (cX + d)^n \left[A_n \left(\frac{-dY + b}{cY - a} \right) Y^n + B_n \left(\frac{-dY + b}{cY - a} \right) \right]$$

$$= (cX + d)^n \frac{\tilde{A}_n(Y)Y^n + \tilde{B}_n(Y)}{\tilde{C}_n(Y)}$$

where $\widetilde{A}_n(Y)$, $\widetilde{B}_n(Y)$, $\widetilde{C}_n(Y)$ are polynomials. There is a 1–1 correspondance between the real zeros of $P_n(X)$ and the real zeros of $\widetilde{A}_n(Y)Y^n + \widetilde{B}_n(Y) = Q_n(Y)$. The number of terms of the polynomials $Q_n(Y)$ is bounded by the number of terms of $\widetilde{A}_n(Y)$ plus those of $\widetilde{B}_n(Y)$. This number is bounded independently of n. By a theorem of Descartes we conclude that the number of zeros of $P_n(X)$ is bounded independently of n.

We now assume that the theorem is true for all values up to $k-1$. Consider

$$P_n(X) = \sum_{i=1}^{k} A_{in}(X)(a_i X + b_i)^n$$

$$= (a_k X + b_k)^n \sum_{i=1}^{k} A_{in}(X) \left(\frac{a_i X + b_i}{a_k X + b_k} \right)^n .$$

Put

$$Y = \frac{a_{k-1}X + b_{k-1}}{a_k X + b_k}.$$

The zeros of $P_n(X)$ are obtained from those of

$$\sum_{i=1}^{k-1} \widetilde{A}_{in}(Y)(\widetilde{a}_i Y + \widetilde{b}_i)^n + \widetilde{A}_k(Y),$$

where the $\widetilde{A}_{in}(Y)$ are polynomials with degrees bounded independently of n. By successive derivations, say d, the last polynomial $\widetilde{A}_k(Y)$ vanishes and we are left with a polynomial

$$\sum_{i=1}^{k-1} \widetilde{\widetilde{A}}_{in}(Y)Y^{n-d}$$

which by induction hypothesis has a number of real zeros bounded independently of n. By successive integrations (d in fact) we conclude that the same is true for $P_n(X)$. ∎

7. APOLOGY

One of the main ideas in the beautiful book of J. Harthong on probability [11] is that basically random variables on bounded sets are always uniformly distributed, and on $\mathbb{R}$ they are centered Gaussian variables with equal standard deviation. But of course they may be submitted to constraints in which case they are as uniform as can be given the extra conditions. In other terms, they adhere to the maximal entropy principle.

In our case

$$\sum_{j=0}^{2^n-1} a_j X^{s(j)}$$

the constraint comes from the fact that the exponents are not distinct. The variance $\binom{n}{k}$ comes out naturally. The random variables

$$A_k = \sum_{s(j)=k} a_j$$

play the role of the main variables whereas the a_j's seem to be "hidden variables"...

In any case, the variances

$$\sigma^2(A_k) = \binom{n}{k}$$

do not appear to be so artificial as first one might think. As for

$$\sigma^2(A_k) = \binom{n}{k} n^{-\beta k}$$

we are afraid we can offer no convincing argument to justify the choice...

REFERENCES

[1] J.-P. Allouche, Automates finis en théorie des nombres, *Expositiones Mathematicae,* **5** (1987), 239–266.

[2] E. Bogomolny, O. Bohigas and P. Lebœuf, Distribution of Roots of Random Polynomials, *Physical Review Letters,* **68** (1992), 2726–2730.

[3]   A. T. Bharuca-Reid and M. Sambandham, *Random polynomials,* Academic Press, 1986.

[4]   G. Christol, T. Kamae, M. Mendès France and G. Rauzy, Suites algébriques, automates et substitutions, *Bull. Soc. Math. France,* **108** (1980), 401–419.

[5]   A. Cobham, On base-dependance of sets of numbers recognizable by finite automata, *Math. Syst. Theor.,* **3** (1969), 186–192.

[6]   A. Cobham, Uniform tag sequences, *Math. Syst. Theor.,* **6** (1972), 164–192.

[7]   C. Davis and D. Knuth, Number representations and dragon curves I and II, *J. Recreational Math.,* **3** (1970), 66–81, 133–149.

[8]   C. Doche and M. Mendès France, Real roots!, in: *Annal. Prob. Methods Number Theory,* A. Dubickas et al (eds.), TEV Vilnius (2002), 42–56.

[9]   M. Dekking, M. Mendès France and A. van der Poorten, Folds!, *Mathematical Intelligencer,* **4** (1982), 130–138, 173–181, 190–195.

[10] A. Edelman and E. Kostlan, How many zeros of a random polynomial are real?, *Bull. (New Series) AMS,* **32** (1995), 1–37.

[11] J. Harthong, *Probalités et statistiques,* Diderot éditeur, 1996.

[12] M. Kac, On the average number of real roots of a random algebraic equation, *Bull. Amer. Math. Soc.,* **49** (1943), 314–320 and 938.

[13] M. Kac, On the average number of real roots of a random algebraic equation (II), *Proc. London Math. Soc.,* **50** (1949), 390–408.

[14] M. Kac, *Probability and related topics in physical sciences,* Lectures in applied mathematics, first printing Interscience publishers 1960, Second printing A.M.S. 1976.

[15] L. Landau and E. Lifchitz, *Physique Statistique,* Éditions de Moscou, MIR, 1967.

Christophe Doche   Michel Mendès France

Laboratoire A2X   *Laboratoire A2X*
Université Bordeaux I   *Université Bordeaux I*
351, cours de la Libération   *351, cours de la Libération*
F-33405 Talence Cedex   *F-33405 Talence Cedex*
France   *France*

cdoche@math.u-bordeaux1.fr   mmf@math.u-bordeaux1.fr

BOLYAI SOCIETY
MATHEMATICAL STUDIES, 15

Conference on Finite
and Infinite Sets
Budapest, pp. 93–141.

EDGE-CONNECTION OF GRAPHS, DIGRAPHS, AND HYPERGRAPHS

A. FRANK*

To the memory of C. St. J. A. Nash-Williams and W. T. Tutte who contributed to the area with fundamental results.

In this work extensions and variations of the notion of edge-connectivity of undirected graphs, directed graphs, and hypergraphs will be considered. We show how classical results concerning orientations and connectivity augmentations may be formulated in this more general setting.

1. INTRODUCTION

A digraph $D = (V, E)$ is called **strongly connected** if there is a directed path from every node to every other node. By an easy exercise, this is equivalent to requiring that $\varrho_D(X) \geq 1$ for every proper non-empty subset X of V, where $\varrho_D(X)$, the **indegree** of X, denotes the number of edges entering X. An undirected graph, (in short, a graph) $G = (V, E)$ is called **2-edge-connected** if there are two edge-disjoint paths from every node to every other. It is not difficult to show that this is equivalent to requiring that $d_G(X) \geq 2$ for every proper non-empty subset X of V, where $d_G(X)$, the **degree** of X, denotes the number of edges connecting X and $V - X$.

---

*The work was started while the author visited the Institute for Discrete Mathematics, University of Bonn, July, 2000. Supported by the Hungarian National Foundation for Scientific Research, OTKA T037547.

The prototypes of theorems we are interested in concern strong-connectivity and 2-edge-connectivity.

1. Augmentation [K. P. Eswaran and R. E. Tarjan] [12]. *A digraph can be made strongly connected by adding at most γ new edges if and only if there are no $\gamma + 1$ disjoint sink-sets (:strongly-connected components with no leaving edges) and there are no $\gamma + 1$ disjoint source-sets (:strongly-connected components with no entering edges). A connected undirected graph can be made 2-edge-connected by adding at most γ new edges if and only if the number of 'leaves' is at most 2γ, where a leaf is a minimal subset X with $d_G(X) = 1$.*

2. Orientation [H. E. Robbins] [52]. *An undirected graph has a strongly connected orientation if and only if it is 2-edge-connected.*

3. Constructive characterization [folklore]. *A digraph is strongly connected if and only if it can be built from a node by the following two operations:* (i) add a new directed edge connecting existing nodes, (ii) subdivide an existing edge by a new node. *A graph is 2-edge-connected if and only if it can be built from a node by the following two operations:* (i) add a new edge connecting existing nodes, (ii) subdivide an existing edge by a new node. In both cases the two operations may be included into one: *add a path (directed, in case of digraphs) connecting two existing nodes (which may be equal),* an operation called **adding an ear.** Therefore these theorems are often formulated in the form: *a graph is 2-edge-connected or a digraph is strongly connected if and only if it can be built from a node by adding ears.* The sequence of ears in such a construction is called an ear-decomposition of the (2-edge-connected) graph or (strongly connected) digraph. *Moreover, such an ear-decomposition exists if the initial (di)graph is an arbitrary 2-edge-connected (respectively, strongly connected) sub(di)graph.*

We survey these types of results concerning higher edge-connection. Here the word 'edge-connection' is used in its informal meaning to describe the intuitive notion of a graph $G = (V, E)$ or a digraph $D = (V, A)$ being 'pretty much connected by edges'. To capture this idea formally, there are (at least) two distinct approaches, and both of them admit several versions.

The first approach requires the (di)graph to be not dismantleable into smaller parts by leaving out only few edges. Here are four possible definitions to make this intuition formal.

(A1) A graph $G = (V, E)$ is k-**edge-connected** if discarding less than k edges leaves a connected graph. (This is easily seen to be equivalent to requiring $d_G(X) \geq k$ whenever $\emptyset \subset X \subset V$.)

(A2) A digraph $D = (V, A)$ is k-**edge-connected** if discarding less than k edges leaves a strongly connected digraph. (This is easily seen to be equivalent to requiring $\varrho_D(X) \geq k$ whenever $\emptyset \subset X \subset V$.) For $k = 1$, k-edge-connectivity is just strong-connectivity.

(A3) G is k-**partition-connected** if discarding less than kq edges leaves a graph with at most q connected components for every $q = 1, 2, \ldots,$ $|V|-1$. Equivalently, there are at least kq edges connecting distinct parts for every partition of V into $q + 1$ non-empty parts for every q, $1 \leq q \leq |V|-1$. Note that for $k = 1$, partition-connectivity is equivalent to connectivity.

(A4) D is **rooted** k-**edge-connected** if there is a root-node s so that after discarding less than k edges every node keeps to be reachable from s. (This is easily seen to be equivalent to requiring $\varrho_D(X) \geq k$ for every non-empty subset X of $V - s$).

The second possible approach to capture the notion of high edge-connection is requiring the graph or digraph to contain several edge-disjoint 'simple' connected constituents. Here are four possibilities.

(B1) In G there are k edge-disjoint paths between every pair u, v of nodes.

(B2) In D there are k edge-disjoint directed paths from every node to every other.

(B3) G contains k edge-disjoint spanning trees (in which case G is called k-**tree-connected**).

(B4) D contains a node s so that there are k edge-disjoint spanning arborescences rooted at s.

Some basic results of graph theory asserts the equivalence of the corresponding definitions. Namely, by the edge-versions of Menger's theorem [15], the definitions **(A1)** and **(B1)** [resp., **(A2)** and **(B2)**] are equivalent:

Theorem 1.1 (Menger). *An undirected graph is k-edge-connected if and only if there are k edge-disjoint paths between every pair of nodes. A digraph is k-edge-connected if and only if there are k edge-disjoint paths from every node to every other.*

The equivalence of **(A3)** and **(B3)** was proved by W. T. Tutte [56].

Theorem 1.2 (Tutte). *A graph contains k edge-disjoint spanning trees if and only if, for every partition $\{V_1, \ldots, V_t\}$ of V, the number of edges connecting distinct parts is at least $k(t-1)$.*

Finally, the equivalence of definitions (**A4**) and (**B4**) was proved by J. Edmonds [9].

Theorem 1.3 (Edmonds). *A digraph D contains k edge-disjoint spanning arborescences rooted at s if and only if $\varrho_D(X) \geq k$ for every non-empty subset X of $V - s$.*

We extend these notions even further. For non-negative integers $l \leq k$, a digraph D is (k,l)-**edge-connected** if D has a node s so that there are k edge-disjoint paths from s to every other node and there are l edge-disjoint paths from every node to s. Equivalently, the digraph is l-edge-connected and rooted k-edge-connected. Note that D is (k,k)-edge-connected exactly if D is k-edge-connected, and $(k,0)$-edge-connectivity is equivalent to rooted k-edge-connectivity. We also remark that, by relying on max-flow min-cut computations, it is possible to decide in polynomial time if a digraph is (k,l)-edge-connected or not.

Another general notion is as follows. For two subsets S, T of nodes, D is said to be k-**edge-connected from S to T** if there are k edge-disjoint paths from every element of S to every element of T. In the special case $S = T$ we briefly say that D is k-**edge-connected in S**. If $S = T = V$ we are back at k-edge-connectivity. If $S = \{s\}$ and $T = V$ we arrive at rooted k-edge-connectivity. Also, for an undirected graph $G = (V, E)$ we say that G is k-**edge-connected in $S \subseteq V$** if there are k edge-disjoint paths in G between any two elements of S. A directed edge st with $s \in S$, $t \in T$ will be called an ST-**edge**.

We say that a partition of V into t non-empty parts is a t-**partition**. For a given partition $\mathcal{P}$ of V, the set of edges in a graph $G = (V, E)$ connecting distinct parts of $\mathcal{P}$ is called the **border** of $\mathcal{P}$. An element of the border is called a **cross-edge** of the partition. The border of a 2-partition is traditionally called a **cut**. For an integer l (which may be negative), we call an undirected graph $G = (V, E)$ (k,l)-**partition-connected** if the border of every t-partition of V ($t \geq 2$) has at least $k(t-1) + l$ elements. For $l \geq 0$, this definition attempts to capture the intuitive notion for higher edge-connection which requires that leaving out only few edges does not result in too many components.

A very first question concerning this notion is whether there exists a polynomially checkable certificate for a graph being (k, l)-partition-connected. The answer depends on whether $l \leq 0$, or $1 \leq l \leq k$, or $k < l$. If $l = 0$, we are back at k-partition-connectivity, and then the certificate (by Tutte's theorem) is a set of k disjoint spanning trees. When $l = -\gamma$ is negative, we will prove (Theorem 2.10) that *a graph is (k, l)-partition-connected if and only if it is possible to add γ new edges so that the resulting graph contains k disjoint spanning trees.* That is, in this case the certificate for (k, l)-partition-connectivity is k disjoint spanning trees whose union may contain γ new edges.

For $l \geq k$, we claim that (k, l)-partition-connectivity is equivalent to $(k + l)$-edge-connectivity. Indeed, if G is (k, l)-partition-connected, then the definition for $t = 2$ implies that every cut contains at least $k(t-1) + l = k+l$ edges, that is, G is $(k + l)$-edge-connected. Conversely, let G be $(k + l)$-edge-connected and let $\mathcal{P} := \{V_1, \ldots, V_t\}$ be a partition. By letting $e_G(\mathcal{P})$ denote the number of cross-edges of $\mathcal{P}$, we have $e_G(\mathcal{P}) = \sum_i^t d_G(V_i)/2 \geq (k+l)t/2 = tk + t(l-k)/2 \geq tk + (l-k) = k(t-1) + l$, and hence we conclude that G is (k, l)-partition-connected. Therefore we will be interested in (k, l)-partition-connectivity only if $l < k$.

Finally, for $0 < l < k$ one has the following characterization (Theorem 4.5): *a graph is (k, l)-partition-connected if and only if it has a (k, l)-edge-connected orientation.* Such an orientation may indeed serve as a certificate for (k, l)-partition-connectivity since a digraph can be tested for (k, l)-edge-connectivity by relying on Menger's theorem.

Given a groundset V, by a **co-partition** (of V) we mean a family of subsets consisting of the complementary sets of a partition of V. A family $\mathcal{F}$ of subsets of V is called a **sub-partition** of V if $\mathcal{F}$ is a partition of a subset of V. For a partition $\mathcal{F}$ of a non-empty proper subset Z of V, the family $\{V - X : X \in \mathcal{F}\}$ is called a **co-partition** of $V - Z$. For a subset X and for two elements x and y, we say that X is an $x\bar{y}$-set if $x \in X$, $y \notin X$.

For non-negative integers k, l, we call an undirected graph G (k, l)-**tree-connected** if deleting any subset of at most l edges leaves a k-tree-connected graph. By Tutte's theorem, G is (k, l)-tree-connected if and only if G is (k, l)-partition-connected.

In a graph $G = (V, E)$ the **local edge-connectivity** $\lambda(x, y; G)$ of nodes x and y is the minimum cardinality of a cut separating x and y. By Menger's theorem, this is equal to the maximum number of edge-disjoint paths connecting x and y. $e_G(X)$ denotes the number of edges with at least one endnode in X.

In a digraph $D = (V, E)$ the **local edge-connectivity** $\lambda(x, y; D)$ from node x to node y is the minimum number of edges entering a $y\bar{x}$-set. By Menger's theorem, this is equal to the maximum number of edge-disjoint paths from x to y. $\varrho(X)$ denotes the number of edges entering X and $\delta(X) := \varrho(V - X)$. For a graph or digraph H, $i_H(X)$ denotes the number of edges induced by X.

Typically we will work with directed or undirected graphs and write (di)graph when either of them is meant. Sometimes mixed graphs are also considered which may contain both directed and undirected edges.

2. Relations between old results

The three motivating theorems mentioned at the beginning of the introduction represent, respectively, the following general problem classes.

1. In a **connectivity augmentation** problem we want to add some new edges to a graph or digraph so that the resulting graph or digraph satisfies a prescribed connectivity property. In a **minimization problem** the number (or, more generally, the total cost) of new edges is to be minimized. In a **degree-specified** problem, in addition to the connectivity requirement, the (di)graph of the newly added edges must meet some (in)degree specification. Another aspect of augmentation problems distinguishes between the type of graphs of usable new edges. In a **restricted** augmentation the new edges must be chosen from a specified graph. We speak of a **free** augmentation if any possible edge is allowed to be added in any number of parallel copies. In the directed case, ST-**free** augmentations will also be considered when the new edges must be ST-edges.

2. In a **connectivity orientation** problem we want to orient the edges of an undirected graph so that the resulting digraph satisfies a prescribed connectivity property. The proof of Robbins' theorem is fairly easy (say, by ear-decomposition) but there are even easier orientation results: (A) *a graph G has a root-connected orientation (:every node is reachable from a root-node) if and only if G is connected*, and (B) *G has an orientation in which a specified node t is reachable from s if and only if s and t belong to the same* *component of G.* These are indeed so trivial that they deserve mentioning only because they serve as a good ground for possible generalizations.

3. In a **constructive characterization** problem we are interested in finding simple operations for a given connectivity property by which every (di)graph with the property may be obtained from a small initial (di)graph. It will turn out that this type of results often help proving connectivity orientation results.

In earlier survey type works ([21] [22], [23]) I endeavored to overview some aspects of connectivity orientations and augmentations with special emphasis on their relationship to sub- and supermodular functions. Therefore in the present paper those results are mentioned only when the overview of the developments of the past decade requires them. Exhibiting this progress is our main goal, with a special emphasis on some known and some newly discovered links connecting the different problems. Some new observations will also be outlined.

By comparing older results, this section is offered to demonstrate how closely the orientation, augmentation, and characterization problems are related to each other. But first a small remark is in order. The augmentation problem may be considered as one of finding a supergraph of a (di)graph with certain connectivity properties. This is naturally related to the sub-graph problem which consists of finding an optimal subgraph of a (di)graph satisfying connectivity requirements (sometimes called generalized Steiner network problem). The minimum cost versions of these problems are actually equivalent, and to explain this we invoke a specific subgraph versus supergraph problem-pair. Subgraph problem: given a digraph $D = (V, A)$ with specified nodes s and t endowed with a cost function c on A, find a minimum cost subdigraph D' of D which is k-edge-connected from s to t. Supergraph (=augmentation) problem: given a digraph $D = (V, A)$ with specified nodes s and t, moreover another digraph $H = (V, F)$ endowed with a cost function c_F on F, find a minimum cost augmentation of D which is k-edge-connected from s to t. Now if the subgraph problem is tractable, then so is the supergraph problem: Let $D_1 = (V, A \cup F)$ be the union of G and H and define a cost function c_1 on $A \cup F$ by $c_1(e) := 0$ if $e \in A$ and $c_1(e) := c_F(e)$ if $e \in F$. Obviously, an optimal solution to the subgraph problem on D_1 determines an optimal solution to the augmentation problem. Conversely, the subgraph problem can be viewed as an augmentation problem because it is equivalent to augment, at a minimum cost, of the empty digraph $(V, \emptyset)$ by using edges of D, (or wording differently, by using arbitrary edges but the ones not in D have cost $+\infty$). Typically we use this equivalence in one direction: when the minimum cost subgraph problem is tractable then so is the augmentation problem. In our concrete case the subgraph problem is indeed solvable with the help of a minimum cost flow algorithm. On the same ground, as the minimum cost connected subgraph problem is solvable with the greedy algorithm, the minimum cost augmentation problem, to make a given graph connected, is also solvable.

We hasten to emphasize however that in several cases the subgraph problem is NP-complete while the corresponding (free) augmentation problem is nicely solvable. A prime example for this phenomenon is the problem of finding a minimum cardinality 2-edge-connected subgraph of a graph G which is known to be NP-complete as it includes the Hamiltonian circuit problem (:the minimum is equal to $|V|$ if and only if G is Hamiltonian). On the other hand, the second introductory problem on the corresponding connectivity augmentation is solvable.

2.1. Splitting and augmentation

The following two splitting lemmas are central to several results. By **splitting off** a pair of undirected edges $e = zu, f = zv$ we mean the operation of replacing e and f by a new edge connecting u and v. In the directed case directed edges uz and zv are replaced by a directed edge uv.

Theorem 2.1 (Lovász's undirected splitting lemma [42]). *Let $k \geq 2$ be an integer and $G = (V + z, E)$ an undirected graph with a special node z of even degree. If G is k-edge-connected in V, then there is a pair of edges $e = zu, f = zv$ which can be split off without destroying k-edge-connectivity in V.*

Theorem 2.2 (Mader's directed splitting lemma [46]). *Let $k \geq 1$ be an integer and $D = (V + z, E)$ a directed graph with a special node z having the same in- and out-degree. If D is k-edge-connected in V, then there is a pair of edges $e = zu, f = vz$ which can be split off without destroying k-edge-connectivity in V.*

Both lemmas may be used repeatedly, as long as there are edges incident to z, and in this case we speak of a **complete splitting**. Sometimes by the splitting lemma this complete version is meant: *Under the same hypotheses,* there is a complete splitting at z so that the resulting (di)graph on node set V is k-edge-connected.

An easy observation shows that the existence of a complete undirected splitting that preserves k-edge-connectivity is equivalent to the following degree-specified augmentation result [19]. Here and throughout the paper, we use the notation $m(X) := \sum [m(v) : v \in X]$.

Theorem 2.3. *We are given an undirected graph* $G = (V, E)$, *a degree-specification* $m : V \to \mathbf{Z}_+$ *with* $m(V)$ *even, and an integer* $k \geq 2$. *There is a graph* $H = (V, F)$ *so that* $d_H(v) = m(v)$ *for every node* $v \in V$ *and* $G + H$ *is* k*-edge-connected if and only if* $m(X) \geq k - d_G(X)$ *for every non-empty subset* $X \subset V$.

This result was used in [19] to exhibit a short derivation of T. Watanabe and A. Nakamura's [57] earlier solution to the minimization form of the undirected edge-connectivity augmentation problem:

Theorem 2.4 (Watanabe and Nakamura). *An undirected graph* G *can be made* k*-edge-connected* ($k \geq 2$) *by adding at most* γ *new edges if and only if* $\sum_i [k - d_G(X_i)] \leq 2\gamma$ *for every subpartition* $\{X_1, \ldots, X_t\}$ *of* V.

Note that the last theorem fails to hold for $k = 1$. On the other hand, for this case, even the minimum cost version is solvable by the greedy algorithm since it is equivalent to the min-cost spanning tree problem (while for $k \geq 2$ the min-cost version is NP-complete.)

Mader's directed splitting lemma is also easily seen to be equivalent to the degree-specified directed edge-connectivity augmentation problem:

Theorem 2.5. *We are given a directed graph* $D = (V, E)$, *in- and out-degree specifications* $m_i : V \to \mathbf{Z}_+$ *and* $m_o : V \to \mathbf{Z}_+$ *so that* $m_i(V) = m_o(V)$. *Let* $k \geq 1$ *be an integer. There is a digraph* $H = (V, F)$ *so that* $\delta_H(v) = m_o(v)$, $\varrho_H(v) = m_i(v)$ *for every node* $v \in V$ *and so that* $D + H$ *is* k*-edge-connected if and only if* $m_i(X) \geq k - \varrho_D(X)$ *and* $m_o(X) \geq k - \delta_D(X)$ *holds for every non-empty subset* $X \subset V$.

This implies the minimization form of directed edge-connectivity augmentation [19]:

Theorem 2.6. *A digraph* $D = (V, E)$ *can be made* k*-edge-connected* ($k \geq 1$) *by adding at most* γ *directed edges if and only if* $\sum_i [k - \varrho_D(X_i)] \leq \gamma$ *and* $\sum_i [k - \delta_D(X_i)] \leq \gamma$ *hold for every subpartition* $\{X_1, \ldots, X_t\}$ *of* V.

2.2. Connectivity orientation and augmentation

The easy orientation results mentioned above concerning strong-connectivity, connectivity from s to t, and s-rooted 1-edge-connectivity naturally raise questions on higher connection: when does a graph G have an orientation which is (a) k-edge-connected from s to t, (b) rooted k-edge-connected, (c) k-edge-connected? Among these, the first one is easy (given Menger's theorem).

Theorem 2.7. *For integers $k_1, k_2 \geq 0$ and specified nodes $s, t \in V$, an undirected graph $G = (V, E)$ has an orientation which is k_1-edge-connected from s to t and k_2-edge-connected from t to s if and only if every cut of G separating s and t has at least $k_1 + k_2$ edges.*

Proof. The necessity of the condition is straightforward. The sufficiency follows by observing that the condition implies, by Menger's theorem, the existence of $k_1 + k_2$ edge-disjoint paths between s and t. One can orient the edges of k_1 paths toward t, the edges of the remaining k_2 paths toward s, and the remaining edges arbitrarily. ■

The first non-trivial result concerning orientation is due to C. St. J. A. Nash-Williams [47]. He proved the following extension of Robbins' theorem (actually in a much stronger form).

Theorem 2.8 (Nash-Williams: weak form). *An undirected graph G has a k-edge-connected orientation if and only if G is $2k$-edge-connected.*

By a straightforward induction, Lovász's undirected splitting lemma implies Nash-Williams' theorem. When rooted k-edge-connectivity is the target in the orientation problem, one has the following result.

Theorem 2.9. *An undirected graph $G = (V, E)$ has a rooted k-edge-connected (that is, $(k, 0)$-edge-connected) orientation if and only if G is k-partition-connected.*

The non-trivial 'if' part is an easy consequence of Theorem 1.2 on disjoint trees since Tutte's theorem implies that a k-partition-connected graph contains k disjoint spanning trees and, by orienting each of these trees away from the root (to become a spanning arborescence) while the remaining edges arbitrarily, one obtains a rooted k-edge-connected orientation of G.

On the other hand, Theorem 2.9, when combined with Edmonds Theorem 1.3, gives rise to Tutte's Theorem 1.2. At this point the question naturally emerges: if the required orientations do not exist, then how many new undirected edges have to be added so that the augmented graph admits an orientation?

The answer is evident when the goal is to augment a graph so as to become k-edge-connected orientable. Namely, by Nash-Williams' theorem this is equivalent to augmenting the graph to make it $2k$-edge-connected, a problem solved in Theorems 2.4 and 2.3. Suppose now we want to augment G to become k-tree-connected (= k-partition-connected). For the special case of free augmentation one has the following:

Theorem 2.10. *Let $G = (V, E)$ be an undirected graph, $s \in V$ a specified node, and γ a nonnegative integer. It is possible to add at most γ new edges to G so that the enlarged graph has an s-rooted k-edge-connected orientation if and only if G is $(k, -\gamma)$-partition-connected. Moreover, all the newly added edges may be chosen to be incident to s.*

Proof. Recall that by definition G is $(k, -\gamma)$-partition-connected if

(1) $$e(\mathcal{F}) \geq k(t - 1) - \gamma$$

holds for every partition $\mathcal{F} := \{V_1, \ldots, V_t\}$ of V, where $e(\mathcal{F})$ denotes the number of cross edges of $\mathcal{F}$. For brevity we call an orientation **good** if it is k-edge-connected from s. If there is a good orientation after adding γ edges, then $\varrho(V_i) \geq k$ holds for every subset $V_i \subset V$ not containing s and hence $e(\mathcal{F}) + \gamma \geq e^+(\mathcal{F}) \geq k(t - 1)$, where e^+ refers to the enlarged graph, proving the necessity of the condition.

To see the sufficiency, add a minimum number of new edges to G, each incident to s so that the enlarged graph has a good orientation and let γ' denote this minimum. Our goal is to prove $\gamma' \leq \gamma$.

Let ϱ denote the in-degree function of the good orientation of the enlarged graph G^+. We may assume that $\varrho(s) = 0$. Let us call a set $X \subseteq V - s$ **tight**, if $\varrho(X) = k$. By standard submodular technique, we see that both the intersection and the union of two tight sets with non-empty intersection are tight. Let T denote the subset of nodes which can be reached from the head of at least one new edge. Clearly, $s \notin T$ and $\varrho(V - T) = 0$.

Lemma 2.11. *If Z is tight and $Z \cap T \neq \emptyset$, then $Z \subseteq T$.*

Proof. Suppose indirectly that $Z \not\subseteq T$. Then for $Y := V - T$ we have
$$k = \varrho(Y) + \varrho(Z) = \varrho(Y \cap Z) + \varrho(Y \cup Z) + d^+(Y, Z) \geq k + 0 + d^+(Y, Z) \geq k,$$
where $d^+(Y, Z)$ denotes the number of edges of G^+ connecting elements of

$Y - Z$ and $Z - Y$. Hence $\varrho(Y \cup Z) = 0$ and $d^+(Y, Z) = 0$. From the first equality there is a new edge $e = st$ for which $t \in Z$ for otherwise no element of $Z \cap T$ would be reachable from the head of any new edge. But then, by the existence of edge e, we have $d^+(Y, Z) > 0$, a contradiction. ∎

There are two cases. If there is a node v in T which does not belong to any tight set, then let st be a new edge for which there is a path P from t to v. Reorient each edge of P and discard e. Since v does not belong to any tight set the revised orientation is good, contradicting the minimality of γ'.

In the second case every element of T belongs to a tight set. Let $V_1, \ldots, V_{t-1}$ be maximal tight sets intersecting T. These are pairwise disjoint and by the lemma they form a partition of T. Let $V_t := V - T$ and $\mathcal{F} := \{V_1, \ldots, V_t\}$. Since $\varrho(V_t) = 0$, and every new edge enters T, we get $k(t - 1) = \sum \left[\varrho(V_i) : i = 1, \ldots, (t - 1) \right] = \sum \left[\varrho(V_i) : i = 1, \ldots, t \right] = i^+(\mathcal{F}) = e(\mathcal{F}) + \gamma'$. This and (1) give rise to $\gamma' = k(t - 1) - e(\mathcal{F}) \le \gamma$, as required. ∎

By combining Theorems 2.10 and 2.9, we obtain the following extension of Tutte's Theorem 1.2 which serves as a characterization of (k, l)-partition-connected graphs in case $l \le 0$.

Theorem 2.12. *An undirected graph $G = (V, E)$ can be augmented by adding $\gamma \ge 0$ new edges so that the enlarged graph is k-tree-connected if and only if G is $(k, -\gamma)$-partition-connected. Moreover, the newly added edges may be chosen to be incident to any given node in V.*

The theorem shows that the free augmentation problem is tractable for k-tree-connectivity as a target. This is, however, not surprising since, by using matroid techniques, even the minimum cost version is solvable in polynomial time. To see this, let $G = (V, E)$ be an undirected graph and let $G_u = (V, E_u)$ be a graph, where E_u is the set of edges usable in the augmentation of G. Let $c_u : E_u \to \mathbf{R}_+$ be a cost function. We want to choose a subset F of edges of G_u of minimum total cost so that the increased graph $G^+ = (V, E + F)$ is k-tree-connected.

To this end, let us define a cost function c' on the edge set of the union $G' = (V, E + E_u)$ of G and G_u so that $c'(e) := 0$ if $e \in E$ and $c'(e) = c(e)$ if $e \in E_u$. Then the problem is equivalent to finding k disjoint spanning trees of G' with minimum total cost. Since the edge-sets which are the union of k disjoint spanning trees form the set of bases of a matroid, this problem is solvable in polynomial time by using Edmonds' matroid partition algorithm and the greedy algorithm. This approach also shows that Edmonds' matroid partition theorem does provide a characterization for the existence of the required augmentation in Theorem 2.10. Our goal has simply been to show a direct, graphical proof.

One may also consider the degree-specified version of the k-tree-connected augmentation problem. This does not seem to be a matroid problem and it does not follow from the previous material either. Section 4 includes an answer even for the more general case of (k, l)-partition-connectivity.

2.3. Constructive characterization and splitting

Let $G' = (V + z, E')$ be an undirected graph with a special node z of even degree and suppose that G' is k-edge-connected in V. By the undirected splitting lemma we know that there is a complete splitting at z so that the resulting graph $G = (V, E)$ is k-edge-connected. In other words, the $d(z)$ edges incident to z can be paired so that splitting off these $j := d(z)/2$ pairs (and discarding z) we obtain a k-edge-connected graph. In a directed graph $D' = (V + z, A')$ a complete splitting at z consists of pairing the edges entering z with those leaving z and then splitting off the pairs. Both in the directed and in the undirected cases the inverse operation of a complete splitting is as follows. *Add a new node z, subdivide j existing edges by new nodes and identify the j subdividing nodes with z.* This will be called **pinching** j **edges (with** z**).** When $j = 0$ this means adding a single new node z, while in case $j = 1$ pinching an edge requires the edge to be subdivided by a node z.

By the operation of adding a new edge to a (di)graph we always mean that the new edge connects existing nodes. Unless otherwise stated, the newly added edge may be a loop or may be parallel to existing edges.

After these definitions, we exhibit how the splitting lemmas give rise to constructive characterizations of $2k$-edge-connected graphs and k-edge-connected digraphs. By using the easy observation that a minimally (with respect to edge-deletion) K-edge-connected undirected graph (with at least two nodes) always contains a node of degree K, one can easily derive from the undirected splitting lemma the following constructive characterization of $2k$-edge-connected graphs.

Theorem 2.13 (Lovász). *An undirected graph $G = (V, E)$ is $2k$-edge-connected if and only if G can be obtained from a single node by the following two operations:* (i) *add a new edge,* (ii) *pinch k existing edges.*

By using a rather difficult theorem of Mader [44], stating that *a minimally (with respect to edge-deletion) k-edge-connected directed graph (with at least two nodes) always contains a node of in-degree and out-degree k*, one can derive from the directed splitting lemma the following constructive characterization of k-edge-connected digraphs.

Theorem 2.14 (Mader). *A directed graph $D = (V, E)$ is k-edge-connected if and only if D can be obtained from a single node by the following two operations: (i) add a new edge, (ii) pinch k existing edges.*

It is useful to observe that Mader's characterizaton in Theorem 2.14 for k-edge-connected digraphs combined with Nash-Williams' orientation result give rise to Theorem 2.13. The same phenomenon will occur later as well: with the help of an orientation result, a constructive characterization for directed graphs may be used to derive its undirected counterpart.

By an easy reduction, Theorem 2.14 provides a constructive characterization of rooted k-edge-connected digraphs.

Theorem 2.15. *A digraph $D = (V, E)$ is rooted k-edge-connected if and only if D can be built up from a root-node s by the following two operations: (j) add a new edge, (jj) pinch i $(0 \leq i \leq k - 1)$ existing edges with a new node z, and add $k - i$ new edges entering z and leaving existing nodes.*

In [46] Mader showed that this characterization, in turn, can be used to derive Edmonds' Theorem 1.3 on disjoint arborescences. Combining Theorems 2.9 and 2.15, one obtains the following constructive characterization.

Theorem 2.16. *An undirected graph $G = (V, E)$ is k-tree-connected ($= k$-partition-connected) if and only if G can be built from a node by the following two operations: (j) add a new edge, (jj) pinch i $(0 \leq i \leq k - 1)$ existing edges with a new node z, and add $k - i$ new edges connecting z with existing nodes.*

3. Splitting and detachment

In this section first we exhibit extensions of the splitting lemmas of section 2 and of their applications. After that the notion of splitting will be extended to detachments.

3.1. Undirected splitting

As a significant generalization of Lovász's undirected splitting lemma, W. Mader [45] proved the following result. Recall (from the introduction) the definition of local edge-connectivity λ.

Theorem 3.1 (Mader). *Let $G = (V + z, E)$ be an undirected graph so that there is no cut-edge incident to z and the degree of z is even. Then there exists a complete splitting at z preserving the local edge-connectivities of all pairs of nodes $u, v \in V$.*

Mader originally formulated his result in a slightly weaker form: *If z is not a cut-node of $G = (V + z, E)$ and $d(z) \geq 4$, then there exists a pair of edges incident to z which can be split off without lowering any local edge-connecivity on V.* However the two forms can be shown to be equivalent. This and a relatively short proof of Mader's theorem was given in [20].

3.1.1. Constructive characterizations. Mader [45] used his result to characterize $(2k + 1)$-edge-connected graphs.

Theorem 3.2 (Mader). *Let $K = 2k + 1 \geq 3$. An undirected graph $G = (U, E)$ is K-edge-connected if and only if G can be constructed from the initial graph of two nodes connected by K parallel edges by the following three operations:*

(i) *add an edge,*

(ii) *pinch k edges with a new node z' and add an edge connecting z' with an existing node,*

(iii) *pinch k edges with a new node z', pinch then again in the resulting graph k edges with another new node z so that not all of these k edges are incident to z', and finally connect z and z' by a new edge.*

The theorem is obviously equivalent to the first part of the following result:

Theorem 3.3. *An undirected graph G with more than two nodes is K-edge-connected (K odd) if and only if G can be obtained from a (smaller) K-edge-connected graph G' by one application of one of the operations (i), (ii), (iii). Moreover, for any node s of G, G' can be chosen so as to contain s.*

Proof. It is not difficult to check that each of these operations preserves K-edge-connectivity. (Note that if all the k edges to be pinched with z' in the second part of (iii) were adjacent to z, then only $K - 1 = 2k$ edges would leave the subset $\{z, z'\}$.)

For a subset $X \subseteq V$, the set of edges connecting X and $V - X$ will be denoted by $[X, V - X]$. We call a cut $[X, V - X]$ **trivial** if $|X| = 1$ or $|V - X| = 1$. By a **minimum cut** we mean one with cardinality K.

Lemma 3.4. *Suppose that X is a minimal subset of nodes of a K-edge-connected graph $G = (U, E)$ for which*

$$(2) \qquad\qquad d_G(X) = K \quad \text{and} \quad |X| \geq 2.$$

Then any minimum cut B containing an edge $e = zz'$ with $z, z' \in X$ is trivial (that is, B is $[z, U - z]$ or $[z', U - z']$).

Proof. Suppose indirectly that there is a subset Y for which $z \in Y$, $z' \in U - Y$, $d(Y) = K$, $|Y| \geq 2$, $|U - Y| \geq 2$. Then by the minimal choice of X we have $Y \not\subseteq X$ and $U - Y \not\subseteq X$. But it is well-known (and an easy exercise anyway to show) that in a K-edge-connected graph with K odd there cannot exist two such crossing sets X, Y. (Indeed, we have $K + K = d(X) + d(Y) = d(X \cap Y) + d(X \cup Y) + 2d(X, Y) \geq K + K + 0$ from which $d(X \cap Y) = K = d(X \cup Y)$ and $d(X, Y) = 0$, where $d(X, Y)$ denotes the number of edges connecting $X - Y$ and $Y - X$. Analogously, we obtain for $\bar{Y} := U - Y$ that $d(X \cap \bar{Y}) = K = d(X \cup \bar{Y})$ and $d(X, \bar{Y}) = 0$. So if $\alpha := d(X \cap Y, Y - X)$, then $d(X \cap Y, X - Y) = K - \alpha = d(Y - X, U - (X \cup Y))$ from which $K = d(Y) = d(X \cap Y, X - Y) + d(Y - X, U - (X \cup Y)) = 2K - 2\alpha$, that is, K is even, a contradiction.) ∎

If there is an edge e so that $G' := G - e$ is K-edge-connected, then G arises from G' by (i). So we may assume that G is minimally K-edge-connected. We may assume that there is no node z which is connected only with s since otherwise, then by the minimality, $d(z) = K$ and then G arises from G' by operation (ii) where G' is a graph arising from G by deleting z and adding k loops at s. (Clearly G' is K-edge-connected.)

If every minimum cut is trivial, then let $e = zz'$ be an arbitrary edge not incident to s. If there are non-trivial minimum cuts, then there is a set X satisfying (2). Since the complement of X also satisfies (2), there exists a minimal set X satisfying (2) so that $s \notin X$.

Let $e = zz'$ be an arbitrary edge induced by X. As X induces a connected subgraph, such an e exists. Now e belongs to at most two minimum cuts, each is trivial. If e belongs to one minimum cut, than exactly one of z and z', say z, is of degree K. Then $G - e$ is K-edge-connected in $U - z$. By Lovász's splitting lemma there is a complete splitting at z resulting in a K-edge-connected digraph G'. Then G arises from G' by operation (ii).

If both z and z' are of degree K, then $G - e$ is K-edge-connected in $U - \{z, z'\}$. It follows from Mader's splitting Theorem 3.1 that there is a complete splitting of $G - e$ at z so that the resulting graph G_1 is K-edge-connected in $U - \{z, z'\}$. By applying the splitting lemma to G_1 (now Lovász's is enough), we obtain that there is a complete splitting at z' so that the resulting graph G' with node set $U - \{z, z'\}$ is K-edge-connected. This construction shows that G arises from G' by operation (iii).

Since in each case z and z' were chosen to be distinct from s, we have also proved the second half of the theorem. ■ ■

Operation (iii) may seem to be a bit too complicated and one's natural wish could be to try to simplify it. For example, a simpler, more symmetric version could be as follows: (iii)′ choose two disjoint subsets F and F' of edges both having k elements, pinch the elements of F with a new node z, pinch the elements of F' with another new node z', and finally connect z and z'. However, Mader in his original paper showed an example which cannot be obtained with operations (i), (ii), (iii)′.

Fortunately, for $K = 3$, operations (iii) and (iii)′ coincide and it is worthwhile to formulate this special case separately:

Corollary 3.5. *An undirected graph G with at least two nodes is 3-edge-connected if and only if G can be built from a node by the following operations:*

(i) *add an edge,*

(ii) *subdivide an existing edge $e = uv$ by a new node z and connect z to an existing node,*

(iii) *subdivide two existing edges by nodes z and z' and connect z and z' by a new edge.*

3.1.2. Orientation.

Lovász's splitting lemma immediately implied Nash-Williams' orientation theorem (:a $2k$-edge-connected graph always has a k-edge-connected orientation). In [29] we observed that Mader's splitting theorem also rather easily gives rise to the following common generalization of theorems 2.8 and 2.7.

Theorem 3.6. *Let k_1, k_2, k be non-negative integers with $k_1 \geq k$, $k_2 \geq k$. An undirected graph $G = (V, E)$ with two specified nodes s and t has a k-edge-connected orientation which is k_1-edge-connected from s to t and k_2-edge-connected from t to s if and only if G is $2k$-edge-connected and G is $(k_1 + k_2)$-edge-connected in $\{s, t\}$.*

This immediately implies a characterization of $(2k + 1)$-edge-connected graphs.

Theorem 3.7. *An undirected graph G is $(2k + 1)$-edge-connected if and only if, for every pair of nodes s and t, G has a k-edge-connected orientation which is $(k + 1)$-edge-connected from s to t.*

Given the easy way how Lovász's splitting lemma implies the weak form of Nash-Williams orientation theorem, one may expect that Mader's stronger splitting result implies immediately the following stronger orientation result of Nash-Williams [47]:

Theorem 3.8 (Nash-Williams: strong form)**.** *Every undirected graph $G = (V, E)$ has an orientation $\vec{G}$ for which $\lambda(x, y; \vec{G}) \geq \lfloor \lambda(x, y; G)/2 \rfloor$ for all $x, y \in V$.*

Mader was indeed able to derive Theorem 3.8 relying on his splitting theorem but the derivation is not at all simple (as neither is Nash-Williams' original proof).

In the introduction of his paper, Nash-Williams [48] remarks that his orientation theorems 'do not seem particularly closely related to much other existing work in graph theory'. These words are painfully true even after 40 years as far as the strong form is concerned, and it remains a major task to find a simple proof of Theorem 3.8 or at least to find some closer link to the body of edge-connectivity problems. Note that by now pretty much is known about the various connections of the weak form along with its numerous strengthenings and extensions. Nash-Williams also remarks that 'these theorems seem to have a somewhat natural character which would suggest that there must ultimately be a place for them in the overall structure of graph theory'. Since then it has turned out that wherever this place is located, it is not a lonely one.

Nash-Williams calls an orientation with the property given in the theorem **well-balanced.** He actually proved the existence of a well-balanced orientation that is, in addition, **near-Eulerian** which means by definition that $\left| \varrho(v) - \delta(v) \right| \leq 1$ for every node v of $\vec{G}$. Nash-Williams also outlined the proof of the following generalization of Theorem 3.8.

Theorem 3.9 [47]. *Let G be a graph and H a subgraph of G. Then G has a well-balanced and near-Eulerian orientation with the additional property that its restriction to H is a well-balanced and near-Eulerian orientation of H.*

Corollary 3.10. *Let $G = (V, E)$ be a $2k$-edge-connected graph and $H = (V, F)$ an Eulerian subgraph of G. For any Eulerian orientation of H, the edges in $E - F$ can be oriented so as to obtain a k-edge-connected orientation of G.*

This implies that in order to find a k-edge-connected orientation of a $2k$-edge-connected graph G one can pick up edge-disjoint circuits one after the other and orient them around. The corollary ensures that the remaining forest can always be oriented to get a k-edge-connected orientation of G. It would be interesting to see a direct constructive proof of this fact which does not rely on Theorem 3.9. We note that there is an easy alternative proof of Corollary 3.10 relying on submodular flows.

3.1.3. Augmentation. Let us turn to the effect of Mader's theorem on connectivity augmentation. The same way as Lovász's splitting lemma could be used for solving (global) connectivity augmentation, Mader's splitting theorem gives rise to a solution of the local edge-connectivity augmentation problem. Let $G = (V, E)$ be an undirected graph and r a non-negative integer-valued function on unordered pairs $\{u, v\}$ of distinct nodes of G, called a **requirement function.** In the local edge-connectivity augmentation problem we want to augment G so that the local edge-connectivity in the increased graph G^+ majorizes r. By Menger's theorem this is equivalent to requiring

$$(3) \qquad d_{G^+}(X) \geq R_r(X) \text{ for every subset } X \subset V,$$

where

$$(4) \qquad R_r(X) := \max \{ r(u, v) : u \in X, v \in V - X \}.$$

The following two results appeared in [19].

Theorem 3.11. *Let $G = (V, E)$ be an undirected graph. Let $m : V \to \mathbf{Z}_+$ be an integer-valued function so that $m(V)$ is even and $m(C) \geq 2$ for each component C of G. There is a set F of new edges so that the local edge-connectivity in $G^+ = (V, E + F)$ is at least r and $d_F(v) = m(v)$ for every node v if and only if*

$$(5.10) \qquad m(X) \geq R_r(X) - d_G(X)$$

for every $X \subseteq V$.

Let $C(\neq V)$ be the node-set of a component of G and call C a **marginal component** (with respect to r) if $R_r(C) \leq 1$ and $R_r(X) \leq d_G(X)$ for every proper subset X of C. Let $q(X) := R_r(X) - d_G(X)$ for $X \subset V$.

Theorem 3.12. *Suppose that there are no marginal components. There is a set F of at most γ edges so that the local edge-connectivity in $G^+ = (V, E + F)$ is at least r if and only if*

$$(5) \qquad\qquad \sum_i q(X_i) \leq 2\gamma$$

holds for every sub-partition $\{X_1, X_2, \ldots, X_t\}$ *of* V.

In [1], J. Bang-Jensen, H. Gabow, T. Jordán and Z. Szigeti investigated the augmentation problem when the possible set of new edges meets a partition constraint. Among their numerous results, we cite here only one:

Theorem 3.13. *Let $G = (V, E)$ be an undirected graph and $\mathcal{P} = \{P_1, \ldots, P_r\}$ a partition of V into at least two non-empty parts. Let $k \geq 2$ be an even integer. It is possible to add at most γ new edges to G each connecting distinct parts of $\mathcal{P}$ so that the resulting graph is k-edge-connected if and only if $\sum_{X \in \mathcal{F}} \left[k - d(X) : X \in \mathcal{F} \right] \leq 2\gamma$ holds for every subpartition $\mathcal{F}$ of V, and $\sum_{X \in \mathcal{F}_i} \left[k - d(X) : X \in \mathcal{F}_i \right] \leq \gamma$ holds for every subpartition $\mathcal{F}_i$ of P_i $(i = 1, \ldots, r)$.*

It is not difficult to check that the conditions in the theorem are necessary for even and odd k, as well. For odd k, however, they are not sufficient. But [1] did provide a characterization even for this more complicated case.

3.2. Directed splitting

Can one extend Mader's directed splitting lemma so as to preserve local edge-connectivities in directed graphs? No such a general result is known but some extensions of the directed splitting lemma are available. The following is a consequence of a result in [22].

Theorem 3.14. *Let $k \geq l \geq 1$ be integers and $D = (V + z, E)$ a directed graph with a special node z having the same in- and out-degree. If D is (k, l)-edge-connected in V, then there is a pair of edges $e = zu$, $f = vz$ which can be split off without destroying (k, l)-edge-connectivity in V.*

This result was proved in [22] in a more general form concerning coverings of crossing supermodular functions by digraphs. It can be used to solve the free- and the degree-specified augmentation problem for digraphs when the target is (k, l)-edge-connectivity. Let $D = (V, E)$ be a digraph with a root-node s and let $0 \leq l \leq k$ be integers. Define $p_{kl}(X) := \left(k - \varrho_D(X) \right)^+$ if $\emptyset \subset X \subset V - s$ and $p_{kl}(X) := \left(l - \varrho_D(X) \right)^+$ if $s \in X \subset V$.

Theorem 3.15. *For in- and out-degree specifications $m_i : V \to \mathbf{Z}_+$ and $m_o : V \to \mathbf{Z}_+$ with $m_i(V) = m_o(V)$, there is a digraph $H = (V, F)$ so that $\delta_H(v) = m_o(v)$, $\varrho_H(v) = m_i(v)$ for every node $v \in V$ and so that $D + H$ is (k, l)-edge-connected with respect to root s if and only if $m_i(X) \geq p_{kl}(X)$ and $m_o(V - X) \geq p_{kl}(X)$ holds for every non-empty subset $X \subset V$.*

Theorem 3.16. *There is a digraph $H = (V, F)$ of at most γ edges so that $D + H$ is (k, l)-edge-connected with respect to root s if and only if $\sum \left[p_{kl}(X) : X \in \mathcal{F} \right] \leq \gamma$ and $\sum \left[p_{kl}(V - X) : X \in \mathcal{F} \right] \leq \gamma$ hold for every partition $\mathcal{F}$ of V.*

3.3. Undirected detachment

Let $G = (V + z, E)$ be an undirected graph. We modify slightly the operation of splitting off a pair of edges $e = uz, f = vz$ as follows. Replace e and f by a new edge $h = uv$ and subdivide then h by a new node z'. More generally, by a **detachment** of node z into p nodes we mean the following operation. Replace z by p new nodes $z_1, \ldots, z_p$ and replace each edge uz by an edge uz_i. If the degree of each new node z_i is required to be a specified number d_i, we speak of a degree-specified detachment of z. In order for this to make sense we assume that $d_1, \ldots, d_p$ add up to $d_G(z)$.

Theorem 3.17 (Nash-Williams, [50]). *Let $G = (U, E)$ be a graph with a given positive integer $p(z)$ at every node z. It is possible to detach each node z into $p(z)$ parts so that the resulting graph is connected if and only if*

$$(6) \qquad\qquad e(X) \geq p(X) + c_G(X) - 1$$

holds for every non-empty subset $X \subseteq V$, where $p(X) := \sum \left[p(v) : v \in X \right]$, $e(X)$ is the number of edges having at least one end-node in X, and $c_G(X)$ denotes the number of components of $G - X$.

Note that Nash-Williams pointed out that this type of detachment can be handled as a matroid partition problem.

Suppose now that we are given at each node z of a graph $G = (U, E)$ a degree specification $d_1(z), \ldots, d_{p(z)}(z)$. Nash-Williams showed that it is possible to detach simultaneously all nodes so that there exists a degree-specified detachment of all nodes so that the resulting graph is connected if and only if (6) holds and $d_i(z) \geq 1$ for each i and $z \in V$.

What if we want a detachment which is k-edge-connected for $k \geq 2$? Clearly, for the existence of such detachment it is necessary that G be k-edge-connected and that each $d_i(z)$ is at least k. This is not always sufficient and we exhibit even two examples to show that. Let k be odd. First, suppose G consists of just two nodes u and v connected with $2k$ parallel edges, and $d_1(u) = d_2(u) = k = d_1(v) = d_2(v)$. Second, suppose that G has a cut node z of degree $2k$ and $d_1(z) = d_2(z) = k$. It is not difficult two check that no k-edge-connected detachment exists in either case. Quite surprisingly, there are no other bad cases:

Theorem 3.18 (Nash-Williams, [50]). *Let $G = (V, E)$ be an undirected graph with a degree specification $d_1(z), \ldots, d_{p(z)}(z)$ at each node z. It is possible to detach each node z into $p(z)$ nodes having specified degrees so that the resulting graph is k-edge-connected if and only if G is k-edge-connected, each requested degree $d_i(z)$ is at least k, except if k is odd and G is one of the two exceptional examples mentioned above.*

How is this result related to Lovász's undirected splitting lemma? They are not really comparable (in the sense that neither implies the other.) The splitting lemma detaches only one node, into nodes of degree two, and is clearly not 'interested' in preserving k-edge-connectivity at the detached nodes. But there is a very nice result of B. Fleiner [13] which is a generalization of Lovász's splitting lemma on one hand and implies easily Theorem 3.18 on the other.

The splitting lemma asserted that if G was k-edge-connected on V then a k-edge-connectivity preserving splitting always existed. If there are odd numbers in the degree-specification of the detachment, then this is not necessarily true. Let G consist of two disjoint triangles plus a node z connected to all the other six nodes. Then G is 3-edge-connected on V (even the whole G is) but it is not possible to detach z into two nodes of degree 3 so that the resulting graph keeps to be 3-edge-connected on V.

Theorem 3.19 (Fleiner). *Let $G = (V + z, E)$ be an undirected graph with a special node z and $k \geq 2$ an integer. Let $d_1, \ldots, d_p$ be integers for which $d_i \geq 2$, $\sum d_i = d_G(z)$. It is possible to detach z into p nodes of degree $d_1, \ldots, d_p$, respectively, so that the resulting graph is k-edge-connected in V if and only if G is k-edge-connected in V and $G - z$ is k'-edge-connected where*

$$(7) \qquad k' := k - \sum_{i=1}^{p} \lfloor d_i/2 \rfloor.$$

Note that if each d_i is even, then $G - z$ is automatically k'-edge-connected so we do not have to explicitly require it, and this special case follows immediately from the undirected splitting lemma. As Lovász's splitting lemma could be used to derive Watanabe and Nakamura's Theorem 2.4 on minimum k-edge-connected augmentation of a graph, Fleiner used his result to prove the following generalization [13].

Theorem 3.20 (Fleiner). *Let $G = (V, E)$ be an undirected graph and $d_1, \ldots, d_p$ and k integers larger than one. It is possible to augment G by adding p new nodes of degree d_i, respectively, so that the enlarged graph G^+ is k-edge-connected on V if and only if*

$$(8) \qquad \sum \left[(k - d_G(X)) \ : \ X \in \mathcal{F} \right] \leq \sum_{i=1}^{p} d_i$$

holds for every sub-partition $\mathcal{F}$ of V, and

$$(9) \qquad \lambda(u, v; G) \geq k - \sum_{i=1}^{p} \lfloor d_i/2 \rfloor$$

holds for every pair of nodes $u, v \in V$, that is, G is k'-edge-connected, where $k' := k - \sum_{i=1}^{p} \lfloor d_i/2 \rfloor$.

So, Fleiner's Theorem 3.19 is one generalization of the undirected splitting lemma while Mader's Theorem 3.1 is another. Does there perhaps exist a common generalization of these difficult theorems? Yes, T. Jordán and Z. Szigeti proved the following theorem [34].

Theorem 3.21 (Jordán and Szigeti). *Let $G = (V + z, E)$ be a graph with a special node z so that there is no cut-edge incident to z. Let $d_1, \ldots, d_p$*

be integers for which $d_i \geq 2$, $\sum d_i = d_G(z)$. Also, we are given a symmetric function $r(u,v)$ on the pairs of nodes in V. There is a detachment of z into p nodes of degree $d_1, \ldots, d_p$, respectively, so that in the resulting graph G' the local edge-connectivity $\lambda(u,v;G')$ is at least $r(u,v)$ for every $u,v \in V$ if and only if

$$(10) \qquad r(u,v) \leq \lambda(u,v;G) \text{ and } \lambda(u,v;G-z) \geq r(u,v) - \sum_{i=1}^{p} \lfloor d_i/2 \rfloor$$

for all $u,v \in V$.

In the augmentation results so far we always added edges to an existing graph $G = (V,E)$. This may be interpreted as adding new nodes of degree two so that the (local) edge-connectivity should attain a certain prescribed value. It is quite natural to investigate an extension of the problem when the newly added nodes are of prescribed degree, not necessarily two. The following result of Jordán and Szigeti [34] is a straight generalization of Theorem 3.12. As in Theorem 3.12, we are given an undirected graph $G = (V,E)$ and a symmetric non-negative integer-valued function $r(u,v)$ on the pair of nodes, called local edge-connectivity requirement. Let $R_r(X) :=$ max$\{r(u,v) : u \in X, v \in V - X\}$ for every $X \subseteq V$ and let $q(X) := R_r(X) - d_G(X)$. Recall the definition from (4) of $R_r(X)$, $q(X)$ and a marginal component of G.

Theorem 3.22 (Jordán and Szigeti [34]). *Let $G = (V,E)$ be an undirected graph, $r(u,v)$ a local edge-connectivity requirement function so that there are no marginal components. Moreover, let $d_1, d_2, \ldots, d_p$ be integers each larger than 1. It is possible to add to G p new nodes of degree d_i, respectively, so that the enlarged graph G^+ satisfies $\lambda(u,v;G^+) \geq r(u,v)$ for every pair of nodes $u,v \in V$ if and only if*

$$(11) \qquad \sum \left[q(X) : X \in \mathcal{F} \right] \leq \sum_{i=1}^{p} d_i$$

holds for every sub-partition $\mathcal{F}$ of V, and

$$(12) \qquad \lambda(u,v;G) \geq r(u,v) - \sum_{i=1}^{p} \lfloor d_i/2 \rfloor$$

holds for every pair of nodes $u,v \in V$.

3.4. Directed detachment

In Mader's directed splitting lemma, it was assumed for the specified node z to have the same in- and outdegree. Without this restriction a splitting at z preserving k-edge-connectivity in V does not necessarily exist. However, Berg, Jackson and Jordán [5] found the following interesting extension of the splitting lemma.

Theorem 3.23 (Berg, Jackson, Jordán). *Let $k \geq 1$ be an integer and $D = (V + z, E)$ a directed graph with a special node z for which $\varrho(z) \geq \delta(z)$. If D is k-edge-connected on V, then for every edge zu there are t edges $v_1 z, \ldots, v_t z$, where $1 \leq t \leq \varrho(z) - \delta(z) + 1$, entering z so that detaching z into two nodes z' and z_1 results in a digraph which is k-edge-connected on V, where z_1 has one outgoing edge $z_1 u$ and t entering edges $v_1 z_1, \ldots, v_t z_1$.*

By repeated applications of the theorem, one easily obtains a complete detachment version: *If k, D, z are the same as before, it is possible to detach the edges at z into $\delta(t)$ nodes so that each contains exactly one edge leaving it and so that the resulting digraph is k-edge-connected in V.*

A directed counter-part of Nash-Williams's detachment theorem was obtained by Berg, Jackson and Jordán [6]. Given a function $r : V \to \mathbf{Z}_+$, by an *r*-**detachment** of a digraph $D = (V, A)$ we mean a digraph arising from D by 'detaching' simultaneously each node v into $r(v)$ pieces so that each edge leaving or entering v would leave or enter one of the pieces.

Theorem 3.24 ([6]). *Let $D = (V, E)$ be a digraph and let $r : V \to \mathbf{Z}_+$. Then D has a k-edge-connected r-detachment if and only if*

(a) *D is k-edge-connected,*

(b) *$\varrho(v) \geq kr(v)$ and $\delta(v) \geq kr(v)$ for every $v \in V$.*

In addition, Berg, Jackson and Jordán proved that the in- and out-degrees of every detached node $v \in V$ can be arbitrarily specified provided that at each node v of D all the values in the indegree specifications are at least k and add up to the indegree of v and similarly for the outdegree specifications.

4. Uncrossing-based results

In the previous two sections we overviewed results evolving from the splitting lemmas. Here some fruits of another fundamental technique, the uncrossing procedure, will be surveyed. The rough idea of this approach is that for a given family of sets with certain properties or parameters one can replace two uncomparable (or intersecting, or crossing) sets by their intersection and union so as to preserve the properties or parameters of the family. By repeating this uncrossing step as long as possible, one arrives in a finite number of steps at a nicer family (chain of sets, laminar, or cross-free), preserving the essential properties or parameters of the initial one. To my best knowledge, the first appearance of this approach that appeared in print [39] was a solution of L. Lovász (a third-grade university student at that time) to Problem 11 (posed by A. Rényi) of the Memorial Mathematical Contest Miklós Schweitzer of the year 1968.

Later Lovász used the technique to provide a simple proof of the Lucchesi-Younger theorem [41] and to prove his theorem on minimum T-joins [40]. Since then the uncrossing method has proved to be an extremely powerful proof technique. In this section we briefly overview some recent results that were obtained this way.

4.0.1. A detour to the origin of uncrossing.

Rényi's Problem 11 was to verify an inequality concerning the probabilities of some events in a finite probability space. In his solution, Lovász first observed that the logarithm of the probability of events is a submodular function (where product and sum of events correspond to intersection and union, respectively), and he then applied the uncrossing technique to derive the requested inequality. Actually, Lovász's proof uses nothing but the submodular property and hence it provides the corresponding inequality for *any* submodular function: we exhibit Lovász's proof in this context. In order to do so, it is useful to introduce the notion of linear extension of a set-function.

Let b be a set-function on a groundset S for which $b(\emptyset) = 0$. For any vector $c \in \mathbf{R}^{|S|}$, arrange the elements of S in such a way that $c(s_1) \geq \cdots \geq c(s_n)$. Let $S_i := \{s_1, \ldots, s_i\}$ and define $\hat{b}(c)$ by $\hat{b}(c) := c(s_n)b(S_n) + \sum_{i=1}^{n-1} \left[c(s_i) - c(s_{i+1}) \right] b(S_i)$. The function $\hat{b} : \mathbf{R}^S \to \mathbf{R}$ defined this way is called the **linear extension** of b. It was introduced also by Lovász in 1983 [43] and therefore often the term *Lovász extension* is used. It should be noted that the correctness of the matroid greedy algorithm is equivalent to stating that the maximum c-weight of bases of a matroid with rank function r equals $\hat{r}(c)$, or more generally, Edmonds' polymatroid greedy algorithm is equivalent to the assertion that, given a fully submodular function b, $\max\{cx : x \in B(b)\} = \hat{b}(c)$, where $B(b) := \{x \in \mathbf{R}^S : x(Z) \le b(Z)$ for every $Z \subset S$ and $x(S) = b(S)\}$ is the so called base-polyhedron.

The solution of Lovász in [39] to Problem 11 contains implicitly the following.

Lemma 4.1. *Let b be a fully submodular function on a ground-set S and $\hat{b}$ its linear extension. Then, for any collection $\{X_1, X_2, \ldots, X_m\}$ of subsets of S,*

$$(13) \qquad \sum_i^m b(X_i) \ge \hat{b}\left(\sum_i^m \chi_{X_i}\right),$$

where χ_X denotes the characteristic function of X.

Proof. Apply the uncrossing procedure to the family $\{X_1, \ldots, X_m\}$, that is, as long as there are two uncomparable sets in the current family, replace them by their intersection and union. Due to the submodularity of b, the sum of the b-values of the members never increases, while the sum of the characteristic vectors of the members stay unchanged.

Since the number of uncomparable sets in the family during an uncrossing step strictly decreases, the uncrossing procedure terminates in a finite number of steps. The final family is a chain $\{Z_1 \subseteq Z_2 \subseteq \cdots \subseteq Z_m\}$ of subsets for which $\sum_i \chi_{X_i} = \sum_i \chi_{Z_i}$, and hence $\sum_i b(X_i) \ge \sum_i b(Z_i) = \hat{b}\left(\sum_i \chi_{Z_i}\right) = \hat{b}\left(\sum_i \chi_{X_i}\right)$. ∎

The inequality in (13) may be called generalized submodular inequality. (We note that the even more general inequality $\sum \hat{b}(c_i) \ge \hat{b}\left(\sum c_i\right)$ also holds true for arbitrary vectors $c_1, \ldots, c_m \in \mathbf{R}^S$.) To see the usefulness of (13), we make a little detour and derive in a few lines the following elegant result on matroids from the partition theorem.

Theorem 4.2 (Greene és Magnanti). *Let B_1 and B_2 be bases of a matroid M and $\{Z_1, Z_2, \ldots, Z_m\}$ a partition of B_1. Then there is a partition $\{Y_1, \ldots, Y_m\}$ of B_2 for which $B_1 - Z_i \cup Y_i$ is a basis for each subscript $i = 1, \ldots, m$.*

Proof. We may assume that B_1 and B_2 are disjoint for otherwise their intersecion can be contracted and the theorem for the contracted matroid implies that for M. Let k denote the rank of M. For each i, consider the matroid $M_i = (B_2, r_i)$ arising from M by contracting first $B_1 - Z_i$ and restricting then the resulting matroid to B_2. For any subset $X \subseteq B_2$, let $X_i := B_1 - Z_i \cup X$. Then $\sum_i \chi_{X_i} = (m-1)\chi_{(B_1 \cup X)} + \chi_X$, and by (13) we have $\sum_i r(X_i) \geq \hat{r}\left(\sum_i \chi_{X_i}\right) = \hat{r}\left[(m-1)\chi_{(B_1 \cup X)} + \chi_X\right] = (m-1)r(B_1 \cup X) + r(X) = (m-1)k + |X|$. From $r_i(X) = r(X_i) - r(B_1 - Z_i) = r(X_i) - |B_1 - Z_i|$, we obtain $\sum_i r_i(X) = \sum_i \left[r(X_i) - |B_1 - Z_i|\right] = \sum_i r(X_i) - (km - k) \geq (m-1)k + |X| - (km - k) = |X|$.

By the matroid partition theorem of Edmonds and Fulkerson [11], B_2 can be partitioned into sets $Y_1, Y_2, \ldots, Y_m$ so that Y_i is independent in M_i. By the definition of M_i, $|Y_i| \leq |Z_i|$ for each i, and hence $\sum |Z_i| = \sum |Y_i|$. Therefore $|Y_i| = |Z_i|$, and then $B_1 - Z_i \cup Y_i$ is a basis of M. ∎

4.1. Orientations and augmentations through submodular flows

A general and flexible framework concerning sub- or supermodular functions is the notion of submodular flow. In [23] a rather exhaustive survey was given to show how basic results on submodular flows can be applied to orientation problems. By an **orientation of a mixed graph** $M = (V, A + E)$, with directed and undirected edge-sets A and E respectively, we mean a directed graph $(V, A + \vec{E})$ arising from M by orienting each undirected edge and leaving alone the directed ones.

Before exhibiting a characterization of mixed graphs having k-edge-connected orientations, let us consider the special case $k = 1$.

4.1.1. Strongly connected orientation of mixed graphs. A straightforward generalization of Robbins' theorem, with a fairly easy proof, is due to F. Boesch and R. Tindell [7].

Theorem 4.3. *A mixed graph $M = (V, A + E)$ has a strongly connected orientation if and only if M has no cut-edge and no subset $\emptyset \subset X \subset V$ of nodes so that neither directed nor undirected edges leave X.*

Proof. We show that the undirected edges can be oriented greedily one by one, taking care only to avoiding the creation of a directed cut. There is nothing to prove if E is empty. Let $e = uv \in E$ be an undirected edge. If orienting e toward v (toward u, respectively) creates a directed cut, then there is a $u\bar{v}$-set X (a $v\bar{u}$-set Y) so that no directed edge leaves X (Y) and e is the only undirected edge leaving X (Y). Then neither $X \cap Y$ nor $V - (X \cup Y)$ admits a leaving edge and hence they must be empty. Therefore X and Y are complementary sets and e is the only edge connecting X and Y, contradicting the assumption on the non-existence of cut-edges. ∎

The simplicity of this result may suggest that Nash-Williams' Theorem 2.8 on k-edge-connected orientability of $2k$-edge-connected undirected graphs can also be extended to mixed graphs in a straightforward way. But this is not the case even for $k = 2$.

4.1.2. An example for $k = 2$. It turns out that in this case the natural cut-type or partition-type necessary conditions are not sufficient anymore. To see this, define a mixed graph $M = (V_4, A + E)$ as follows. Let $V_4 = \{v_1, v_2, v_3, v_4\}$, let E consist of two edges $e_1 = v_1 v_2$, $e_2 = v_3 v_4$, and let A consist of the following nine edges: $v_1 v_3$, $v_1 v_3$, $v_3 v_1$, $v_2 v_3$, $v_2 v_3$, $v_3 v_2$, $v_2 v_4$, $v_2 v_4$, $v_4 v_2$.

The digraph $D = (V_4, A)$ is strongly connected, that is, every in-deficient set (with respect to 2-edge-connectivity) is of indegree one, and there are exactly three such sets:

$$X_1 := \{v_1\}, \quad X_2 := \{v_1, v_2, v_3\}, \quad X_3 := \{v_2, v_4\}.$$

Let $\mathcal{A}_3 := \{X_1, X_2, X_3\}$. In order to have a 2-edge-connected orientation of M, one has two orient the two edges of $G = (V_4, E)$ so that each member of $\mathcal{A}_3$ admits at least one newly oriented entering edge. An easy case checking shows that no such orientation may exist. Note, however, that for every two members of $\mathcal{A}_3$, there is an orientation of G in which the indegree of these two members is at least 1. This implies that any certificate of the nonexistence of a 2-edge-conneced orientation of M which consists of in-deficient sets must include all the three members of $\mathcal{A}_3$.

Note that $\mathcal{A}_3$ is neither a partition nor a co-partition of any subset of V. The example therefore indicates why one needs more general families of sets in the characterization of k-edge-connected orientable mixed graphs. The result will also show that the use of submodular functions is unavoidable in the solution of this purely graph-theoretic problem. The approach easily extends to (k, l)-edge-connected orientability.

4.1.3. Tree-compositions. For a proper non-empty subset S of V we introduce the notion of a tree-composition of S. Let $\{S_1, \ldots, S_\alpha\}$ be a partition of S and $\{Z_1, \ldots, Z_\beta\}$ a partition of $V - S$ ($\alpha, \beta \geq 1$). Let $T = (U, F)$ be a directed tree such that $U := \{s_1, \ldots, s_\alpha, z_1, \ldots, z_\beta\}$ and each directed edge goes from a z_j to an s_i. For each edge f of the tree, let T_f denote the set of nodes of that component of $T - f$ which is entered by f. The family $\mathcal{A} := \{\varphi^{-1}(T_f) : f \in F\}$ is called a **tree-composition of** S where $\varphi(v) = s_i$ if $v \in S_i$ and $\varphi(v) = z_j$ if $v \in Z_j$. We will also say that a partition or a co-partition of V is a **tree-composition of** V. Note that a tree-composition $\mathcal{A}$ of S is cross-free and every element of S belongs to the same number t of members of $\mathcal{A}$ and every element of $V - S$ belongs to $t - 1$ members. (If $\alpha = \beta = 1$, then $\mathcal{A}$ consists of the single set S. If $\beta = 1 < \alpha$, then $\mathcal{A}$ is a partition of S. If $\alpha = 1 < \beta$, then $\mathcal{A}$ is a co-partition of S.)

Let us consider the subset $S := \{v_1, v_2\}$ in the example above. We claim that the family $\mathcal{A}_3$ forms a tree-composition of S. This can be seen by defining $S_1 := \{v_1\}$, $S_2 := \{v_2\}$, $Z_1 := \{v_3\}$, $Z_4 := \{v_4\}$ and by letting T be a directed tree on node set $\{s_1, s_2, z_1, z_2\}$ having three edges: $f_1 = v_3v_1$, $f_2 = v_3v_2$, $f_3 = v_4v_2$. Now $T_{f_1} = s_1$, $T_{f_2} = \{z_2, s_2\}$ and $T_{f_3} = \{s_1, s_2, z_1\}$. Let $\varphi(v_1) = s_1$, $\varphi(v_2) = s_2$, $\varphi(v_3) = z_1$, $\varphi(v_4) = z_2$. Then $\mathcal{A}_3$ indeed arises in the form described in the definition of tree-composition.

Suppose now that $G = (V, E)$ is an arbitrary undirected graph. Let $\mathcal{A}$ be a tree-composition of a subset $S \subseteq V$ and $j = uv$ an edge of G. Let $e_{u\bar{v}}(\mathcal{A})$ denote the number of $u\bar{v}$-sets in $\mathcal{A}$. That is, $e_{u\bar{v}}(\mathcal{A})$ is the number of sets in $\mathcal{A}$ entered by the directed edge with tail v and head u. Let $e_j(\mathcal{A}) := \max\{e_{u\bar{v}}(\mathcal{A}), e_{\bar{u}v}(\mathcal{A})\}$ and

$$(14) \qquad e_G(\mathcal{A}) := \sum_{j \in E} e_j(\mathcal{A}).$$

Note that $|e_{u\bar{v}}(\mathcal{A}) - e_{v\bar{u}}(\mathcal{A})| \leq 1$ with equality if and only if $|S \cap \{u, v\}| = 1$. The quantity $e_j(\mathcal{A})$ indicates the (maximal) possible contribution of an edge $j = uv$ to the sum $\sum[\varrho_{\vec{G}}(X) : X \in \mathcal{A}]$ for any orientation $\vec{G}$ of G. Hence $e_G(\mathcal{A})$ measures the total of these contributions and we have

$$(15) \qquad \sum_{X \in \mathcal{A}} \varrho_{\vec{G}}(X) \leq e_G(\mathcal{A})$$

for any orientation $\vec{G}$ of G. Let $D = (V, A)$ be a digraph and $M = (V, A + E)$ a mixed graph. Let s be a root-node of M. For integers $0 \leq l \leq k$ define $p_{kl}(X) := (k - \varrho_D(X))^+$ if $\emptyset \subset X \subset V - s$ and $p_{kl}(X) := (l - \varrho_D(X))^+$ if $s \in X \subset V$.

Theorem 4.4 [23]. *A mixed graph M has a (k, l)-edge-connected orientation (with respect to root-node s) if and only if*

(16) $$\sum \left[p_{kl}(X) : X \in \mathcal{A} \right] \leq \sum \left[e_G(\mathcal{A}) : e \in E \right]$$

holds for every tree-composition $\mathcal{A}$.

In the example above, where $k = l = 2$, $\mathcal{A}_3$ violates (16) since $p_{kl}(X) = 1$ for each $X \in \mathcal{A}_3$ while $e_G(\mathcal{A}_3) = 2$ since each of the two edges of G can contribute to the indegree of the sets in $\mathcal{A}_3$ by one.

4.1.4. Special cases. While tree-compositions are inevitable in general, in some important special cases they are not, as we have already seen in Theorems 2.8 and 2.9. We now exhibit a common generalization of these last two results when partition type conditions turn out to be sufficient. We investigate the orientation problem when l-edge-connectivity and rooted k-edge-connectivity are simultaneously required (that is, we want a (k, l)-edge-connected orientation).

Theorem 4.5 [18]. *Let $0 \leq l \leq k$ be integers. An undirected graph $G = (V, E)$ has a (k, l)-edge-connected orientation if and only if G is (k, l)-partition-connected.*

Another special case of the mixed graph (k, l)-edge-connected orientation problem when only partition type conditions are required is the case of $l \leq 1$. The case $l = 0$, which is a generalization of Theorem 2.9, appeared in [16].

Theorem 4.6. *A mixed graph $D + G = (V, A + E)$ with a root-node s has a $(k, 0)$-edge-connected (that is, s-rooted k-edge-connected) orientation if and only if the number of cross-edges of G is at least*

(17) $$\sum_{i=1}^{t} \left[k - \varrho_D(V_i) \right]$$

for every partition $\{V_0, V_1, \ldots, V_t\}$ of V into non-empty parts with $s \in V_0$.

The case $l = 1$ appeared in [23].

Theorem 4.7. *A mixed graph $D + G = (V, A + E)$ with a root-node s has a $(k, 1)$-edge-connected orientation (that is, strongly connected and s-rooted k-edge-connected) if and only if the number of cross-edges of G is at least $\sum_{i=1}^{t} \left[k - \varrho_D(V_i) \right] + 1$ for every partition $\{V_0, V_1, \ldots, V_t\}$ of V into non-empty parts with $s \in V_0$.*

4.1.5. An augmentation result. The rooted edge-connectivity augmentation problem (in digraphs) behaves nicely in the sense that even the minimum cost version is tractable. Suppose that we are given a digraph with a special root-node s and we want to augment the digraph by adding a minimum cost of new edges so as to have a rooted k-edge-connected digraph. At the beginning of section 2, we mentioned that the minimum cost subgraph problem is equivalent to the minimum cost augmentation problem, and in this case the subgraph problem (:find in a digraph a minimum cost rooted k-edge-connected subgraph) can be solved with the help of submodular flows, see [17] and [54]. Here we mention only one consequence of this:

Theorem 4.8. *Let $D = (V, E)$ and $H = (V, A)$ be two digraphs so that their union $D + H = (V, E \cup A)$ is k-edge-connected from a root-node s. The minimum number of edges of H whose addition to D results in a s-rooted k-edge-connected digraph is equal to the maximum of $\sum [k - \varrho_D(X) : X \in \mathcal{F}]$, where the maximum is taken over all laminar families $\mathcal{F}$ of nonempty subsets of $V - s$ for which no edge of H enters more than one member of $\mathcal{F}$.*

4.2. Connectivity orientation and augmentation combined

Now comes an account on some new developments making possible to combine certain orientation and augmentation problems. In subsection 2.2 we have already mentioned this type of results: Theorem 2.10 characterized undirected graphs which can be augmented by adding at most γ edges so as to have a $(k, 0)$-edge-connected orientation. We also remarked that even the minimum cost augmentation was tractable by using matroid techniques. Here we consider the same problem for mixed graphs (where those matroid techniques do not work.) Let us consider Theorem 4.6 and suppose that the required orientation does not exists, that is, the necessary and sufficient condition in (17) fails to hold. How many new undirected edges should be added to M so as to have a $(k, 0)$-edge-connected orientation. Or more generally, what is the minimum cost of required new edges? By considering the existing undirected edges having zero cost, this latter problem is equivalent to the following.

Given a mixed graph with a root node s endowed with a non-negative cost function on the set of undirected edges, delete a maximum cost of edges so that the resulting mixed graph has a $(k, 0)$-edge-connected orientation.

S. Khanna, J. Naor and F. B. Shepherd [35] solved this problem in an even more general form when the directed edges may also have costs and the two possible directions $e' = uv$ and $e'' = vu$ of an undirected edge uv may have different costs.

To be more specific, let $M = (V, A + E)$ be a mixed graph consisting of a digraph $D = (V, A)$ and an undirected graph (V, E). Let s be a root-node of M and let $A_1 := A \cup \{e', e'' : e \in E\}$. Furthermore we are given a nonnegative cost function $c : A_1 \to \mathbf{R}_+$. We say that a subset $F \subseteq A_1$ of directed edges (or the subdigraph $D' := (V, F)$) is **orientation-constrained** if F may contain at most one of the two possible directions e' and e'' of any undirected edge $e \in E$.

The $(k, 0)$-**orientable subgraph problem** consists of finding a minimum cost $(k, 0)$-edge-connected orientation-constrained subdigraph $D' = (V, F)$ of $D_1 := (V, A_1)$.

Khanna, Naor and Shepherd considered the following linear program:

(18) $$\min \sum \left[c(f)x(f) : f \in A_1 \right]$$

subject to

(19) $$0 \leq x(f) \leq 1 \quad \text{for every directed edge} \quad f \in A_1$$

(20) $$x(e') + x(e'') \leq 1 \quad \text{for every edge} \quad e \in E$$

(21)
$$\sum \left[x(f) : f \in A_1, \ f \text{ enters } Z \right] \geq k \quad \text{for every subset} \quad \emptyset \subset Z \subseteq V - s.$$

Let P denote the polytope described by the three constraints. Clearly, an integer vector in P is actually $0 - 1$-valued and the $0 - 1$ vectors of P are precisely the characteristic vectors of orientation constrained $(k, 0)$-edge-connected subdigraphs of D_1.

The main result of [35] is as follows:

Theorem 4.9 (Khanna, Naor, and Shepherd). *The vertices of polytope P are $0 - 1$ vectors, or equivalently, P is the convex hull of (characteristic vectors) of orientation-constrained $(k, 0)$-edge-connected subdigraphs of D_1.*

By relying on linear programming duality, this theorem provides a min-max formula for the minimum cost of a solution. We avoid formulating this since the result can be even further improved [29]. We emphasize, however, that the improvement uses only known ideas, and the main point here is the recognition of Khanna, Naor, and Shepherd that even this general framework is tractable by standard techniques.

Theorem 4.10. *The linear inequality system of* (19), (20), *and* (21) *is totally dual integral* (*implying the integrality of* P). *Moreover,* P *is a submodular flow polyhedron.*

This theorem enables us to solve the problem algorithmically by invoking a submodular flow algorithm. Furthermore, one has a better structured duality theorem. For the sake of simplicity we formulate it only for $0-1$-valued cost functions.

Theorem 4.11. *Let* $M = (V, A + E)$ *be a mixed graph with a root-node* s *endowed with a* $0-1$ *valued cost function* $c : A \cup E \rightarrow \{0, 1\}$. *The minimum cost of a mixed subgraph of* M *which has a* $(k, 0)$-*edge-connected orientation is equal to the maximum of*

$$tk - e_G(\mathcal{F}) - \sum \left[\varrho_D(X) : X \in \mathcal{F} \right] + q(\mathcal{F}),$$

where the maximum is taken over all laminar families $\mathcal{F}$ *of* t ($t \geq 0$) *subsets of* $V - s$. *Here* $G = (V, E)$ *is the undirected part of* M, $e_G(\mathcal{F})$ *is defined in* (14), *and* $q(\mathcal{F})$ *denotes the number of* (*directed or undirected*) *edges of cost* 1 *which enter at least one member of* $\mathcal{F}$.

This is a common generalization of Theorems 4.6 and 4.8. When c is zero on all directed edges, we are back at our starting problem of finding a smallest set of new undirected edges to be added to a mixed graph to have a $(k, 0)$-edge-connected orientation.

So, we can solve quite reassuringly the combined orientation/augmentation problem in mixed graphs when the target is $(k, 0)$-edge-connectivity. Wouldn't it be natural to lift our horizon to (k, l)-edge-connectivity? The directed (k, l)-edge-connectivity augmentation problem is solved by Theorem 3.16. The (k, l)-edge-connectivity orientation problem is solved for undirected graphs by Theorem 4.5 (and even for mixed graphs by Theorem 4.4). We show now how to solve the problem of augmenting an undirected graph by adding undirected edges so that the resulting graph has a (k, l)-edge-connected orientation. Due to the relatively complicated nature of tree-compositions in Theorem 4.4, so far we have not taken courage to try to attack the corresponding augmentation problem for mixed graphs. And even for undirected graphs the minimum cost version is out of question because the NP-complete problem of finding a Hamiltonian circuit problem is a special case. We consider the degree-specified and the minimum augmentation problems as well. The following results are taken from [28].

Theorem 4.12. *Let* $G = (V, E)$ *be an undirected graph,* $k \geq l \geq 0$ *integers, and* $m := V \to \mathbf{Z}_+$ *a degree-specification for which* $m(V)$ *is even. There exists a graph* $H = (V, A)$ *so that* $d_H(v) = m(v)$ *for every* $v \in V$ *and so that* $G + H$ *is* (k, l)-*tree-connected* $(= (k, l)$-*partition-connected* $= (k, l)$-*edge-connected orientable) if and only if*

$$(22) \qquad\qquad m(V)/2 \geq (t - 1)k + l - e_G(\mathcal{F})$$

and

$$(23) \qquad\qquad \min_{X \in \mathcal{F}} m(V - X) \geq (t - 1)k + l - e_G(\mathcal{F})$$

hold for every partition $\mathcal{F}$ *of* V *into* $t \geq 2$ *non-empty parts.*

Let us indicate briefly the proof of necessity. If $G + H$ has a (k, l)-edge-connected orientation, then it is (k, l)-partition-connected, that is, $e_{G+H}(\mathcal{F}) \geq k(t - 1) + l$ and hence $e_H(\mathcal{F}) \geq k(t - 1) + l - e_G(\mathcal{F})$. If H satisfies the degree-specification, then $m(V)/2 = |A| \geq e_H(\mathcal{F})$ and $m(V - X) \geq e_H(\mathcal{F})$ for every $X \in \mathcal{F}$ from which both (22) and (23) follow.

This result might be interesting even in the special case of $l = 0$:

Corollary 4.13. *Let* $G = (V, E)$ *be an undirected graph,* $k \geq 1$ *an integer, and* $m := V \to \mathbf{Z}_+$ *a degree-specification for which* $m(V)$ *is even. There exists a graph* $H = (V, A)$ *so that* $d_H(v) = m(v)$ *for every* $v \in V$ *and so that* $G + H$ *is* k-*tree-connected if and only if*

$$(24) \qquad\qquad m(V)/2 \geq (t - 1)k - e_G(\mathcal{F})$$

and

$$(25) \qquad\qquad \min_{X \in \mathcal{F}} m(V - X) \geq (t - 1)k - e_G(\mathcal{F})$$

hold for every partition $\mathcal{F}$ *of* V *into* $t \geq 2$ *non-empty parts.*

The following theorem is a bit out of the main line of the paper since the target of the augmentation is not a connectivity property. As a counterpart to tree-packing in corollary 4.13, here our target is tree-covering:

Theorem 4.14 [28]. *Let $G = (V, E)$ be an undirected graph, $k \geq 1$ an integer, and $m := V \to \mathbf{Z}_+$ a degree-specification for which $m(V)$ is even. There exists a graph $H = (V, A)$ so that $d_H(v) = m(v)$ for every $v \in V$ and so that $G + H$ is the union of k forests if and only if*

$$(26) \qquad m(X) - m(V)/2 \leq k(|X| - 1) - i_G(X)$$

for every $\emptyset \subset X \subseteq V$, where $i_G(X)$ denotes the number of edges of G induced by X.

Again it is useful to prove the necessity. If H is a graph for which $G + H$ is the union of k forests, then $e_{G+H} \leq k(|X| - 1)$ holds for every subset $X \subseteq V$, that is, $i_H(X) \leq k(|X| - 1) - i_G(X)$. If H satisfies the degree-specification, then $|A| = m(V)/2$ and at most $m(V - X)$ edges may be incident with an element of $V - X$. So at least $m(V)/2 - m(V - X)$ edges are induced by X in H and hence $m(X) - m(V)/2 = m(V)/2 - m(V - X) \leq i_H(X) \leq k(|X| - 1) - i_G(X)$.

To conclude this subsection, we cite a result from [28] on the minimization form of (k, l)-tree-connectivity augmentation.

Theorem 4.15. *Let $G = (V, E)$ be an undirected graph. It is possible to add at most γ new edges to G so that the resulting graph G^+ is (k, l)-tree-connected (that is, G^+ has a (k, l)-edge-connected orientation) if and only if*

$$(27) \qquad \gamma \geq k(t - 1) + l - e_G(\mathcal{F})$$

holds for every partition $\mathcal{F}$ of V with t members, and

$$(28) \qquad 2\gamma \geq t_1 k + t_2 l - e_G(\mathcal{F})$$

holds whenever $\mathcal{F}$ is the union a partition $\mathcal{F}_1$ of a subset $Z \subseteq V$ and a co-partition $\mathcal{F}_2$ of Z so that $|\mathcal{F}_i| = t_i$ $(i = 1, 2)$ and so that $\mathcal{F}_1$ is a finer partition of Z than partition $\{X : V - X \in \mathcal{F}_2\}$.

4.3. Directed edge-connectivity augmentation

In [25] we proved a general min-max formula concerning minimum coverings of a so-called bi-supermodular function by directed graphs. This result implies Theorem 3.16 (which has had an independent and simpler proof) and implies the following, as well.

Theorem 4.16. *Let $D = (V, A)$ be a directed graph and S, T two (not necessarily disjoint) non-empty subsets. It is possible to add at most γ ST-edges so that the resulting digraph is k-edge-connected from S to T if and only if*

$$(29) \qquad \sum \left[k - \varrho_D(X) : X \in \mathcal{F} \right] \leq \gamma$$

holds for every family $\mathcal{F}$ of pairwise ST-independent sets, where two sets X, Y are ST-independent if $X \cap Y \cap T = \emptyset$ or $S - (X \cup T) = \emptyset$.

In sharp contrast with the existence of a good characterization in Theorem 3.12 concerning local edge-connectivity augmentations of undirected graphs, the directed counterpart of this problem is NP-complete [19] even in the special case when the requirement is one between the nodes of a specified subset T of nodes and zero otherwise. (That is, given a digraph, add a minimum number of new edges so that there is a path from every element of T to every other element of T.) Recently, however, I found the following characterization for $|T| = 2$ [24]. (This result seems to be independent of the rather general main theorem of [25].)

Theorem 4.17. *Let $D = (V, E)$ be a digraph with two specified nodes s, t and let k, l be two non-negative integers. Let S, T be non-empty subsets of V so that every $s\bar{t}$-set X with $\varrho_D(X) < k$ and every $t\bar{s}$-set X with $\varrho_D(X) < l$ is entered by an ST-edge. D can be augmented by adding at most γ (possibly parallel) ST-edges so that in the resulting digraph there are k edge-disjoint paths from s to t and there are l edge-disjoint paths from t to s if and only if $\gamma \geq k - \varrho_D(X)$ whenever $t \in X \subseteq V - s$, $\gamma \geq l - \varrho_D(X)$ whenever $s \in X \subseteq V - t$, and $\gamma \geq \left(l - \varrho_D(X) \right) + \left(k - \varrho_D(Y) \right)$ holds whenever $s \in X, t \in Y$ and $X \cap Y \cap T = \emptyset$ or $X \cup Y \supseteq S$.*

5. CONSTRUCTIVE CHARACTERIZATIONS

We have already seen constructive characterizations of k-edge-connected graphs and digraphs (Theorems 2.13, 3.2, 2.14), of $(k, 0)$-edge-connected digraphs (2.15) and k-tree-connected graphs (2.16). For integers $0 \le l < k$ we offer the following:

Conjecture 5.1. *A directed graph D is (k, l)-edge-connected if and only if it can be built from a node by the following two operations: (j) add a new edge, (jj) pinch i $(l \le i < k)$ existing edges with a new node z, and add $k - i$ new edges entering z and leaving existing nodes. An undirected graph is (k, l)-tree-connected $(= (k, l)$-partition-connected) if and only if it can be built from a node by the following two operations: (j) add a new edge, (jj) pinch i $(l \le i < k)$ existing edges with a new node z, and add $k - i$ new edges connecting z with existing nodes.*

Note that by Theorem 4.5 the undirected version of the conjecture follows from the directed one. As mentioned above, the case $l = 0$ is settled by Theorem 2.15. Jointly with Zoltán Király [27], we characterized $(k, k-1)$-edge-connected digraphs (and hence $(k, k - 1)$-partition-connected graphs, as well). At the other end of the range of l, recently in [31] we proved the case $l = 1$. All other cases of the conjecture are open (for example, when $k = 4$, $l = 2$).

The theorem in [27] concerning the case $l = k - 1$, in turn, can be used to derive the following orientation result. Let $G = (V, E)$ be an undirected graph. A subset T of nodes is called G-**even** if $|T| + |E|$ is even. We call an orientation of G T-**odd** if the indegree of a node v is odd precisely when v belongs to T. The following is taken from [27].

Theorem 5.2. *An undirected graph G has a k-edge-connected and T-odd orientation for every G-even subset T if and only if G is $(k+1, k)$-partition-connected.*

Corollary 5.3. *A $(2k + 2)$-edge-connected graph always admits a k-edge-connected orientation in which the indegree of all nodes but possibly one are odd.*

As mentioned above, the proof is based on the constructive characterization of $(k+1, k)$-partition-connected graphs. It would be interesting to have a simple direct proof of the corollary, even for the special case $k = 1$ when it asserts that a 4-edge-connected graph has a strongly connected orientation in which every node but possibly one is of odd indegree.

The motivation behind such a theorem is the natural attempt to have a better understanding of problems where both parity and connectivity are involved. In Theorem 5.2 we charaterized graphs having a certain orientation for *every* G-even subset T. It would be interesting to know the necessary and sufficient condition of the existence of a k-edge-connected T-odd orientation of a graph G for *one* specified G-even subset T. This is open. However, the analogous question concerning k-tree-connectivity has been settled in [26].

Theorem 5.4. *Let $G = (V, E)$ be a graph with a root-node s. Let T be a G-even subset of $V - s$. G has a $(k, 0)$-edge-connected ($= s$-rooted k-edge-connected) T-odd orientation if and only if the number of cross edges of every partition $\mathcal{P} := \{V_1, \ldots, V_t\}$ of V into at least two non-empty parts is at least*

$$k(t - 1) + o(\mathcal{P}),$$

where $o(\mathcal{P})$ (which depends also on G, k, and T) denotes the number of those parts X of $\mathcal{P}$ for which $|X \cap T| - i_G(X) - k$ is odd.

As a possible counterpart to Corollary 5.3, we can derive:

Corollary 5.5. *Let $G = (V, E)$ be an undirected graph with $|E| + |V|$ even. If G is $(k + 1)$-tree-connected, then G has a $(k, 0)$-edge-connected V-odd orientation.*

But this is straightforward anyway since we can take $k + 1$ edge-disjoint trees, orient the edges of k of these away from a root node s, orient the remaining edges not in the last tree F_{k+1} arbitrarily, and finally, orient the edges of F_{k+1} so as to meet the parity prescription.

A problem related to the constructive characterization of k-edge-connected digraphs is to find a characterization of (acyclic) digraphs whose all directed cuts admit at least k edges. Such an approach could perhaps be used to prove D. Woodall's long-standing conjecture:

Conjecture 5.6. *If every directed cut of a digraph D has at least k edges, then the edge-set of D can be partitioned into k parts so that each part has at least one edge from every directed cut.*

Woodall's conjecture can easily be seen to be true for $k = 2$ but no answer is known even for $k = 3$ and for planar digraphs. (In which case, after planar dualization, the conjecture reads as follows: *in a simple planar digraph, the edge-set can be coloured by three colours so that every directed triangle contains each colour.*) A straightforward generalization of Woodall's conjecture concerning a crossing family of directed cuts was disproved by A. Schrijver [53] even for $k = 2$.

We call a graph $G = (V, E)$ **nearly k-tree-connected** if $G + e$ is the union of k edge-disjoint spanning trees for every possible new edge $e = uv$ ($u, v \in V$). It follows that such a graph has exactly $k(|V|-1) - 1$ edges and that every subset $X \subseteq V$ with $|X| \geq 2$ induces at most $k(|X|-1) - 1$ edges. A theorem of Nash-Williams [49] implies that these properties actually characterize nearly k-tree-connected graphs.

This notion for $k = 2$ (under different name) has been introduced in the theory of graph rigidity. By combining theorems of L. Henneberg [33] and of G. Laman [37], one obtains the following constructive characterization of nearly 2-tree-connected graphs.

Theorem 5.7 (Henneberg and Laman). *A graph G is nearly 2-tree-connected if and only if G can be constructed from one (non-loop) edge by the following two operations: (i) add a new node z and connect z to two distinct existing nodes, (ii) subdivide an existing edge uv by a node z and connect z to an existing node distinct from u and v.*

Jointly with László Szegő [31], we were able to extend this result for general k.

Theorem 5.8. *A graph G is nearly k-tree-connected if and only if G can be constructed from an initial graph, consisting of two nodes and $k - 1$ parallel edges, by the following operation: choose a subset F of j existing edges ($0 \leq j \leq k - 1$), pinch the elements of F with a new node z, and add $k - j$ new edges connecting z with other nodes so that there are no k parallel edges among these new edges.*

$(k, 1)$-tree-connectivity has meant that the graph has k disjoint spanning trees even after *deleting* any edge. What can be said about graphs which can be covered by k forests even after *adding* any new edge? We call such a graph k-**sparse**. By a theorem of Nash-Williams, we know that a graph $G = (V, E)$ is k-sparse if and only if every subset X of nodes with at least two elements induces at most $k(|X| - 1) - 1$ edges. Note that k-sparse graphs with $k(|V| - 1) - 1$ edges are exactly the nearly k-tree-connected graphs.

Theorem 5.9 [31]. *An undirected graph $G = (V, E)$ is k-sparse if and only if G can be built from a single node by applying the following operations. (i) add a new node z and at most k new edges ending at z so that no k parallel edges can arise, (ii) choose a subset F of i existing edges ($1 \leq i \leq k - 1$), pinch the elements of F with a new node z, and add $k - i$ new edges connecting z with other existing nodes so that there are no k parallel edges in the resulting graph.*

6. Hypergraphs

So far our interest has been fully occupied by graphs and digraphs. In this last section we let hypergraphs take over the center stage. A hypergraph $H = (V, \mathcal{F})$ consists of a ground-set V and a family $\mathcal{F}$ of (not necessarily distinct) subsets of V, called hyperedges. The cardinality $|Z|$ of a hyperedge Z is called its **size**. We are naturally back at undirected graphs when each hyperedge is of size two. Such a hyperedge will be referred as a graph-edge. The maximum size of a hyperedge is called the **rank** of H. Throughout we will assume that the size of every hyperedge is at least two.

It is often useful to associate a bipartite graph $B = B_H = (V, U_{\mathcal{F}}; E)$ with hypergraph H as follows. The elements of $U_{\mathcal{F}}$ correspond to the hyperedges of H and a node $v \in V$ is connected to a node $u_X \in U_{\mathcal{F}}$ precisely if $u \in X$. In this correspondence the size of a hyperedge Z will be the degree of its corresponding node u_Z in B.

For a subset $X \subseteq V$ let $d_H(V)$ denote the number of hyperedges of H intersecting both X and $V - X$. For a specified subset $R \subseteq V$, a hypergraph H is called k**-edge-connected in** R if $d_H(X) \geq k$ for every subset $X \subset V$ separating R. (X is said to **separate** R if $X \cap R \neq \emptyset, R - X \neq \emptyset$.) If $R = V$, the hypergraph itself is called k**-edge-connected.** When $k = 1$ we simply say that H is connected.

From the definitions it follows that H is k-edge-connected in R if and only if the elements of R belong to one component of the graph arising from the associated bipartite graph $(V, U_{\mathcal{F}}; E)$ by deleting at most $k - 1$ elements of $U_{\mathcal{F}}$. By a version of Menger's theorem, it follows that B has this property if and only if there are k paths between any pair of nodes u, v of R so that each node of $U_{\mathcal{F}}$ belongs to at most one of these paths (but the paths may share freely elements of V).

This implies that a hypergraph H is k-edge-connected in R if and only if there are k hyperedge-disjoint hyperpaths between every pair of nodes $u, v \in R$. Here a hyperpath means a sequence $\{u_1 := u, F_1, u_2, F_2, \ldots, u_t, F_t, u_{t+1} := v\}$ so that $u_i, u_{i+1} \in F_i \in \mathcal{F}$ for $i = 1, \ldots, t$.

Theorem 2.4 has been extended by J. Bang-Jensen and B. Jackson to hypergraphs [2].

Theorem 6.1 (Bang-Jensen and Jackson). *A hypergraph $H = (V, A)$ can be made k-edge-connected by adding at most γ new graph-edges if and only if $\sum(k - d_H(X) : X \in \mathcal{P}) \leq 2\gamma$ holds for every sub-partition $\mathcal{P}$ of V and $c(H') - 1 \leq \gamma$ for every hypergraph $H' = (V, A')$ arising from H by leaving out $k - 1$ hyperedges where $c(H')$ denotes the number of components of H'.*

In [4] we extended this to the case when the target is k-edge-connectivity in a specified subset $R \subseteq V$.

For $q \geq 3$, T. Király [36] recently to characterized hypergraphs which can be made k-edge-connected by adding at most γ hyperedges of size at most q. The special case, when H is already $(k - 1)$-edge-connected, was solved by T. Fleiner and T. Jordán [14].

Let r be again a requirement function on the set of unordered pairs of nodes, We say that H is **r**-edge-connected if there are at least $r(u, v)$ edge-disjoint hyperpaths between every pair of nodes u, v. Again by Menger's theorem, this is equivalent to requiring $d_H(X) \geq R_r(X)$ for every non-empty subset $X \subset V$.

Since local edge-connectivity augmentation is nicely tractable for undirected graphs, one may want to extend this to hypergraphs and determine the minimum number of new graph edges whose addition to H results in an **r**-edge-connected hypergraph. However, B. Cosh, B. Jackson and Z. Király [8] pointed out that this problem is NP-complete even if r is $(1 - 2)$-valued. For 3-uniform hypergraphs, however, the local edge-connectivity augmentation problem is tractable in the case when the newly added hyperedges are of size three or size two and for both types the number of new hyperedges are specified. This follows from Theorem 3.22 of Jordán and Szigeti and is based on the observation that intuitively says that the contribution of a hyperedge $\{a, b, c\}$ of size three to the edge-connectivity is the same as that of a star graph with three edges, that is, a graph with node set $\{z, a, b, c\}$ and edge set $\{za, zb, zc\}$.

Another interesting version of the local edge-connectivity augmentation of hypergraphs was solved nicely by Z. Szigeti [55].

Theorem 6.2 (Szigeti). *Given a requirement function r, a hypergraph H can be made **r**-edge-connected by adding hyperedges with total size at most γ if and only if $\sum_i \left(R_r(X_i) - d_H(X_i) \right) \leq \gamma$ holds for every subpartition $X_1, \ldots, X_t$ of V.*

The material below is taken from [30]. A hypergraph $H = (V, \mathcal{E})$ is called **connected** if there is a hyperedge intersecting both X and $V - X$ for every non-empty, proper subset X of V. The hypergraph is **partition-connected** if there are at least $t - 1$ hyperedges intersecting at least two parts for every t-partition of V. For graphs these two notions coincide but for hypergraphs they do not (consider the hypergraph on three elements a, b, c having a single hyperedge $\{a, b, c\}$).

The connectivity of a hypergraph is equivalent to the connectivity of the bipartite graph associated with H. Therefore deciding whether a hypergraph is connected is an easy task. Testing a hypergraph for partition-connectivity is not so straightforward. To this end we call a hypergraph $H = (V, \mathcal{F})$ **wooded** if it is possible to select two elements from each hyperdege of H so that the selected pairs, as graph edges, form a forest.

Theorem 6.3 (Lovász). *A hypergraph $H = (V, \mathcal{F})$ is wooded if and only if H satisfies the strong form of the Hall condition, that is, the union of any j hyperedges ($j \geq 1$) has at least $j + 1$ nodes.*

Proof. (outline) The necessity is staightforward. To see the sufficiency, consider the bipartite graph $B = (V, U; E)$ associated with H. Since the Hall condition is satisfied, there is a matching M of B covering the elements of U. Let S denote the set of nodes not covered by M. Orient the elements of M toward V while all other edges toward U. It follows from the strong form of the Hall condition that each node of B is reachable from S. Hence there is a spanning branching of B rooted at S and this determines the required forest. ∎

Theorem 6.4 (Lorea, [38]). *Given a hypergraph $H = (V, \mathcal{E})$, the wooded subhypergraphs of H form a family of independent sets of a matroid on ground-set $\mathcal{E}$.*

Theorem 6.5 [30]. *A hypergraph $H = (V, \mathcal{E})$ is partition-connected if and only if H contains a wooded subhypergraph $(V, \mathcal{F})$ with $|V| - 1$ hyperedges.*

A hypergraph is k-**partition-connected** if there are at least $k(t-1)$ hyperedges intersecting at least two parts for every t-partition of V.

Tutte's Theorem 1.2 characterizes those graphs that can be decomposed into k edge-disjoint connected (or equivalently, partition-connected) spanning subgraphs, asserting that exactly the k-partition-connected graphs have this property. The problems of decomposing a hypergraph into k connected or into k partition-connected spanning subhypergraphs are not equivalent anymore. The first one can be shown to be NP-complete, while the second one is tractable.

Theorem 6.6 [30]. *A hypergraph $H = (V, \mathcal{F})$ can be decomposed into k partition-connected subhypergraphs if and only if H is k-partition-connected.*

The following corollary is well-known for graphs (case $q = 2$).

Corollary 6.7. *If a hypergraph H of rank at most q is (kq)-edge-connected, then H can be decomposed into k partition-connected (and thus connected) spanning subhypergraphs.*

Proof. By Theorem 6.6 it suffices to show that H is k-partition-connected. Let $\mathcal{P} = \{V_1, \dots, V_t\}$ be a partition of V. There are at least kq hyperedges intersecting both V_i and its complement for each i. Since every hyperedge is of cardinality at most q, the total number of hyperedges intersecting at least two members of $\mathcal{P}$ is at least $kqt/q = kt \geq k(t-1)$. Therefore H is indeed k-partition-connected and Theorem 6.6 applies. ∎

6.1. Directed hypergraphs

There may be several choices to define directed hypergraphs, we work with the following definition. A **directed hyperedge** (Z, z) is a pair of a subset Z of the ground-set V and an element z of Z. The element z is called the **head** of Z. By a **directed hypergraph** we mean a collection of directed hyperedges. This obviously generalizes the notion of directed graphs. A disadvantage of this definition is that the symmetry between the head and the tail of a directed graph edge is lost. On the positive side of this definition is that several results concerning edge-connectivity of directed graphs can be carried over nicely to directed hypergraphs.

We say that a directed hyperedge (Z, z) **enters** a subset $X \subseteq V$ if the head z is in X but $Z - X \neq \emptyset$. A directed hypergraph is called

k-**edge-connected** if there are at least k hyperedges entering each non-empty proper subset of V. More generally, for integers $0 \leq l \leq k$, a directed hypergraph is called (k, l)-**edge-connected** if there is a node $s \in V$ so that each non-empty subset $X \subseteq V - s$ is entered by at least k hyperedges and each subset $X \subset V$ containing s is entered by at least l hyperedges.

By orienting an (undirected) hypergraph we mean the operation that consists of assigning a head to every hyperedge.

Theorem 6.8 [29]. *A hypergraph has a (k, l)-edge-connected orientation if and only if there are at least $kt - k + l$ hyperedges intersecting more than one part of every t-partite partition of V.*

Finally we mention that Edmonds' Theorem 1.3 can also be carried over to hypergraphs. To this end we say that a directed hypergraph H is a spanning hyper-arborescence of root s if H has $|V| - 1$ hyperedges whose heads are distinct elements of $V - s$ and H is $(1, 0)$-edge-connected.

Theorem 6.9 [29]. *A directed hypergraph contains k disjoint spanning hyper-arborescences of root s if and only if H is $(k, 0)$-edge-connected (with respect to s).*

Note that the special case $l = 0$ of Theorem 6.8 combined with Theorem 6.9 immediately implies Theorem 6.6 (without using matroids).

The paper [5] of Berg, Jackson and Jordán contains extensions of Mader's directed splitting lemma and of the directed augmentation Theorem 2.6 to directed hypergraphs.

REFERENCES

[1] J. Bang-Jensen, H. Gabow, T. Jordán and Z. Szigeti, Edge-connectivity augmentation with partition constraints, *SIAM J. Discrete Mathematics,* **12** No. 2 (1999), 160–207.

[2] J. Bang-Jensen and B. Jackson, Augmenting hypergraphs by edges of size two, in: *Connectivity Augmentation of Networks: Structures and Algorithms, Mathematical Programming,* (ed. A. Frank), Ser. B **84** No. 3 (1999), pp. 467–481.

[3] J. Bang-Jensen, A. Frank and B. Jackson, Preserving and increasing local edge-connectivity in mixed graph, *SIAM J. Discrete Math.,* **8** (1995 May), No. 2, pp. 155–178.

[4]   A. Benczúr and A. Frank, Covering symmetric supermodular functions by graphs, in: Connectivity Augmentation of Networks: Structures and Algorithms, *Mathematical Programming,* (ed. A. Frank), Ser. B **84** No. 3 (1999), pp. 483–503.

[5]   A. Berg, B. Jackson and T. Jordán, Edge-splitting and connectivity augmentation in directed hypergraphs, *Discrete Mathematics,* **273** (2003), pp. 71–84.

[6]   A. Berg, B. Jackson and T. Jordán, Highly edge-connected detachments of graphs and digraphs, *J. Graph Theory,* **43** (2003), pp. 67–77.

[7]   F. Boesch and R. Tindell, Robbins's theorem for mixed multigraphs, *Am. Math. Monthly,* **87** (1980), 716–719.

[8]   B. Cosh, B. Jackson and Z. Király, Local connectivity augmentation in hypergraphs is NP-complete, submitted.

[9]   J. Edmonds, Edge-disjoint branchings, in: *Combinatorial Algorithms,* Academic Press, New York (1973), 91–96.

[10]   J. Edmonds, Minimum partition of a matroid into independent sets, *J. Res. Nat. Bur. Standards Sect.,* **869** (1965), 67–72.

[11]   J. Edmonds and D. R. Fulkerson, Transversal and matroid partition, *Journal of Research of the National Bureau of Standards* (B), **69** (1965), 147–153.

[12]   K. P. Eswaran and R. E. Tarjan, Augmentation problems, *SIAM J. Computing,* **5** No. 4 (1976), 653–665.

[13]   B. Fleiner, Detachment of vertices preserving edge-connectivity, *SIAM J. on Discrete Mathematics,* **3** No. 3. (2005), pp. 581–591.

[14]   T. Fleiner and T. Jordán, Covering and structure of crossing families, in: *Connectivity Augmentation of Networks: Structures and Algorithms, Mathematical Programming,* (ed. A. Frank), Ser. B **84** No. 3 (1999), pp. 505–518.

[15]   L. R. Ford and D. R. Fulkerson, Flows in Networks, Princeton Univ. Press, Princeton NJ., 1962.

[16]   A. Frank, On disjoint trees and arborescences, in: Algebraic Methods in Graph Theory, *Colloquia Mathematica, Soc. J. Bolyai,* North-Holland **25** (1978), 159–169.

[17]   A. Frank, Kernel systems of directed graphs, *Acta Scientiarum Mathematicarum* (Szeged), **41** No. 1–2 (1979), 63–76.

[18]   A. Frank, On the orientation of graphs *J. Combinatorial Theory,* Ser. B **28** No. 3 (1980), 251–261.

[19]   A. Frank, Augmenting graphs to meet edge-connectivity requirements, *SIAM J. on Discrete Mathematics,* **5** No. 1. (1992 February), pp. 22–53.

[20]   A. Frank, On a theorem of Mader, *Annals of Discrete Mathematics,* **101** (1992), 49–57.

[21]   A. Frank, Applications of submodular functions, in: Surveys in Combinatorics, *London Mathematical Society Lecture Note Series 187,* Cambridge Univ. Press (Ed. K. Walker), 1993, 85–136.

[22] A. Frank, Connectivity augmentation problems in network design, in: *Mathematical Programming: State of the Art* 1994 (eds.: J. R. Birge and K. G. Murty), The University of Michigan, pp. 34–63.

[23] A. Frank, Orientations of graphs and submodular flows, *Congressus Numerantium,* **113** (1996) (A. J. W. Hilton, ed.), 111–142.

[24] A. Frank, An intersection theorem for supermodular functions, preliminary draft (2004).

[25] A. Frank and T. Jordán, Minimal edge-coverings of pairs of sets, *J. Combinatorial Theory,* Ser. B, **65** No. 1 (1995, September), pp. 73–110.

[26] A. Frank, T. Jordán and Z. Szigeti, An orientation theorem with parity conditions, *Discrete Applied Mathematics,* **115** (2001), pp. 37-47.

[27] A. Frank and Z. Király, Graph orientations with edge-connection and parity constraints, *Combinatorica,* **22** No. 1. (2002), pp. 47–70.

[28] A. Frank and T. Király, Combined connectivity augmentation and orientation problems, in: Submodularity, *Discrete Applied Mathematics,* guest ed. S. Fujishige, **131** No. 2. (September 2003), pp. 401–419.

[29] A. Frank, T. Király and Z. Király, On the orientation of graphs and hypergraphs, in: Submodularity, *Discrete Applied Mathematics,* guest ed.: S. Fujishige, 131, No. 2. (September 2003), pp. 385–400.

[30] A. Frank, T. Király and M. Kriesell, On decomposing a hypergraph into k connected sub-hypergraphs, in: Submodularity, *Discrete Applied Mathematics,* (guest ed. S. Fujishige), **131** No. 2. (September 2003), pp. 373–383.

[31] A. Frank and L. Szegő, Constructive characterizations for packing and covering with trees, in: Submodularity, *Discrete Applied Mathematics,* (guest ed. S. Fujishige), **131** No. 2. (September 2003), pp. 347–371.

[32] C. Greene and T. L. Magnanti, Some abstract pivot algorithms, *SIAM Journal on Applied Mathematics,* **29** (1975), 530–539.

[33] L. Henneberg, Die graphische Statik der starren Systeme, Leipzig 1911.

[34] T. Jordán and Z. Szigeti, Detachments preserving local edge-connectivity of graphs, *SIAM J. Discrete Mathematics,* **17** No. 1. (2003), pp. 72–87.

[35] S. Khanna, J. Naor and F. B. Shepherd, Directed network design with orientation constraints, *SIAM J. Discrete Mathematics,* to appear in 2005, a preliminary version appeared in: *Proceedings of the Eleventh Annual ACM-SIAM Symposium on Discrete Algorithms,* San Francisco, California, Jan. 9–11 (2000), 663–671.

[36] T. Király, Covering symmetric supermodular functions by uniform hypergraphs, *J. Combinatorial Theory,* Ser. B, **91** (2004), pp. 185–200.

[37] G. Laman, On graphs and rigidity of plane skeletal structures, *J. Engineering Mathematics,* **4** (1970), pp. 331–340.

[38] M. Lorea, Hypergraphes et matroides, *Cahiers Centre Etud. Rech. Oper.,* **17** (1975), pp. 289–291.

[39] L. Lovász, Solution to Problem 11, see pp. 168–169, in: Report on the Memorial Mathematical Contest Miklós Schweitzer of the year 1968 (in Hungarian), *Matematikai Lapok,* **20** (1969), pp. 145–171.

[40] L. Lovász, 2-matchings and 2-covers of hypergraphs, *Acta Mathematica Academiae Scientiarium Hungaricae,* **26** (1975), 433–444.

[41] L. Lovász, On two minimax theorems in graph theory, *J. Combinatorial Theory,* Ser. B **21** (1976), 96–103.

[42] L. Lovász, Combinatorial Problems and Exercises, North-Holland 1979.

[43] L. Lovász, Submodular functions and convexity, in: *Mathematical programming – The state of the art,* (eds. A. Bachem, M. Grötschel and B. Korte), Springer 1983, 235–257.

[44] W. Mader, Ecken vom Innen- und Aussengrad k in minimal n-fach kantenzusammenhängenden Digraphen, *Arch. Math.,* **25** (1974), 107–112.

[45] W. Mader, A reduction method for edge-connectivity in graphs, *Ann. Discrete Math.,* **3** (1978), 145–164.

[46] W. Mader, Konstruktion aller n-fach kantenzusammenhängenden Digraphen, *Europ. J. Combinatorics,* **3** (1982), 63–67.

[47] C. St. J. A. Nash-Williams, On orientations, connectivity and odd vertex pairings in finite graphs, *Canad. J. Math.,* **12** (1960), 555–567.

[48] C. St. J. A. Nash-Williams, Well-balanced orientations of finite graphs and unobtrusive odd-vertex-pairings in: *Recent Progress in Combinatorics* ed. W. T. Tutte (1969), Academic Press, pp. 133–149.

[49] C. St. J. A. Nash-Williams, Decomposition of finite graphs into forests, *J. London Math. Soc.,* **39** (1964), 12.

[50] C. St. J. A. Nash-Williams, Connected detachments of graphs and generalized Euler trails, *J. London Math. Soc.,* **31** No. 2 (1985), 17–19.

[51] C. St. J. A. Nash-Williams, Strongly connected mixed graphs and connected detachments of graphs, *Journal of Combinatorial Mathematics and Combinatorial Computing,* **19** (1995), 33–47.

[52] H. E. Robbins, A theorem on graphs with an application to a problem of traffic control, *American Math. Monthly,* **46** (1939), 281–283.

[53] A. Schrijver, A counterexample to a conjecture of Edmonds and Giles, *Discrete Mathematics,* **32** (1980), 213–214.

[54] A. Schrijver, Total dual integrality from directed graphs, crossing families and sub- and supermodular functions, in: *Progress in Combinatorial Optimization,* (ed. W. R. Pulleyblank), Academic Press (1984), 315–361.

[55] Z. Szigeti, Hypergraph connectivity augmentation, in: Connectivity Augmentation of Networks: Structures and Algorithms, *Mathematical Programming,* (ed. A. Frank), Ser. B, **84** No. 3 (1999), pp. 519–527.

[56] W. T. Tutte, On the problem of decomposing a graph into n connected factors, *J. London Math. Soc.,* **36** (1961), 221–230.

[57]   T. Watanabe and A. Nakamura, Edge-connectivity augmentation problems, *Computer and System Sciences,* **35** No. 1 (1987), 96–144.

András Frank

Department of Operations Research
Eötvös University
Egerváry Research Group of the
Hungarian Academy of Sciences and
Eötvös University Budapest
Pázmány P. s. 1/c
Budapest, Hungary, H-1117
and Ericsson Hungary
Laborc u. 1
Budapest, Hungary H-1037

`frank@cs.elte.hu`

BOLYAI SOCIETY
MATHEMATICAL STUDIES, 15

Conference on Finite
and Infinite Sets
Budapest, pp. 143–155.

PERFECT POWERS IN PRODUCTS WITH CONSECUTIVE TERMS FROM ARITHMETIC PROGRESSIONS

K. GYŐRY*

I. INTRODUCTION

There is an extensive literature on perfect powers and "almost" perfect powers in products of the form

$$(1) \qquad\qquad n(n+d)\ldots\bigl(n+(k-1)d\bigr)$$

where n, d, k are positive integers with $\gcd(n,d)=1$ and $k \geq 3$. By an "almost" perfect power we mean a number of the shape b times a perfect power, where b is a positive integer having no prime factor greater than a given number, say k. The classical case $d=1$ has been completely settled. Further, for $d>1$, a lot of interesting partial results have been published. For survey papers on results obtained before 1999 we refer to Tijdeman [41], [42], Shorey and Tijdeman [37, 38], Shorey [33, 34] and Győry [19].

Since 1999, considerable progress has been made in the case $d > 1$. Several results have been established on squares and "almost" squares of the form (1), on those d for which (1) can be a perfect or an "almost" perfect power, and on the situation when at least one of the factors $n + id$ is omitted from the product (1). For an account of these results we refer to Shorey [35, 36].

Recently, it has been proved by Győry [19, case $k = 3$], Győry, Hajdu and Saradha [20, case $k = 4, 5$] and Bennett, Győry and Hajdu [2, case

*Research supported in part by the Hungarian Academy of Sciences, by the Netherlands Organization for Scientific Research, and by Grant 29330 from the Hungarian National Foundation for Scientific Research.

$6 \leq k \leq 11$] that apart from some exceptions, (1) cannot be a perfect or an "almost" perfect power whenever $k \leq 11$. Further, Győry, Hajdu and Saradha [20] showed, for each k, the finiteness of the numbers n, d for which (1) is an "almost" perfect power. The purpose of the present paper is to give an overview of the above-mentioned results of Bennett, Hajdu, Saradha and the author. This article may be considered as a continuation of Sections 1 to 4 of Győry [19].

In the second section a brief survey is given on the most important results obtained in the case $d = 1$. In the first part of Section III general finiteness theorems are presented. The second part of Section III is devoted to recent results which, for $k \leq 11$, provide all perfect or "almost" perfect powers of the form (1). In Section IV we deal with an application to rational solutions of a related superelliptic equation. Finally, in the last section some methods will be discussed which were needed in our proofs. It will be pointed out that ternary diophantine equations and the theory of Galois representations and modular forms play a crucial rôle in recent investigations concerning (1).

II. PRODUCTS OF CONSECUTIVE INTEGERS

The case $d = 1$

It was an old **conjecture** from the 1820's that equation

$$(2) \qquad n(n+1)\ldots(n+k-1) = y^l \quad \text{in integers} \quad n \geq 1, \ k, y, l \geq 2$$

has no solution. After many special results, Erdős [9] and Rigge [28] confirmed the conjecture for $l = 2$. Their proof was elementary and ingenious. Erdős [10] and, independently, Rigge showed that for every $l > 2$ there is a $k_0 = k_0(l)$ such that for $k \geq k_0$, (2) is impossible. By means of the Thue-Siegel method Erdős and Siegel proved in 1940 the conjecture for all sufficiently large k. Their proof remained unpublished. Later, Erdős [12] gave another, elementary proof for this theorem.

Using Erdős' method, Erdős and Selfridge [13] proved the conjecture in full generality.

Theorem A (Erdős and Selfridge, [13]). *Equation (2) has no solution.*

Saradha and Shorey [31] recently showed that omitting one of the factors $n+i$ on the left hand side of (2), all the solutions of the equation so obtained are

$$2 \cdot (2 + 2 \cdot 1) = 2^3 \quad \text{and} \quad 1 \cdot (1 + 1) \cdot (1 + 3 \cdot 1) = 2^3.$$

The binomial equation

Consider now the equation

$$(3) \qquad \binom{n + k - 1}{k} = y^l \quad \text{in integers} \quad k, l, y \geq 2, n \geq k + 1.$$

For $k = l = 2$, this leads to a Pell equation and hence it has infinitely many solutions (n, y). For $k = 3, l = 2$, Meyl [23, n odd], and Watson [43, n even] proved that

$$\binom{50}{3} = 140^2$$

is the only solution of (3).

Erdős [10] **conjectured** that for $l > 2$, equation (3) has no solution. In the same article he proved this for $l = 3$ and for $k \geq 2^l$. The cases $l = 4, 5$ were settled by Oblàth [24].

Using his elementary method applied earlier to (2), Erdős [11] proved that for $k \geq 4$, equation (3) has no solution.

For $k < 4$, the approach of Erdős does not work. By means of Baker's method Tijdeman [40] proved in an effective form that for $k = 2$ and 3, equation (3) has only finitely many solutions. Later Terai [39] showed that in this case $l < 4250$. We note that recently Terai derived a bound for l also in the case when in (3) y^l is replaced by py^l with an odd prime p.

Finally, in Győry [17] I succeeded to prove Erdős' conjecture for the remaining cases $k = 2, 3$ and $l > 2$. The proof is based on a combination of some results of Győry [15] and Darmon and Merel [6] on generalized Fermat equations with a theorem of Bennett and de Weger [4] on binomial Thue equations. In fact this was the first time that generalized Fermat equations were used in the study of equations (3) and (4) and their generalizations.

Summing up the above results, we have the following.

Theorem B (Erdős, [11, case $k \geq 4$]; Győry, [17, case $k < 4$]). *Apart from the cases $(k, l) = (2, 2), (3, 2)$, equation (3) has no solution.*

A common generalization of equations (2) and (3)

Denote by $P(b)$ the greatest prime factor of any integer $b > 1$, and let $P(1) = 1$. The equation
(4)
$$n(n + 1) \ldots (n + k - 1) = by^l \text{ in integers } n, b, y \geq 1, k, l \geq 2 \text{ with } P(b) \leq k$$

is a common generalization of equations (2) and (3). For $b = 1$ this is just equation (2), while for $b = k!$ it gives equation (3).

For $k = b = l = 2$, this is again a Pell equation, having infinitely many solutions.

Let $p^{(k)}$ denote the least prime with $p^{(k)} > k$. As was pointed out in Győry [19], n, k yield a solution of (4) with $P(y) \leq k$ if and only if $n \in \{1, 2, \ldots, p^{(k)} - k\}$. This means that for given k, equation (4) has only finitely many solutions with $P(y) \leq k$ and all these can be easily determined. Hence, in what follows, we are interested only in those solutions for which $P(y) > k$.

It was proved by Erdős and Selfridge [13] that under the restriction $P(b) < k$, equation (4) has no solution with $P(y) > k$. However, this result cannot be applied to equation (4) if k is prime.

The following theorem gives the complete solution of equation (4).

Theorem C (Saradha [30, case $k \geq 4$]; Győry [18, case $k < 4$]). *Apart from the case $(k, b, l) = (2, 2, 2)$, equation (4) has the only solution*

$$48 \cdot 49 \cdot 50 = 6 \cdot 140^2$$

with $P(y) > k$.

To prove this theorem for $k \geq 4$, Saradha [30] combined Erdős' method with a result of Shorey and Tijdeman and with some computations. Her method of proof cannot be applied to the case $k < 4$. In Győry [18], the results of Wiles [44], Ribet [27] and Darmon and Merel [6] concerning generalized Fermat equations were used to resolve (4) for $k = 2$ and 3.

Theorems A and B are consequences of Theorem C; cf. Győry [19]. Further, it is clear that Theorem C is valid also with $P(b) \leq k$ replaced by $P(b) < p^{(k)}$.

In (4), $P(y) > k$ implies that $n > k^l$. Recently, Theorem C has been refined by Saradha [30] for $k \geq 9$, Hanrot, Saradha and Shorey [22] for

$6 \leq k \leq 8$ and Bennett [1] for $3 \leq k \leq 5$. They proved that under the assumptions $l \geq 3$, $n > k^l$ and $P(b) \leq p^{(k)}$, equation (4) does not hold.

For $3 \leq k \leq 5$, a further refinement has been recently obtained by Győry and Pintér [21]. They showed that (4) is impossible even if $l \geq 3$, $n > k^l$ and $P(b) \leq p_k$, where p_k denotes the k-th prime. It is clear that $p_k > p^{(k)}$ if $k \geq 4$.

III. PRODUCTS OF CONSECUTIVE TERMS IN ARITHMETIC PROGRESSION

In this section we deal with the equations

$$(5) \qquad n(n + d) \ldots \big(n + (k - 1)d\big) = y^l$$

and

$$(6) \qquad n(n + d) \ldots \big(n + (k - 1)d\big) = by^l,$$

where $n, d, b, y \geq 1$ and $k \geq 3$, $l \geq 2$ are unknown integers such that $\gcd(n, d) = 1$ and $P(b) \leq k$.

Finiteness results

It is easy to see that both equations (5) and (6) have infinitely many solutions if $k = 2$ or if $(k, l) = (3, 2)$. Tijdeman [41] showed that (6) has infinitely many solutions with $P(y) > k$ for $(k, l) = (3, 3)$ and $(4, 2)$, too.

Erdős **conjectured** that in (5) k must be bounded. Further, by a **conjecture** of Tijdeman [41], the total number of solutions of (6) with $P(y) > k$ and $k + l > 6$ is finite.

Using Faltings' theorem [14] on rational points of curves of genus > 1, Darmon and Granville [5] proved that for given $k \geq 3$, $l \geq 4$, equation (5) has only finitely many solutions. The following theorem refines this and extends it to the case $b > 1$.

Theorem 1 (Győry, Hajdu, Saradha [20]). *For given $k \geq 3$, $l \geq 2$ with $k + l > 6$, equation (6) has only finitely many solutions (n, d, b, y).*

In view of Tijdeman's result (6) has infinitely many solutions for each choice of $k \geq 3$, $l \geq 2$ with $k + l \leq 6$.

Shorey [33] proved that for $l \geq 4$, the abc conjecture implies Erdős' conjecture on the boundedness of k. In fact, from the abc conjecture one can deduce a more precise result.

Theorem 2 (Győry, Hajdu, Saradha [20]). *The abc conjecture implies that (6) has only finitely many solutions* (n, d, k, b, y, l) *with* $k \geq 3$, $l \geq 4$ *and* $d > 1$.

We note that the assumption $d > 1$ is necessary. For $d = 1$, (6) has the solution $n = y = 1$, $b = k!$ for each $k \geq 3$.

On the resolution of equations (5) and (6)

First consider equation (5). As was mentioned above, (5) has infinitely many solutions both for $k = 2$ and for $(k, l) = (3, 2)$. Euler proved that for $(k, l) = (4, 2)$, equation (5) has no solution. The same result was proved by Obláth [25, 26] for $(k, l) = (5, 2), (3, 3), (3, 4)$ and $(3, 5)$.

Using results of Wiles, Ribet and Darmon and Merel on generalized Fermat equations, Győry [19] showed that equation (5) is impossible for $k = 3$ and $l > 2$.

Recently, the following theorem has been established for $k \leq 11$.

Theorem 3 (Győry, Hajdu, Saradha [20, case $k \leq 5$]; Bennett, Győry, Hajdu [2, case $k \geq 6$]). *For* $4 \leq k \leq 11$, *equation (5) has no solution.*

We note that for $6 \leq k \leq 11$ and $l = 2$, Theorem 3 was independently proved by Hirata–Kohno and Shorey.

Summarizing the above results on equation (5) we have the following.

Theorem D. *Apart from the case* $(k, l) = (3, 2)$, *equation (5) is impossible for* $3 \leq k \leq 11$.

Conjecture 1. *For* $k \geq 3$ *and* $(k, l) \neq (3, 2)$, *(5) has no solution.*

Concerning equation (6), Győry [19] proved more generally that for $k = 3$, $l > 2$ and $P(b) \leq 2$, (6) is not solvable. As Tijdeman's result [41] concerning the case $(k, l) = (3, 3)$ shows, the assumption on b cannot be relaxed to $P(b) \leq 3$.

Recently Győry's theorem [19] has been extended to the case $k \leq 11$.

Theorem 4 (Bennett, Győry, Hajdu [2]). *For $4 \leq k \leq 6$, $P(b) \leq 2$ and for $6 < k \leq 11$, $P(b) \leq 3$, equation (6) has no solution.*

For $k = 4$ and 5, this was proved in a less precise form by Győry, Hajdu and Saradha [20]. When $b = 1$, Theorem 4 gives back Theorem 3.

The above results on (6) can be summarized as follows.

Theorem E. *If $3 \leq k \leq 11$, $(k, l) \neq (3, 2)$ and $P(b) \leq 2$, then (6) has no solution.*

We note that in Győry, Hajdu and Saradha [20] and Bennett, Győry and Hajdu [2], the results were extended to the case when n and b are not necessarily positive integers. As will be seen in Section IV, this extension is important for certain applications.

Conjecture 2. *For $k \geq 3$, $(k, l) \neq (3, 2)$ and $P(b) \leq 2$, (6) has no solution.*

The examples

$$2 \cdot 9 \cdot 16 = 2^5 \cdot 3^2 \cdot 1^l \quad \text{and} \quad 1 \cdot 2 \cdot 3 \cdot 4 = 2^3 \cdot 3 \cdot 1^l$$

show that for $k = 3$ and 4, the assumption $P(b) \leq 2$ cannot be replaced by $P(b) \leq 3$. It is likely that for $k \geq 5$, the assumption on the greatest prime factor of b can be relaxed in Theorems 4, E and in Conjecture 2.

IV. AN APPLICATION OF THEOREMS 3 AND 4

The results concerning equations (5) and (6) can be applied to the superelliptic equation

$$(7) \qquad\qquad x(x + 1) \ldots (x + k - 1) = z^l,$$

where the unknowns are now k, l, x, z with $k, l \geq 2$ and *rational* x, z. It is clear that $(x, z) = (-i, 0)$ are solutions of (7) for $i = 0, \ldots, k - 1$. These solutions are called *trivial*.

It follows from Faltings' theorem [14] that for fixed k, l with $k + l > 6$, equation (7) has only finitely many solutions.

Sander [29] proved that if $2 \leq k \leq 4$ and $(k, l) \neq (2, 2)$, then (7) has only trivial solutions. Further he **conjectured** that except for the case $(k, l) = (2, 2)$, (7) has no non-trivial solution.

Putting $x = n/d$, $z = y/u$, with integers n, d, y, u such that $d, u > 0$, $n, y \neq 0$ and $\gcd(n, d) = \gcd(y, u) = 1$, (7) leads to the equation

$$n(n + d) \ldots (n + (k - 1)d) = y^l, \qquad u^l = d^k.$$

By applying now the extended version of Theorem 3 with not necessarily positive n, the following theorem follows.

Theorem 5 (Győry, Hajdu, Saradha [20, case $2 \leq k \leq 5$, $l \geq 3$], Bennett, Győry, Hajdu [2, the other cases]). *For $2 \leq k \leq 11$, the only non-trivial solutions of (7) are given by*

$$(k, l, x, z)$$

$$= \left(3, 3, -\frac{2}{3}, -\frac{2}{3}\right), \left(3, 3, -\frac{4}{3}, \frac{2}{3}\right), \left(4, 2, -\frac{3}{2}, \pm\frac{3}{4}\right), \left(8, 2, -\frac{7}{2}, \pm\frac{105}{16}\right).$$

As is seen, for $(k, l) = (3, 3)$ and $(4, 2)$ there exist non-trivial solutions which are missing from the theorem and the conjecture of Sander. Hence Sander's conjecture should be modified accordingly.

We note that we proved Theorem 5 in a more general form, we solved equation (7) with z^l replaced by $\pm 2^\alpha \cdot z^l$, where $\alpha \in \mathbb{Z}$ is also unknown. Further, under the assumptions $l \geq 3$ and $\gcd(k, l) = 1$, Theorem 5 has been extended in Győry, Hajdu, and Saradha [20] to the case $k \leq 18$.

V. THE METHOD OF PROOFS OF THEOREMS 1 TO 4

We briefly present the basic ideas and the main tools used in the proofs of Theorems 1, 2 and 4.

From the equation

$$(6) \qquad\qquad n(n + d) \ldots (n + (k - 1)d) = by^l,$$

where $n, d, b, y \geq 1$ and $k \geq 3$, $l \geq 2$ are unknown integers with $\gcd(n, d) = 1$ and $P(b) \leq k$, one can deduce that

$$(8) \qquad\qquad n + id = A_i X_i^l, \qquad i = 0, \ldots, k - 1,$$

where $A_i, X_i \geq 1$ are unknown integers with $P(A_i) \leq k$. It is obvious that conversely, (8) implies (6). Depending on the situation investigated, we can choose the A_i, X_i such that either A_i is l-th power free (when, for fixed k, there are only finitely many possibilities for A_i) or X_i is free of prime factors $\leq k$.

Using (8), (6) can be reduced to systems of equations consisting of generalized Fermat equations. There are two possibilities:

1. For distinct integers $0 \leq p, q, r \leq k - 1$, one can easily find non-zero integers $\lambda_p, \lambda_q, \lambda_r$ with absolute values $\leq k$ such that

$$\lambda_p(n + pd) + \lambda_q(n + qd) = \lambda_r(n + rd).$$

Hence, in view of (8), we get an equation of the form

(9) $\quad AX^l + BY^l = CZ^l \quad$ in coprime non-zero integers $\quad X, Y, Z,$

where A, B, C are relatively prime non-zero integers with $P(ABC) \leq k$.

2. For integers $0 \leq p < q \leq r < s \leq k - 1$ with $p + s = q + r$, we deduce that

$$(n + qd)(n + rd) - (n + pd)(n + sd) = (qr - ps)d^2.$$

Thus, by (8), we obtain an equation of the shape

(10) $\quad AX^l + BY^l = CZ^2 \quad$ in coprime non-zero integers $\quad X, Y, Z,$

where A, B, C are relatively prime non-zero integers with $P(AB) \leq k$ and $|C| \leq (k - 1)^2$.

The basic ideas of the proofs of Theorems 1 and 2

To prove Theorem 1, we choose in (8) the A_i to be l-th power free. Then we arrived at equations of the form (9) with coefficients which can be taken fixed. For $k = 3$ and $l \geq 4$, one can use Falting's theorem to prove the finiteness of the number of solutions of the equation (9) so obtained, whence Theorem 1 follows. If $k \geq 4$, the situation is more complicated but a similar argument can be applied in that case, too.

In the proof of Theorem 2 we may assume that $(n, d, k) \neq (2, 7, 3)$. Then a theorem of Shorey and Tijdeman [37] gives that $P(y) > k$. Under this assumption Shorey [33] used the abc conjecture to prove that k is bounded. So we can fix k. Then we reduce equation (6) to equations of the form (9), where A, B, C are not fixed, but $P(ABC) \leq k$. The abc conjecture can be applied to (9) in a well-known way to show that X, Y, Z and l are bounded. So we may assume that X^l, Y^l and Z^l are fixed. Now (9) becomes an S-unit equation in A, B, C for the set of primes $S = \{p \mid p \leq k\}$, hence, by a theorem of Győry [16], $\max\{|A|, |B|, |C|\}$ is bounded. This implies that $\max\{n, d, b, y\}$ is also bounded.

We note that using an effective version of the abc conjecture, the above proof provides an effective upper bound for $\max\{n, d, k, b, y, l\}$.

The main tools in the proof of Theorem 4

The proof of Theorem 4 is long and complicated.

The case $l = 2$. In the case $l = 2$ one can reduce equation (6) to finding rational points on some elliptic curves of rank 0. Then one can use the program package MAGMA to find all rational points on the curves in question.

The case $l > 2$. In this case we may assume that $l > 2$ is a prime. After having reduced equation (6) to (8), we have to distinguish several subcases, according to the possible choices of the A_i. If there are $0 \leq i, j \leq k - 1$ such that $P(A_i A_{i+1} \ldots A_{i+j}) \leq j + 1$, then (6) reduces to the case when k is replaced by $j + 1 < k$. However, this is not the case in general. Then, as remarked above, equation (6) can be reduced to systems of equations consisting of ternary equations of the form (9) and (10).

We applied different methods to deal with *non-trivial* solutions X, Y, Z of (9) and (10), i.e. with solutions for which $XYZ \neq 0, \pm 1$.

When $3 \leq l \leq 7$, for certain choices of the A_i we used local methods and showed that at least one of the equations (9) and (10) involved is not solvable (mod p) for some appropriate prime p.

For $l = 3$, classical results of Selmer [32] and others can be used to prove that some equations (9) coming from (6) have no non-trivial solutions.

For $l = 5, 7$, one can use some results of Dirichlet, Lebesgue, Maillet (cf. [8]), Dénes [7], Győry [15] and Bennett, Győry and Hajdu [2] on the equations of the form (9) with $A = B = 1$.

For $l \geq 7$, the main ingredients are some recent results on ternary equations of the form (9) and (10) whose proofs are based upon the theory of Frey curves, Galois representations and modular forms.

It should be mentioned here the celebrated results of Wiles [44], Ribet [27], Darmon and Merel [6] and others on equations of the form (9). For example, we utilize the fact that for $A = B = 1$, $C = 2^{\alpha}$, $\alpha \geq 0$ integer, equation (9) has no non-trivial solutions.

Some results of Bennett and Skinner [3] and Bennett, Győry and Hajdu [2] play also an important rôle in the proof of Theorem 4. For example, we showed with Bennett and Hajdu that for $P(AB) \leq 3$ and $C = 1$, equation (10) has no solutions with $5 \mid XY$ if $l \geq 7$, and with $7 \mid XY$ if $l \geq 11$.

Acknowledgements. The author is indebted to Professor A. Pethő and Dr. L. Hajdu for pointing out some typist's errors in the manuscript.

REFERENCES

[1] M. A. Bennett, Products of consecutive integers, *Bull. London Math. Soc.*, **36** (2004), 683–694.

[2] M. A. Bennett, K. Győry and L. Hajdu, Powers from products of consecutive terms in arithmetic progression, *Proc. London Math. Soc.*, to appear.

[3] M. A. Bennett and C. Skinner, Ternary Diophantine equations via Galois representations and modular forms, *Canad. J. Math.*, **56** (2004), 23–54.

[4] M. A. Bennett and B. M. M. de Weger, On the Diophantine equation $|ax^n - by^n| = 1$, *Math. Comp.*, **67** (1998), 413–438.

[5] H. Darmon and A. Granville, On the equations $z^m = F(x, y)$ and $Ax^p + By^q = Cz^r$, *Bull. London Math. Soc.*, **27** (1995), 513–543.

[6] H. Darmon and L. Merel, Winding quotients and some variants of Fermat's last theorem, *J. Reine Angew. Math.*, **490** (1997), 81–100.

[7] P. Dénes, Über die diophantische Gleichung $x^l + y^l = cz^l$, *Acta Math.*, **88** (1952), 241–251.

[8] L. E. Dickson, *History of the Theory of Numbers*, Vol. II, Carnegie Inst., Washington DC. (1919).

[9] P. Erdős, Note on products of consecutive integers, *J. London Math. Soc.*, **14** (1939), 194–198.

[10] P. Erdős, Note on the product of consecutive integers (II), *J. London Math. Soc.* **14** (1939), 245–249.

[11] P. Erdős, On a diophantine equation, *J. London Math. Soc.* **26** (1951), 176–178.

[12] P. Erdős, On the product of consecutive integers III, *Indag. Math.*, **17** (1955), 85–90.

[13] P. Erdős and J. L. Selfridge, The product of consecutive integers is never a power, *Illinois J. Math.*, **19** (1975), 292–301.

[14] G. Faltings, Endlichkeitssätze für abelsche Varietäten über Zahlkörpern, *Invent. Math.*, **73** (1983), 349–366.

[15] K. Győry, Über die diophantische Gleichung $x^p + y^p = cz^p$, *Publ. Math. Debrecen*, **13** (1966), 301–305.

[16] K. Győry, On the number of solutions of linear equations in units of an algebraic number field, *Comment. Math. Helv.*, **54** (1979), 583–600.

[17] K. Győry, On the diophantine equation $\binom{n}{k} = x^l$, *Acta Arith.*, **80** (1997), 289–295.

[18] K. Győry, On the diophantine equation $n(n + 1) \ldots (n + k - 1) = bx^l$, *Acta Arith.*, **83** (1998), 87–92.

[19] K. Győry, Power values of products of consecutive integers and binomial coefficients, in: *Number Theory and Its Applications*, Kluwer Acad. Publ. (1999), pp. 145–156.

[20] K. Győry, L. Hajdu and N. Saradha, On the diphantine equation $n(n + d) \ldots (n + (k - 1)d) = by^l$, *Canad. Math. Bull.*, **47** (2004), 373–388.

[21] K. Győry and Á. Pintér, On products of consecutive integers, *Monatshefte Math.*, **145** (2005), 19–33.

[22] G. Hanrot, N. Saradha and T. N. Shorey, Almost perfect powers in consecutive integers, *Acta Arith.*, **99** (2001), 13–25.

[23] A. J. J. Meyl, Question 1194, *Nouv. Ann. Math.*, **17** (1878), 464–467.

[24] R. Obláth, Note on the binomial coefficients, *J. London Math. Soc.*, **23** (1948), 252–253.

[25] R. Obláth, Über das Produkt fünf aufeinander folgender Zahlen in einer arithmetischen Reihe, *Publ. Math. Debrecen*, **1** (1950), 222–226.

[26] R. Obláth, Eine Bemerkung über Produkte aufeinander folgender Zahlen, *J. Indian Math. Soc.*, **15** (1951), 135–139.

[27] K. A. Ribet, On the equation $a^p + 2^\alpha b^p + c^p = 0$, *Acta Arith.*, **79** (1997), 7–16.

[28] O. Rigge, Über ein diophantisches Problem, in: *9th Congress Math. Scand., Helsingfors*, Mercator (1939), 155–160.

[29] J. W. Sander, Rational points on a class of superelliptic curves, *J. London Math. Soc.*, **59** (1999), 422–434.

[30] N. Saradha, On perfect powers in products with terms from arithmetic progressions, *Acta Arith.*, **82** (1997), 147–172.

[31] N. Saradha and T. N. Shorey, Almost perfect powers in arithmetic progression, *Acta Arith.*, **99** (2001), 363–388.

[32] E. Selmer, The diophantine equation $ax^3 + by^3 + cz^3 = 0$, *Acta Math.*, **85** (1951), 203–362.

[33] T. N. Shorey, Exponential diphantine equations involving products of consecutive integers and related equations, in: *Number Theory,* Hindustan Book Agency (1999), pp. 463–495.

[34] T. N. Shorey, Mathematical contributions, *Bombay Math Colloquium,* **15** (1999), 1–19.

[35] T. N. Shorey, Powers in arithmetic progression, in: *A Panorama in Number Theory,* Cambridge (2002), pp. 325–336.

[36] T. N. Shorey, Powers in arithmetic progression (II), in: *New Aspects of Analytic Number Theory,* Kyoto (2002b), pp. 202–214.

[37] T. N. Shorey and R. Tijdeman, On the greatest prime factor of an arithmetic progression, *A Tribute to Paul Erdős,* Cambridge (1990), pp. 385–389.

[38] T. N. Shorey and R. Tijdeman, Some methods of Erdős applied to finite arithmetic progressions, in: *The Mathematics of Paul Erdős,* I, Springer (1997), pp. 251–267.

[39] N. Terai, On a diophantine equation of Erdős, *Proc. Japan Acad. Ser. A,* **70** (1994), 213–217.

[40] R. Tijdeman, Applications of the Gelfond–Baker method to rational number theory, in: *Topics in Number Theory,* North-Holland (1976), pp. 399–416.

[41] R. Tijdeman, Diophantine equations and Diophantine approximations, in: *Number Theory and Applications,* Kluwer Acad. Press (1989), pp. 215–243.

[42] R. Tijdeman, Exponential diophantine equations 1986–1996, in: *Number Theory,* de Gruyter (1998), pp. 523–539.

[43] G. N. Watson, The problem of the square pyramid, *Messenger Math.,* **48** (1919), 1–22.

[44] A. Wiles, Modular elliptic curves and Fermat's last theorem, *Ann. of Math.,* **141** (1995), 443–551.

Kálmán Győry

Institute of Mathematics
University of Debrecen
4010 Debrecen, P.O. Box 12
Hungary

gyory@math.klte.hu

BOLYAI SOCIETY
MATHEMATICAL STUDIES, 15

Conference on Finite
and Infinite Sets
Budapest, pp. 157–174.

THE TOPOLOGICAL VERSION OF FODOR'S THEOREM

I. JUHÁSZ* and A. SZYMANSKI[†]

The following purely topological generalization is given of Fodor's theorem from [3] (also known as the "pressing down lemma"):

Let X be a locally compact, non-compact T_2 space such that any two closed unbounded (c u b) subsets of X intersect [of course, a set is bounded if it has compact closure]; call $S \subset X$ stationary if it meets every c u b in X. Then for every neighbourhood assignment U defined on a stationary set S there is a stationary subset $T \subset S$ such that

$$\bigcap \{U(x) : x \in T\} \neq \emptyset.$$

Just like the "modern" proof of Fodor's theorem, our proof hinges on a notion of diagonal intersection of c u b's, definable under some additional conditions.

We also use these results to present an (alas, only partial) generalization to this framework of Solovay's celebrated stationary set decomposition theorem.

1. INTRODUCTION

One of the most frequently used results in set-theory is Fodor's theorem (also known as the pressing down lemma) from [3]:

Theorem 1. *Let α be an ordinal of uncountable cofinality. If $S \subset \alpha$ is stationary in α [i.e. $S \cap C \neq \emptyset$ for every closed unbounded (in short: c u b) subset C of α] and $f : S \to \alpha$ is a regressive function on S [i.e. $f(\xi) < \xi$ whenever $\xi \in S \setminus \{0\}$] then there is a stationary subset $T \subset S$ and an ordinal*

*Research supported by OTKA grant no. 37758.
[†]Research supported by Charles University and the Czech Academy of Sciences.

$\zeta \in \alpha$ with $f(\xi) \le \zeta$ for all $\xi \in T$. In particular, if α is (an uncountable) regular cardinal then T and ζ above may be chosen in such a way that $f(\xi) = \zeta$ for all $\xi \in T$.

A precursor of Fodor's result was Neumer's theorem from [8] that, under the same assumptions, yields the same conclusion with only an unbounded $T \subset S$, instead of a stationary one.

Since a regressive function f defined on $S \subset \alpha$ is equivalent to the neighbourhood assignment $\xi \mapsto (f(\xi), \xi]$ in the ordinal space α [i.e. α considered with its natural order topology], and the conclusion of the above results can be reformulated to state that the neighbourhoods assigned to all elements of T have non-empty intersection, both Fodor's and Neumer's theorems can be viewed as purely topological statements about the ordinal space α. This was clear to Fodor himself, and raises naturally the question if these results could be generalized to a purely topological setting.

In [5] and [10] such generalizations were successfully achieved for the case of Neumer's result, but not for Fodor's. In fact, both authors explicitly stated the problem of finding a purely topological generalization of Fodor's theorem.

It should be mentioned that the authors of [2] took a completely different approach to viewing Fodor's and Neumer's results as topological: they viewed the regressive function f as one that assigns to the open set $\xi = [0, \xi)$ the compact subset $[0, f(\xi)]$. Still, for them the same phenomenon occurred: they found a satisfactory generalization of Neumer's theorem but not that of Fodor's.

Finally, we should like to note that the generalization of Fodor's theorem to finite products of uncountable regular cardinals given in [1], contrary to the title of that paper, is a generalization towards partial orders rather than topological spaces.

We hope to convince the reader that our generalization of Fodor's theorem, formulated and proved below, does provide a/the satisfactory solution to the above problem.

2. THE THEOREM FOR LOCALLY COMPACT SPACES

Let us start with the basic definitions.

Definition 1. Let X be a locally compact but non-compact T_2 space. The set $A \subset X$ is called bounded in X if its closure $\overline{A}$ is compact, unbounded if it is not bounded. It follows from our assumptions that X itself is unbounded. We say that X is *good* if the intersection of any two closed unbounded (in short: c u b) subsets of X is non-empty. We shall denote by $\mathcal{C}(X)$ the family of all c u b sets in X.

It is easy to see that an ordinal space α is good exactly if it has uncountable cofinality. A more general statement is formulated below.

Lemma 1. *If X is good then X is countably compact.*

Proof. Assume, indirectly that X is not countably compact, hence we have an infinite set $A \subset X$ with no accumulation point. But then every infinite subset of A is c u b, contradicting that X is good. ∎

Another easy but frequently used result is the following.

Lemma 2. *Let X be a good, non-compact, locally compact T_2 space. Then $C_1, C_2 \in \mathcal{C}(X)$ implies $C_1 \cap C_2 \in \mathcal{C}(X)$.*

Proof. Assume that $C_1 \cap C_2 \notin \mathcal{C}(X)$, hence actually $C_1 \cap C_2$ is compact. By the local compactness of X then there is a bounded open set U in X with $C_1 \cap C_2 \subset U$. However, then $C_1 \setminus U$ and $C_2 \setminus U$ would be two disjoint members of $\mathcal{C}(X)$, contradicting that X is good. ∎

Definition 2. Let X be a locally compact, noncompact T_2 space. We say that a set $S \subset X$ is stationary in X if it meets every c u b, i.e. every member of $\mathcal{C}(X)$. We denote the family of all stationary subsets of X by $\mathrm{St}\,(X)$.

Now, the following are immediate from Lemma 2 and the definitions.

Lemma 3. *Let X be as in Lemma 2. Then*

  (i) $\mathcal{C}(X) \subset \mathrm{St}\,(X)$;

  (ii) *if $C \in \mathcal{C}(X)$ and $S \in \mathrm{St}\,(X)$ then $C \cap S \in \mathrm{St}\,(X)$;*

(iii) *every stationary set is unbounded.*

The following definition introduces two concepts that, for good spaces, serve as generalizations of cofinality for ordinal spaces.

Definition 3. Let X be as above. We set

$$cf(X) = \min\big\{\,|A| \,:\, A \text{ is unbounded}\big\}$$

and

$$\varrho(X) = \min\big\{\,|\mathcal{C}| \,:\, \mathcal{C} \subset \mathcal{C}(X) \text{ and } \cap\mathcal{C} = \emptyset\big\}.$$

It is easy to see that $cf(X) \leq \varrho(X)$ holds for any good space X, moreover for a good ordinal space α we have $\varrho(\alpha) = cf(\alpha)$. If X is the well-known Ostaszewski space from [9] then clearly X is good and we have $cf(X) = \omega < \varrho(X) = \omega_1$.

The next result on $\varrho(X)$ is a strengthening of Lemma 2.

Lemma 4. Let X be as above and let $\mathcal{D} \subset \mathcal{C}(X)$ with $|\mathcal{D}| < \varrho(X)$. Then $\cap\mathcal{D} \in \mathcal{C}(X)$. Consequently, the union of fewer than $\varrho(X)$ non-stationary sets is non-stationary.

Proof. Assume that $\cap\mathcal{D} \notin \mathcal{C}(X)$ then $\cap\mathcal{D}$ is compact and hence can be covered by a bounded open set U. But then the family $\{D \setminus U \,:\, D \in \mathcal{D}\} \subset \mathcal{C}(X)$ and has empty intersection, and is of cardinality less than $\varrho(X)$, a contradiction. ∎

From this and Lemma 1 we now can directly conclude

Lemma 5. If X is as above then $\varrho(X)$ is an uncountable regular cardinal.

The Ostaszewski space, mentioned above, is an example showing that this is not necessarily true for $cf(X)$ for a good space X. However, we do not have an example of a good space X for which $cf(X)$ is a singular cardinal.

If κ is an uncountable regular cardinal and $\{C_\alpha \,:\, \alpha \in \kappa\}$ is a κ-sequence of c u b's in κ then their diagonal intersection is defined by

$$\Delta\{C_\alpha \,:\, \alpha \in \kappa\} = \Big\{\delta \in \kappa \,:\, \delta \in \bigcap\{C_\alpha \,:\, \alpha \in \delta\}\Big\},$$

and is known to be c u b. Our next goal is to generalize this concept to our general setting for sequences of c u b's in X of length $\varrho(X)$. However, this will only be possible under special circumstances.

To this end, let us recall (see e.g. [6]) that a free sequence in a space X is a transfinite sequence $P = \{p_\alpha \,:\, \alpha \in \eta\}$ with the property that

$$\overline{\{p_\beta \,:\, \beta \in \alpha\}} \cap \overline{\{p_\beta \,:\, \beta \in \eta \setminus \alpha\}} = \emptyset$$

for every $\alpha \in \eta$. Clearly, if P is a free sequence in X then, as a subspace, P is discrete.

Definition 4. Let X be as above and assume, in addition, that $P = \{p_\alpha : \alpha \in \varrho = \varrho(X)\}$ is a free sequence in X such that P has no complete accumulation point in X. In this case we say that P is a *spine* of the good space X.

For $\{H_\alpha : \alpha \in \varrho\} \subset \mathcal{C}(X)$ we set

$$\Delta_P\{H_\alpha : \alpha \in \varrho\} = \bigcup\left\{P'_\alpha \cap \bigcap\{H_\beta : \beta \in \alpha\} : \alpha \in \varrho\right\},$$

where $P_\alpha = \{p_\beta : \beta \in \alpha\}$ and P'_α is the derived set of P_α, i.e. the set of all limit points of P_α.

Before we prove that this P-diagonal intersection $\Delta_P\{H_\alpha : \alpha \in \varrho\}$ is c u b, let us first show that it is very closely related to the ordinary diagonal intersection of c u b's on an uncountable regular cardinal κ. Indeed, then the sequence of all successor ordinals $P = \{p_\alpha = \alpha + 1 : \alpha \in \kappa\}$ is clearly a free sequence in κ with no complete accumulation point. Moreover, for any limit ordinal $\delta \in \kappa$ we have $\delta \in P'_\alpha$ if and only if $\alpha \geq \delta$, hence clearly

$$\delta \in \Delta_P\{H_\alpha : \alpha \in \kappa\} \leftrightarrow \delta \in \bigcap\{H_\beta : \beta \in \delta\} \leftrightarrow \delta \in \Delta\{H_\alpha : \alpha \in \kappa\}.$$

Lemma 6. *Under the conditions of Definition 4, we have*

$$H = \Delta_P\{H_\alpha : \alpha \in \varrho\} \in \mathcal{C}(X).$$

Proof. Let us note first of all that, as P is a discrete subspace in X, we have $P' = \overline{P} \setminus P$ and $P'_\alpha = \overline{P}_\alpha \setminus P_\alpha$ for every $\alpha \in \varrho$. The fact that P has no complete accumulation point implies that every final segment $P \setminus P_\alpha$ of P is unbounded and so, as P is free and X is good, $\overline{P}_\alpha$ and consequently P'_α as well are compact, moreover $P' = \bigcup\{P'_\alpha : \alpha \in \varrho\}$.

To show that H is closed, consider any point $y \in X \setminus H$. If $y \notin P'$, then, as P' is closed and $H \subset P'$, we have $y \notin \overline{H}$. If $y \in P'$ then let $\alpha \in \varrho$ be minimal such that $y \in P'_\alpha$. Then, however, $y \notin H$ implies that $y \notin \bigcap\{H_\beta : \beta \in \alpha\}$, so we can choose an ordinal $\gamma \in \alpha$ with $y \notin H_\gamma$. Since H_γ is closed and by the minimality of α then $y \notin P'_\gamma$ as well, there is a neighbourhood V of y such that both $V \cap H_\gamma = \emptyset$ and $V \cap P_\gamma = \emptyset$. But then for every $\beta \leq \gamma$ we have $V \cap P'_\beta = \emptyset$ and for every $\beta \in \varrho \setminus (\gamma + 1)$ we have $V \cap \bigcap\{H_\nu : \nu \in \beta\} = \emptyset$, consequently $V \cap H = \emptyset$, showing again that $y \notin \overline{H}$.

Finally, to conclude that $H \in \mathcal{C}(X)$, it suffices to show that H cannot be covered by any bounded open set. So let us fix U open with $\overline{U}$ compact.

Then, as P has no complete accumulation point, there is a final segment $P \setminus P_{\alpha_0}$ of P disjoint from U (or even of $\overline{U}$).

Let us note now that for every unbounded set $A \subset X$ we have A' unbounded as well. Indeed, otherwise A' would be compact and if V is open with $\overline{V}$ compact and $A' \subset V$, then $A \setminus V$ is infinite being unbounded and $(A \setminus V)' = \emptyset$, contradicting the countable compactness of X.

Applying this remark to $A = P \setminus P_{\alpha_0}$ and using Lemma 4, we can pick a point y_0 in

$$\left(P \setminus P_{\alpha_0} \right)' \cap \bigcap \{ H_\beta \, : \, \beta \in \alpha_0 \}.$$

Clearly, then there is an ordinal $\alpha_1 \in \varrho$ with $\alpha_1 > \alpha_0$ such that actually $y_0 \in \left(P_{\alpha_1} \setminus P_{\alpha_0} \right)'$. Repeating the above procedure then by a straightforward recursion we may pick an increasing sequence of ordinals $\alpha_n \in \varrho$ and points y_n such that

$$y_n \in \left(P_{\alpha_{n+1}} \setminus P_{\alpha_n} \right)' \cap \bigcap \{ H_\beta \, : \, \beta \in \alpha_n \}$$

for all $n \in \omega$. Since $\left(P \setminus P_{\alpha_0} \right) \cap U = \emptyset$ we clearly have $y_n \in X \setminus U$ for every $n \in \omega$.

Now let $\alpha = \sup \{ \alpha_n \, : \, n \in \omega \} \in \varrho$ and y be any limit point of the set $\{ y_n \, : \, n \in \omega \}$; y exists because X is countably compact. But then we clearly have $y \in P'_\alpha$ as well, moreover $y \in \bigcap \{ H_\beta \, : \, \beta \in \alpha \}$ because the sequence $\{ y_n \, : \, n \in \omega \}$ is contained eventually in H_β for every $\beta \in \alpha$. Consequently, we have

$$y \in P'_\alpha \cap \bigcap \{ H_\beta \, : \, \beta \in \alpha \} \subset H,$$

moreover $y \notin U$ because $\{ y_n \, : \, n \in \omega \} \subset X \setminus U$, hence $H \setminus U \neq 0$, as required. ∎

After all this preparation, we are now ready to prove the main result of this section.

Theorem 2. *Let X be a locally compact, non-compact T_2 space that is good. If $S \subset X$ is stationary and U is any neighbourhood assignment on S [i.e. $U(x)$ is an open set containing x for each $x \in S$] then there is a stationary subset $T \subset S$ such that*

$$\cap \{ U(x) \, : \, x \in T \} \neq \emptyset.$$

Proof. We distinguish two cases:

Case 1. There is an unbounded subset A of X with $|A| < \varrho = \varrho(X)$, i.e. $cf(X) < \varrho(X)$. Then $S \cap \overline{A}$ is stationary and for every point $x \in S \cap \overline{A}$ we have $U(x) \cap A \neq \emptyset$. But then, in view of Lemma 4, there is a point $q \in A$ for which the set

$$T = \{x \in S \cap \overline{A} : q \in U(x)\}$$

is stationary in X, and we are done.

Case 2. For every set $A \subset X$ with $|A| < \varrho$ its closure $\overline{A}$ is compact, i.e. $cf(X) = \varrho(X)$. Then we can define a spine for X, i.e. a free sequence $P = \{p_\alpha : \alpha \in \varrho\}$ as in Definition 4, as follows.

Let us first fix a sequence $\{F_\nu : \nu \in \varrho\} \subset \mathcal{C}(X)$ such that $\bigcap\{F_\nu : \nu \in \varrho\} = \emptyset$ and $\nu < \mu$ implies $F_\nu \supset F_\mu$, using the definition of $\varrho(X)$ and Lemma 4. We then define points p_α and ordinals $\nu_\alpha \in \varrho$ inductively as follows. If $\alpha \in \varrho$ and $\{p_\beta : \beta \in \alpha\}$, $\{\nu_\beta : \beta \in \alpha\}$ have already been defined then, by assumption, $\overline{\{p_\beta : \beta \in \alpha\}}$ is compact, hence we can pick $\nu_\alpha \in \varrho \setminus \bigcup\{\nu_\beta : \beta \in \alpha\}$ such that $F_{\nu_\alpha} \cap \overline{\{p_\beta : \beta \in \alpha\}} = \emptyset$. Then $p_\alpha \in F_{\nu_\alpha}$ is picked arbitrarily. $P = \{p_\alpha : \alpha \in \varrho\}$ is a free sequence because for every $\alpha \in \varrho$ we have $\overline{\{p_\beta : \beta \geq \alpha\}} \subset F_{\nu_\alpha}$. Moreover, P has no complete accumulation point because $\bigcap\{F_{\nu_\alpha} : \alpha \in \varrho\} = \emptyset$ and if $x \notin F_{\nu_\alpha}$ then

$$|P \setminus F_{\nu_\alpha}| \leq |\alpha| < \varrho,$$

i.e. the complement of F_{ν_α} in X is a neighbourhood of the point x in X that meets P in a set of size smaller than ϱ.

Now, we claim that there is a point $p_\alpha \in P$ such that $T = \{x \in S : p_\alpha \in U(x)\}$ is stationary in X. Assume, indirectly, that for every $\alpha \in \varrho$ there is a c u b $H_\alpha \in \mathcal{C}(X)$ such that $p_\alpha \notin U(x)$ for every $x \in S \cap H_\alpha$. Set $H = \Delta_P\{H_\alpha : \alpha \in \varrho\}$, then $H \in \mathcal{C}(X)$ by Lemma 5. Consequently $H \cap S \neq \emptyset$ as S is stationary, so let $q \in H \cap S$. But then we have $q \in P'_\alpha \cap \bigcap\{H_\beta : \beta \in \alpha\}$ for some $\alpha \in \varrho$, hence on the one hand there are (infinitely many) $\beta \in \alpha$ such that $p_\beta \in U(q)$, while on the other hand $q \in H_\beta \cap S$ implies $p_\beta \notin U(q)$ for all $\beta \in \alpha$. This contradiction completes the proof. ∎

For later use, let us note that the above proof actually established the following somewhat stronger result: Using the notation

$$M(U) = \{q \in X : \{x \in S : q \in U(x)\} \in \mathrm{St}\,(X)\},$$

we have $M(U) \cap A \neq \emptyset$ if $cf(X) < \varrho(X)$ and A is any unbounded set of size less than $\varrho(X)$, moreover $M(U) \cap P \neq \emptyset$ if $cf(X) = \varrho(X)$ and P is any spine of X.

The following strengthening of Theorem 2 is now obtained as an easy consequence.

Corollary. *Let X, S, and U be as in Theorem 2. Then the set $M(U)$ is stationary in X.*

Proof. Let C be any c u b in X, then it is obvious that C as a subspace of X is locally compact, non-compact, and good. Also, $S \cap C$ is stationary in C and thus we may apply Theorem 2 to the neighbourhood assignment $V(x) = U(x) \cap C$ defined on $S \cap C$. Thus there is a set $T \subset S \cap C$ stationary in C and therefore also in X and a point $q \in C$ with

$$q \in \cap \{ U(x) : x \in T \},$$

hence $q \in C \cap M(U)$. Since $C \in \mathcal{C}(X)$ was arbitrary, we indeed have $M(U) \in \mathrm{St}\,(X)$. ∎

We conclude this section by a result which shows that the condition of goodness for assuring the general Fodor-type result is not only natural, in some sense it is also necessary. If X is a locally compact, non-compact T_2 space then let us denote by $PDL(X)$ the statement that Fodor's theorem holds true for X, i.e. whenever S is stationary in X and U is a neighbourhood assignment on S then there is a stationary $T \subset S$ with $\cap \{ U(x) : x \in T \} \neq \emptyset$, or more concisely, $M(U) \neq \emptyset$. Also, we denote by $SPDL(X)$ the statement that for S and U as before, $M(U) = \{ q \in X : \{ x \in S : q \in U(x) \} \in \mathrm{St}\,(X) \}$ is even stationary in X.

Theorem 3. *Let X be a locally compact, non-compact T_2 space. Then the following three statements (i)–(iii) are equivalent.*

(i) *X is good;*

(ii) *X is normal and $PDL(X)$;*

(iii) *$SPDL(X)$.*

Proof. If X is good and K, L are disjoint closed sets in X then one of them must be compact, hence X is clearly normal. The implications (i) $\Longrightarrow$ (ii) and (i) $\Longrightarrow$ (iii) now follow from Theorem 2 and its corollary, respectively.

To see (ii) $\implies$ (i) assume that X is normal but not good. Then we have two disjoint $c\,u\,b$'s, say K and L, and by normality we have disjoint open sets V and W with $K \subset V$ and $L \subset W$. Consider the neighbourhood assignment U defined on X as follows:

$$U(x) = \begin{cases} V & \text{if } x \in K; \\ W & \text{if } x \in L; \\ X \setminus (K \cup L) & \text{if } x \notin K \cup L. \end{cases}$$

Then U witnesses the failure of $PDL(X)$: indeed, if $q \in V$, then $\{x : q \in U(x)\} \cap L = \emptyset$, if $q \in W$ then $\{x : q \in U(x)\} \cap K = \emptyset$, and if $q \in X \setminus (K \cup L)$ $\{x : q \in U(x)\} \cap (K \cup L) = \emptyset$, hence all three types of sets are non-stationary in X.

Finally, to see (iii) $\implies$ (i) assume that K and L are disjoint $c\,u\,b$'s in X and then define a neighbourhood assignment U on X with the following stipulations:

$$U(x) = \begin{cases} X \setminus K & \text{if } x \in L, \\ X \setminus L & \text{if } x \notin L. \end{cases}$$

We claim that U is a witness for the failure of $SPDL(X)$, i.e. $M(U) = \{q \in X : \{x \in X : q \in U(x)\} \in \mathrm{St}\,(X)\}$ is non-stationary in X. Indeed, we have $L \cap M(U) = \emptyset$ since otherwise there is some $q \in L \cap M(U)$ which, by the definition of U implies $\{x \in X : q \in U(x)\} = X \setminus K \in \mathrm{St}\,(X)$, clearly a contradiction. $\blacksquare$

3. A possible generalization

In this section we present a further generalization of our topological framework for Fodor's theorem, where local compactness is omitted and the notion of boundedness is extended from the fixed ideal of sets having compact closure. The precise definition is given below.

Definition 5. Let X be any topological space and I be any proper ideal of subsets of X containing all finite sets. We shall call the elements of I *bounded* and, of course, the subsets of X not in I *unbounded*. We say that I is a *good ideal* on X if it satisfies the following four conditions:

(i) If $A \in I$ then $\overline{A} \in I$ as well (i.e. the closure of any bounded set is bounded).

(ii) For every $A \in I$ there is an open set $U \in I$ with $A \subset U$ (i.e. every bounded set has a bounded neighbourhood).

(iii) Any two closed unbounded (in short: c u b) sets have non-empty intersection. (We shall use $C(X, I)$ to denote the family of all c u b sets in this case.)

(iv) Setting

$$\varrho(X, I) = \min \{ |\mathcal{C}| : \mathcal{C} \subset C(X, I) \text{ and } \cap \mathcal{C} = \emptyset \},$$

we have

$$\varrho(X, I) = \min \{ |\mathcal{C}| : \mathcal{C} \subset C(X, I) \text{ and for every }$$
$$A \in I \text{ there is } C \in \mathcal{C} \text{ with } A \cap C = \emptyset \}.$$

In other words, this says that whenever we have a subfamily of $C(X, I)$ such that every point in X is missed by an element of this family then there is a (possibly different) subfamily of the *same size* for which every member of I is missed by an element. Condition (ii) insures that a subfamily with the latter property does exist.

In addition to $\varrho(X, I)$ we may again define as another generalization of cofinality

$$cf(X, I) = \text{non} - I = \min \{ |A| : A \text{ is unbounded} \}.$$

Clearly, if a set A meets every every member of a family $\mathcal{C} \subset C(X, I)$ having the property that every bounded set is avoided by some member of $\mathcal{C}$, then A must be unbounded, hence by condition (iv) we have

$$cf(X, I) \leq \varrho(X, I).$$

Now, a subset of X is called I-stationary (or simply stationary, if I is understood) if it meets every member of $C(X, I)$ and St (X, I) denotes the family of all I-stationary sets in X. From condition (ii) it is obvious that the family $C(X, I)$ is $\varrho(X, I)$-complete, i.e. it is closed under intersections subfamilies of size less than $\varrho(X, I)$, hence the union of fewer than $\varrho(X, I)$ non-stationary sets is always non-stationary.

Finally, $PDL(X, I)$ denotes the statement that Fodor's theorem holds in this setting, i.e. for every neighbourhood assignment U defined on an I-stationary subset S of X the set

$$M_I(U) = \{q \in X : \{x \in S : q \in U(x)\} \in \mathrm{St}\,(X, I)\}$$

is non-empty. Moreover, $SPDL(X, I)$ stands for the statement that for any such neighbourhood assignment U the set $M_I(U)$ is even I-stationary.

Examples. (1) Of course, the motivating example for a space with a good ideal is given by any good locally compact, non-compact space together with the ideal of its subsets having compact closure.

(2) Now, let α be any ordinal number with $cf(\alpha) > \omega_1$ and consider the subspace

$$X = \{\beta \in \alpha : cf(\beta) > \omega\}$$

of the ordinal space α. Then X is neither locally compact, nor countably compact, and the ideal I of all bounded subsets of X (in the sense of order) is easily seen to be good.

(3) With α as in (2), let us now consider its subspace

$$Y = \{\beta \in \alpha : cf(\beta) = \omega\}.$$

Then Y is countably compact but not locally compact, while the ideal of order-bounded subsets of Y is again good.

(4) If I is a good ideal on a space X and $S \in \mathrm{St}\,(X, I)$ is I-stationary then obviously

$$I \upharpoonright S = \{A \in I : A \subset S\}$$

(that is the restriction of I to S) is a good ideal on the subspace S of X. Clearly, both (2) and (3) are particular cases of this.

We may now formulate a result of which theorem 2 is clearly a special case by example (1) above. The proof will closely parallel that of Theorem 2, however the lack of countable compactness causes a bit of a complication. The following lemma shows that a certain amount of "compactness" is still present and this will be sufficient for the proof to go through.

Lemma 7. *Let I be a good ideal on the T_1 space X. Then for every unbounded set A its derived set A' is also unbounded and hence c u b.*

Proof. Let us start by fixing a family $\mathcal{C} = \{C_\xi : \xi \in \varrho = \varrho(X, I)\} \subset \mathcal{C}(X, I)$ such that every bounded set misses some member of $\mathcal{C}$, with the additional property that $\mathcal{C}$ is strictly decreasing, i.e. $\xi < \eta$ implies that C_η is a proper subset of C_ξ.

Now assume, indirectly, that A' is bounded and fix a bounded open set G with $A' \subset G$. Then $B = A \setminus G$ is again unbounded with $B' = \emptyset$, hence B is a closed discrete set in X. In particular, then B is c u b, hence we have $B \cap C_\xi \neq \emptyset$ for all $\xi \in \varrho$. Thus we can easily select two disjoint subsets D and E of B such that both D and E intersect every member of $\mathcal{C}$. But this contradicts property (iii) of I because both D and E are c u b. ∎

Now the promised generalization of Fodor's theorem reads as follows.

Theorem 4. *Let X be a T_1 topological space carrying a good ideal I. Then $(S)PDL(X, I)$ holds.*

Proof. Let U be a neighbourhood assignment defined on the I-stationary set S. If there is an unbounded set A with $|A| < \varrho(X, I)$ then clearly we even have $A \cap M_I(U) \neq \emptyset$. Otherwise, that is if $cf(X, I) = \varrho(X, I)$, one can easily construct an I-spine for X, that is a free sequence $P = \{p_\alpha : \alpha \in \varrho = \varrho(X, I)\}$ which is unbounded, all its proper initial segments P_α are bounded, and has no complete accumulation point in X. We may also assume without any loss of generality that $p_\alpha \in C_\alpha$ holds for all $\alpha \in \varrho$ where the c u b's C_α are chosen as in the proof of lemma 7.

Then exactly as in definition 4 we can define the P-diagonal intersection of any ϱ-sequence

$$\{H_\alpha : \alpha \in \varrho\}$$

of c u b's, and with the help of lemma 7 we will show that this P-diagonal intersection

$$H = \Delta_P\{H_\alpha : \alpha \in \varrho\} = \bigcup \left\{ P'_\alpha \cap \bigcap \{H_\beta : \beta \in \alpha\} : \alpha \in \varrho \right\},$$

is again a c u b. The proof that H is closed is the same as above. To prove that it is unbounded we first need some notation.

For any $\alpha \in \varrho$ let us set $P_\alpha = \{p_\beta : \beta \in \alpha\}$ and $Q_\alpha = P \setminus P_\alpha$. Recall that for any $\alpha \in \varrho$ we have $P'_\alpha \cap Q'_\alpha = \emptyset$, moreover

$$P' = \bigcup\{P'_\alpha : \alpha \in \varrho\}.$$

For any point $x \in P'$ we set

$$\varphi(x) = \min\{\alpha : x \in P'_\alpha\}.$$

Clearly we have $Q_\alpha \subset C_\alpha$, consequently also $Q'_\alpha \subset C_\alpha$ for every $\alpha \in \varrho$, moreover $\varphi(x) > \alpha$ implies $x \in Q'_\alpha$, hence if $\sup \{ \varphi(x) : x \in A \} = \varrho$ holds for a set $A \subset P'$ then A meets every C_α and so is unbounded.

Let us now define by transfinite recursion on α points $y_\alpha \in P'$ and ordinals $\nu_\alpha \in \varrho$ as follows. Assume that $\alpha \in \varrho$ and that both $Y_\alpha = \{y_\beta : \beta \in \alpha\}$ and $\{\nu_\beta : \beta \in \alpha\}$ have already been defined. Then we set

$$\nu_\alpha = \sup \{ \varphi(y_\beta) : \beta \in \alpha \}$$

and then choose

$$y_\alpha \in Q'_{\nu_\alpha} \cap \bigcap \{ H_\beta : \beta \in \nu_\alpha \}.$$

The latter is possible because Q'_{ν_α} is c u b by lemma 7.

It is easy to see from the construction that

$$\varphi(y_\beta) \leq \nu_\alpha < \varphi(y_\alpha)$$

whenever $\beta < \alpha$, consequently the set $Y = \{y_\alpha : \alpha \in \varrho\}$ is unbounded because $\sup \{ \varphi(y) : y \in Y \} = \varrho$. Thus to prove that H is unbounded, by lemma 7, it suffices to show that $Y' \subset H$. So let $y \in Y'$ be any accumulation point of Y. Then $Y \subset P'$ implies that $y \in P'$, hence $\varphi(y) = \nu$ is defined. But then $y \notin Q'_\nu$, hence $y \notin \{ y_\alpha : \varphi(y_\alpha) > \nu \}'$ and so $y \in \{ y_\alpha : \varphi(y_\alpha) < \nu \}'$. Now let δ be the smallest ordinal with $y \in Y'_\delta$, where $Y_\delta = \{y_\alpha : \alpha < \delta\}$. Clearly δ must be a limit ordinal because X is T_1.

Note that then we have

$$\nu_\delta = \sup \{ \varphi(y_\alpha) : \alpha < \delta \} = \sup\{\nu_\alpha : \alpha < \delta\},$$

and thus clearly $y \in P'_{\nu_\delta}$. Moreover, the minimality of δ implies that we have

$$y \in \{y_\alpha : \beta < \alpha < \delta\}'$$

for every $\beta < \delta$, consequently, as the final segment $\{y_\alpha : \beta < \alpha < \delta\}$ of Y_δ is contained in $\bigcap\{H_\xi : \xi < \nu_\beta\}$, we also have $y \in \bigcap\{H_\xi : \xi < \nu_\beta\}$. Putting all this together we indeed have

$$y \in P'_{\nu_\delta} \cap \bigcap \{ H_\xi : \xi < \nu_\delta \} \subset H.$$

We can now complete the proof of $PDL(X, I)$ by repeating the analogous argument given in the finishing part of the proof of Theorem 2 showing that actually $P \cap M_I(U) \neq \emptyset$. Finally, $SPDL(X, I)$ is then obtained as an easy corollary again, by restricting both U and I to any c u b C. We make use here of the fact that the restriction of the ideal I to the subspace C is also good. ∎

Examples (2) and (3) show that, at least formally, Theorem 4 is a genuine extension of Theorem 2. However, we must admit that we don't as yet have an application of Theorem 4 which cannot be easily reduced to Theorem 2. We emphasize, on the other hand, that the stationary decomposition results of the next section work just as easily in the general framework of spaces with a good ideal of bounded sets as in the restricted case of good locally compact spaces.

4. STATIONARY SET DECOMPOSITION

Solovay proved that for every uncountable regular cardinal κ if S is any stationary subset of κ then S can be decomposed into κ disjoint stationary subsets (see e.g. [7], theorem 85). As an immediate corollary, it follows that if α is an ordinal with $cf(\alpha) > \omega$ then every stationary set in α is $cf(\alpha)$-decomposable. Our aim in this section is to generalize this result to our topological setting. We start with a lemma that shows the relevance of our Fodor-type results to this.

Lemma 8. *Let X be a space and I be an ideal (of bounded sets) on X such that for some set $A \subset X$ we have*

$$A \cap M_I(U) \neq \emptyset$$

whenever U is any neighbourhood assignment defined on an I-stationary set. Assume moreover that for some $S \in \mathrm{St}\,(X, I)$ there are a cardinal κ and a neighbourhood assignment V on S such that for every set $B \in [A]^{<\kappa}$ we have

$$\{\, x \in S \, : \, B \cap V(x) = \emptyset \,\} \in \mathrm{St}\,(X, I).$$

Then S is κ-decomposable, i.e. it splits into κ disjoint I-stationary sets.

Proof. By transfinite recursion on $\alpha \in \kappa$ we define points $q_\alpha \in A$ as follows: If $Q_\alpha = \{q_\beta \, : \, \beta \in \alpha\}$ has already been defined, by our second assumption the set

$$S_\alpha = \{\, x \in S \, : \, Q_\alpha \cap V(x) = \emptyset \,\}$$

is I-stationary. Thus applying our first assumption to the restriction of V to S_α we can pick q_α from the non-empty set $A \cap M_I(V \restriction S_\alpha)$. Now it is obvious from our construction that the sets

$$T_\alpha = \{\, x \in S_\alpha \, : \, q_\alpha \in V(x) \,\}$$

are pairwise disjoint I-stationary subsets of S (for $\alpha \in \kappa$), hence S is indeed κ-decomposable. ∎

The main result of this section is the following theorem that implies Solovay's decomposition theorem in many cases. Roughly, it says that if I is a good ideal on X and all points in an I-stationary set S have small character in X, then S is $cf(X, I)$-decomposable. We recall that the character $\chi(x, X)$ of the point x in the space X is defined as the smallest size of a neighbourhood base for x in X. Note also that if α is an ordinal space then for any $\beta \in \alpha$ the character of the point β in α is equal to its cofinality $cf(\beta)$.

Theorem 5. *Let I be a good ideal of bounded sets on a T_1 space X, moreover set $\gamma = cf(X, I) = non - I$. Then every I-stationary set S satisfying*

$$\sup \left\{ \chi(x, X) : x \in S \right\} = \mu < \gamma$$

is γ-decomposable.

Proof. For every point $x \in S$ let us fix first of all a neighbourhood base of the form $\left\{ U_\alpha(x) : \alpha \in \mu \right\}$ (repetitions are permitted). Then we prove the following claim: For every regular cardinal $\kappa \leq \gamma$ there is an $\alpha \in \mu$ such that for every set $Q \in [X]^{<\kappa}$ we have

$$\left\{ x \in S : Q \cap U_\alpha(x) = \emptyset \right\} \in St\,(X, I).$$

Indeed, assume indirectly that $\kappa = cf(\kappa) \leq \gamma$ but for every $\alpha \in \mu$ there is a set $Q_\alpha \in [X]^{<\kappa}$ for which

$$A_\alpha = \left\{ x \in S : Q_\alpha \cap U_\alpha(x) = \emptyset \right\}$$

is non-stationary. Set $Q = \bigcup \{Q_\alpha : \alpha \in \mu\}$, then $|Q| < \gamma$ follows from the regularity of $\kappa \leq \gamma$ if $\mu < \kappa$ and from $\mu < \gamma$ otherwise. Hence, by definition, Q is a bounded set. On the other hand, since $\mu < \gamma \leq \varrho$ and so the union of μ non-stationary sets is again non-stationary, we have that the set

$$T = S \setminus \bigcup \{A_\alpha : \alpha \in \mu\}$$

is stationary and so unbounded. But for any point $x \in T$ we have $Q \cap U_\alpha(x) \neq \emptyset$ for all $\alpha \in \mu$, which leads to the absurd conclusion that the closure of the bounded set Q, that is also bounded, contains the unbounded set T.

Now, using lemma 8 we immediately conclude from this claim that S is κ-decomposable for every regular cardinal $\kappa \leq \gamma$. In particular, this finishes the proof if γ is regular. Otherwise, if γ is singular and so $cf(\gamma) = \kappa < \gamma$, then we may first decompose S into κ many disjoint stationary subsets $\{S_\alpha : \alpha \in \kappa\}$ and note that the condition of the theorem is trivially inherited by (stationary) subsets. Thus if we have

$$\gamma = \sum \{\gamma_\alpha : \alpha \in \kappa\}$$

with $\gamma_\alpha < \gamma$ for each $\alpha \in \kappa$ then every set S_α is γ_α-decomposable and hence S itself is γ-decomposable again. ∎

Note that if $\alpha = \mu^+$ is a successor cardinal then for every point $\beta \in \alpha$ the cofinality (i.e. the character) of β is at most μ, hence in this case our theorem applies. This particular case of Solovay's decomposition result was obtained by Fodor in [4].

We can get a decomposition result for our topological setting that yields a complete generalization of Solovay's theorem if "a certain amount of compactness" is assumed for our underlying space. However, the problem with this result is that, unlike in the case of the generalized Fodor theorem or Theorem 5, in its proof we have to make use of Solovay's original theorem! Still, for the sake of completeness, we present this result below. We recall that a space X is said to be *initially* $< \varrho$-*compact* iff every infinite set $A \in [X]^{<\varrho}$ has a complete accumulation point in X.

Theorem 6. *Assume that I is a good ideal of bounded sets on a T_1 space X that is initially $< \varrho = \varrho(X, I)$-compact, moreover there is an I-spine $P = \{p_\alpha : \alpha \in \varrho\}$ in X with the property that for every bounded subset A of P' there is an $\alpha \in \varrho$ with $A \subset P'_\alpha$. Then every I-stationary subset of X is ϱ-decomposable.*

Proof. Since, by lemma 7, the set P' is c u b, it suffices to show that any stationary $S \subset P'$ is ϱ-decomposable. To accomplish this we shall again consider the map $\varphi : P' \to \varrho$ defined (in the proof of Theorem 4) for $x \in P'$ by

$$\varphi(x) = \min \{\alpha : x \in P'_\alpha\}.$$

It is easy to see that this map φ is continuous because each P'_α is clopen in P'. Next we show that φ is also a *closed map*, i.e. the φ-image of any closed set is closed.

Indeed, assume indirectly that $A \subset P'$ is closed but $B = \varphi[A]$ is not. Let $\alpha \in B'$ be the smallest limit point of B that is not in B. Then we may choose a set $Z \subset \alpha \cap B$ that has order type $cf(\alpha)$ and is cofinal in α. For each ordinal $\zeta \in Z$ pick a point $y_\zeta \in A$ with $\varphi(y_\zeta) = \zeta$ and set

$$Y = \{y_\zeta : \zeta \in Z\}.$$

Then $|Y| = cf(\alpha) < \varrho$, hence by our assumption there is a complete accumulation point y of Y and $y \in A$ because A is closed. So if we can show that $\varphi(y) = \alpha$ then we get a contradiction because $\varphi(y) \in B$.

But the continuity of φ immediately implies that $\varphi(y)$ is in the closure of Z, hence $\varphi(y) \leq \alpha$. On the other hand, we cannot have $\varphi(y) = \beta < \alpha$ because then P'_β would be a clopen neighbourhood of y in P' such that

$$P'_\beta \cap Y = \{y_\zeta : \zeta \leq \beta\},$$

contradicting that y is a complete accumulation point of Y.

Now, let S be any I-stationary subset of P'. We claim that its image $T = \varphi[S]$ is stationary in ϱ. Indeed, let C be any c u b in ϱ. Then by continuity $\varphi^{-1}(C)$ is closed in P' and it must also be unbounded by our assumption on the I-spine P. Thus we have $S \cap \varphi^{-1}(C) \neq \emptyset$ and consequently $T \cap C \neq \emptyset$ as well.

But we also have the converse of this statement, i.e. for any stationary subset T of ϱ its inverse image $S = \varphi^{-1}(T)$ is I-stationary; to prove this we use that φ is a closed map. Indeed, for any c u b $C \in \mathcal{C}(X, I)$ with $C \subset P'$ its image $\varphi[C]$ is clearly c u b in ϱ. Therefore we have $T \cap \varphi[C] \neq \emptyset$, consequently $\varphi^{-1}(T) \cap C \neq \emptyset$ as well.

To complete our proof, let us consider any I-stationary subset S of P'. We may then apply Solovay's theorem to the stationary set $T = \varphi[S]$ in ϱ and decompose T into the disjoint stationary sets $\{T_\alpha : \alpha \in \varrho\}$. But then the family $\{\varphi^{-1}(T_\alpha) : \alpha \in \varrho\}$ forms a ϱ-decomposition of S. ∎

Note that the existence of an I-spine as required in the above theorem is insured by the very natural assumption $cf(X, I) = \varrho(X, I)$. This assumption also occurs in the following conjecture that, if true, would provide us with a purely topological version of Solovay's theorem in the spirit of Theorem 5.

Conjecture. Let I be a good ideal of bounded sets on a T_1 space X such that $cf(X, I) = \varrho(X, I)$. Then every I-stationary set S with the property that $\chi(x, X) < \varrho(X, I)$ holds for all points $x \in S$ is $\varrho(X, I)$-decomposable.

REFERENCES

[1]   M. Bekkali and R. Bonnet, On Fodor's theorem: a topological version, *Discrete Mathematics*, **53** (1985), 3–19.

[2]   L. Babai and A. Máté, *Inner set mappings on locally compact spaces*, Colloq. Math. Soc. J. Bolyai, vol. 8, North-Holland, Amsterdam, 1974; pp. 77–95.

[3]   G. Fodor, Eine Bemerkung zur Theorie der regressiven Funktionen, *Acta Sci. Math. (Szeged)*, **17** (1956), 139–142.

[4]   G. Fodor, On stationary sets and regressive functions, *Acta Sci. Math. (Szeged)*, **27** (1966), 105–110.

[5]   I. Juhász, On Neumer's Theorem, *Proc. Amer. Math. Soc.*, **54** (1976), 453–454.

[6]   I. Juhász, *Cardinal Functions – Ten Years Later,* Math. Centre Tract no. 123, Amsterdam, 1980.

[7]   T. Jech, *Set Theory*, 2nd edition, Springer, 1997

[8]   W. Neumer, Verallgemeinerung eines Satzes von Alexandrov und Urysohn, *Math. Z.*, **54** (1951), 254–261.

[9]   A. Ostaszewski, On countably compact, perfectly normal spaces, *J. London Math. Soc.*, **14** (1976), 501–516.

[10]   B. Scott, A Generalized Pressing-Down Lemma and Isocompactness, *Proc. Amer. Math. Soc.*, **81** (1981), 316–320.

I. Juhász

A. Rényi Mathematical Institute
Hungarian Academy of Sciences
Budapest, P.O.B. 127.
H-1364, Hungary

juhasz@renyi.hu

A. Szymanski

Slippery Rock University
Slippery Rock
PA 16057
U.S.A.

andrzej.szymanski@sru.edu

BOLYAI SOCIETY
MATHEMATICAL STUDIES, 15

Conference on Finite
and Infinite Sets
Budapest, pp. 175–197.

COLOR-CRITICAL GRAPHS AND HYPERGRAPHS WITH FEW EDGES: A SURVEY

A. KOSTOCHKA*

The current situation with bounds on the smallest number of edges in color-critical graphs and hypergraphs is discussed.

1. INTRODUCTION

The theory of graph and hypergraph coloring plays a central role in discrete mathematics. It has applications in areas with seemingly little connection to coloring. Coloring deals with the fundamental problem of partitioning a set of objects into classes that avoid certain conflicts. Many timetabling, sequencing, and scheduling problems are of this nature.

A hypergraph is *color-critical* if deleting any edge or vertex reduces the chromatic number; a color-critical hypergraph with chromatic number k is k-*critical*. Every k-chromatic hypergraph contains a k-critical hypergraph, so one can study chromatic number by studying the structure of k-critical (hyper)graphs. There is vast literature on k-critical graphs and hypergraphs. Many references can be found in [23, Chapters 5 and 1].

In this survey we concentrate on k-critical graphs and hypergraphs with few edges. Lower bounds on, say, average degree of k-critical graphs can be applied as follows. If we know that the average degree of every k-critical graph in a family $\mathcal{H}$ is at least x, and the average degree of every subgraph H' of a graph $H \in \mathcal{H}$ is less than x, then we know that H is

*This work was partially supported by the NSF grant DMS-0099608.

$(k-1)$-colorable. For example, a theorem of Gallai (described in Section 4 below) says that $|E(G)|/|V(G)| \geq 0.5(k-1+\frac{k-3}{k^2-3})$ when G is a k-critical graph other than K_k. For $k \geq 6$ and arbitrary g, this implies that the problem of testing k-colorability is solvable in polynomial time for graphs that embed on the orientable surface of genus g. If such a graph is not k-colorable, then it has a $(k+1)$-critical subgraph G'. Euler's Formula yields $|E(G')| \leq 3(|V(G')|+2g-2)$, but Gallai's Theorem requires $|E(G')| \geq 0.5(k+\frac{k-2}{k^2+2k-2})|V(G')|$. For $k \geq 6$, this requires $|V(G')| \leq 138(g-1)$. Therefore, it suffices to test the k-colorability of every subgraph of G having at most $138(g-1)$ vertices.

Another application of such bounds to coloring of graphs on surfaces appears in [7], and Krivelevich [36] presents interesting applications to random graphs.

In connection with *list coloring* originated by Vizing [54] and Erdős, Rubin, and Taylor [19], one can study also *list-k-critical* (hyper)graphs. Given a (hyper)graph G and a list assignment L for the vertices of G, G is *L-critical* if there is no proper coloring of vertices of G from their lists, but after deleting any edge or vertex, such a coloring does exist. A list assignment L for the vertices of a (hyper)graph G is called *t-uniform* if $|L(v)| = t$ for every $v \in V(G)$.

Two basic questions will be discussed:

(a) what is the minimum possible number of edges in a k-critical (hyper)graph in a given class $\mathcal{G}$?

(b) what is the minimum possible number of edges in a k-critical (hyper)graph on n vertices in a given class $\mathcal{G}$? In particular what is the inf $\frac{|E(G)|}{|V(G)|}$ taken over k-critical (hyper)graph in a given class $\mathcal{G}$?

In the next section we give proofs of a few basic facts. Then graph questions are discussed in Sections 3, 4, and 5, and hypergraph questions in Sections 6, 7, and 8.

2. Preliminaries

It is well known that color-critical graphs and hypergraphs do not have vertices of small degree. This is true also for list-critical hypergraphs. We state this folklore observation as a proposition because of its importance.

Proposition 1. *Let L be a $(k-1)$-uniform list for a hypergraph G and let G be L-critical. Then $\deg_G(v) \geq k - 1$ for every $v \in V(G)$. Moreover, for every $v \in V(G)$, there exist some $k - 1$ edges $e_1, \ldots, e_{k-1}$ such that $e_i \cap e_j = \{v\}$ for every $1 \leq i < j \leq k - 1$.*

Proof. Let $v \in V(G)$. By definition, there is an L-coloring f of $G - v$, but we cannot extend f to v. This means that for every $\alpha \in L(v)$, there exists an edge e_α containing v such that all vertices of $e_\alpha - v$ are colored with α. And for distinct $\alpha, \beta \in L(v)$, the sets $e_\alpha - v$ and $e_\beta - v$ must be disjoint. This proves the proposition. ∎

Dirac [12] observed that k-critical hypergraphs have not only the minimum degree at least $k - 1$, but also the edge-connectivity at least $k - 1$.

Proposition 2 [12]. *Let G be a k-critical hypergraph. Then G is $(k-1)$-edge-connected.*

Proof.[1] Assume that $V(G) = W \cup U$, $W \cap U = \emptyset$, and the only edges intersecting both W and U are $e_1, \ldots, e_s$, where $s \leq k - 2$. For $j = 1, \ldots, s$, let w_j be some vertex in $W \cap e_j$ and u_j be some vertex in $U \cap e_j$. Since G is k-critical, there exists a $(k-1)$-coloring f_W of $G(W)$ and a $(k-1)$-coloring f_U of $G(U)$, both using colors $1, \ldots, k-1$. We can change the names of colors in f_U in $(k-1)!$ ways, keeping the same partition of vertices. For a given j, in at most $(k-2)!$ ways we will get the colors of w_j and u_j the same. Thus, there are at least $(k-1)! - (k-2)!s = (k-1-s)(k-2)! > 0$ ways to choose the names of colors in f_U so that the resulting $(k-1)$-coloring of G will be proper. ∎

In view of the simple proof of Proposition 2 above, it is a bit surprising that for list colorings, this proposition does not hold.

Example 1. Let $H(k)$ denote the graph with $V(H(k)) = W \cup U$, where $W = \{w_1, \ldots, w_k\}$, $U = \{u_1, \ldots, u_k\}$, such that the subgraphs of $H(k)$ induced by W and U are complete graphs and there is exactly one edge, namely $w_k u_k$, connecting W with U.

Define the list L for the $H(k)$ by

$$L(v) = \begin{cases} \{1, \ldots, k-1\}, & \text{if } v \in V(G) - \{w_k, u_k\}; \\ \{2, \ldots, k\}, & \text{if } v \in \{w_k, u_k\}. \end{cases}$$

---

[1]I've learned this proof from Jacent Tokaz via Douglas West. Another short proof the reader can find in [52].

Assume that $H(k)$ is L-colorable. Then all colors $1, \ldots, k-1$ should be used on $w_1, \ldots, w_{k-1}$, and the same holds for $u_1, \ldots, u_{k-1}$. Thus, both w_k and u_k must be colored with k, a contradiction. It is also easy to check that after deleting any edge, we get an L-colorable graph. Hence, $H(k)$ is L-critical and has connectivity 1.

3. Dirac–type bounds

Critical graphs were first defined and used by Dirac [11] in 1951. Dirac was interested in

$$F(k, n) = \min \left\{ |E(G)| \ : \ G \text{ is } k\text{-critical and } |V(G)| = n \right\}.$$

In view of Proposition 1, for every k-critical graph G on n vertices,

$$\sum_{v \in V(G)} \deg(v) \geq (k-1)n.$$

Thus $2F(k, n) \geq (k-1)n$. This motivates introducing the *excess*

$$\varepsilon(k, G) = \sum_{v \in V(G)} (\deg(v) - k + 1)$$

and

(1) $$\varepsilon(k, n) = \min \left\{ \varepsilon(k, G) \mid G \text{ is } k\text{-critical and } |V(G)| = n \right\}$$

$$= 2F(k, n) - (k-1)n.$$

Brooks' Theorem yields that $\varepsilon(k, n) \geq 1$ for $k \geq 4$ and $n \geq k+1$. Dirac [13] proved the following.

Theorem 3 [13]. *Let $k \geq 4$ and G be a k-critical graph. If G is not a K_k, then $\varepsilon(k, G) \geq k - 3$.*

Shorter and more elegant proofs of this result were given by Kronk and Mitchem [37] and Weinstein [56]. We present here a proof using ideas from [10, 29].

Proof of Dirac's Theorem. Assume that G is a vertex minimum k-critical graph distinct from K_k, with $\varepsilon(k, G) \leq k - 4$. For every $v \in V(G)$, define $\varepsilon_G(k, v) = \deg_G(v) - k + 1$. Then $\varepsilon(k, G) = \sum_{v \in V(G)} \varepsilon_G(k, v)$.

Let $w \in V(G)$ and $\deg_G(v) = k - 1$. Since $G \neq K_k$, there are non-adjacent vertices $x_1, x_2 \in N_G(w)$. The graph G^* obtained from G by merging x_1 and x_2 into a new vertex x^* is not $(k - 1)$-colorable, since every its $(k - 1)$-coloring generates a $(k - 1)$-coloring of G. Hence G^* contains a k-critical subgraph G_1^*. Note that $x^* \in V(G_1^*)$ (otherwise, G_1^* would be a subgraph of G). Since $\deg_{G^*}(w) = k - 2$, $w \neq V(G_1^*)$.

Let $V_1^* = V(G_1^*)$, $V_1 = V_1^* - x^* + x_1 + x_2$, and $V_2 = V(G^*) - V_1^* = V(G) - V_1$. Let $E_{1,2}$ be the set of edges connecting V_1 with V_2 in G. Assume that $G_1 \neq K_k$. Then by the minimality of G, $\varepsilon(k, G_1) \geq k - 3$. Since every edge in $E_{1,2}$ contributes 1 to $\sum_{v \in V_1^*} \left(\varepsilon_{G^*}(k, v) - \varepsilon_{G_1^*}(k, v) \right)$, we have

$$(2) \qquad \varepsilon(k, G) \geq \sum_{v \in V_2} \varepsilon_G(k, v) + \varepsilon(k, G_1^*) + |E_{1,2}| - (k - 1),$$

where the last $-(k - 1)$ reflects merging x_1 with x_2. By Proposition 2, $|E_{1,2}| \geq k - 1$. Hence (2) yields $\varepsilon(k, G) \geq k - 3$, a contradiction.

Thus G^* contains a K_k one of whose vertices is x^*. In other words, G contains a triple (M, y_1, y_2), where M is a clique of size $k - 1$ and y_1 and y_2 are non-adjacent vertices with $N_G(\{y_1, y_2\}) \supset M$.

Among all such triples, choose a triple (M, y_1, y_2) with maximum $|N_G(y_1) \cap M|$.

Now, let $x_1 = y_1$ if $N_G(y_1) \cap M$ contains a vertex w of degree $k - 1$, and let $x_1 = y_2$ otherwise. Since $\varepsilon(k, G) \leq k - 4$, in both cases there is $w \in N_G(x_1) \cap M$ with $\deg_G(w) = k - 1$. Let x_2 be a non-adjacent to x_1 vertex in M of the smallest degree. Define graphs G^* and G_1^* and sets V_1^*, V_1, V_2, and $E_{1,2}$ as above. Then again $G_1^* = K_k$. Let $M' = V(G_1^*) - x^*$, $M^* = M - x_2 + x^*$, $M_1^* = M^* \cap V(G_1^*)$ and $m_1 = |M_1^*|$. Since every $v \in M_1^*$ has at least $k - 1$ neighbors in G_1^* and at least $k - 1 - m_1$ neighbors outside of G_1^*, $\varepsilon(k, G) \geq (m_1 - 1)(k - 1 - m_1)$. Since $w \notin M^*$, in order to have $(m_1 - 1)(k - 1 - m_1) \leq k - 4$, we need $m_1 = 1$, which means $M_1^* = \{x^*\}$, i.e. $M' \cap M^* = \emptyset$.

Let $|N_G(y_1) \cap M| = m$ and $|N_G(x_1) \cap M| = m'$. Then $|N_G(x_2) \cap M| \geq k - 1 - m$ by the choice of y_1. Hence $\deg_G(x_2) \geq (k - 2) + (k - 1 - m)$. By the choice of x_2, $\varepsilon_G(k, v) \geq k - 2 - m$ for every $v \in M - N_G(x_1)$. Taking into account that all vertices in $M' \cup M - x_2$ are adjacent to x_1 or x_2, and at least m' of them to both, we have

$$(3) \qquad \varepsilon(k, G) \geq (k - 2 - m)(k - 2 - m') + m' - 1.$$

If $x_1 = y_1$, then $m' = m$ and the minimum of $(k - 2 - m)^2 + m - 1$ over integers $m, 1 \leq m \leq k - 2$ is exactly $k - 3$. This contradicts the choice of G. So, let $x_1 = y_2$. In this case, y_1 is not adjacent to at least 3 vertices of degree $k - 1$ in M, and hence $k - 2 - m \geq 2$. Then (3) yields $\varepsilon(k, G) \geq 2(k - 2 - m') + m' - 1 \geq k - 3$, again. This proves the theorem. ■

For $k \geq 3$, let $\mathcal{D}_k$ denote the family of all graphs G whose vertex set consists of three non-empty pairwise disjoint sets A, B_1, B_2 with $|B_1| + |B_2| = |A| + 1 = k - 1$ and two additional vertices a, b such that A and $B_1 \cup B_2$ are cliques in G not joined by any edge, $N_G(a) = A \cup B_1$ and $N_G(b) = A \cup B_2$. Obviously, such a graph G has $2k - 1$ vertices, $\deg_G(x) = k - 1$ for all vertices $x \neq a, b$, and $\varepsilon(k, G) = \deg_G(a) + \deg_G(b) - 2(k - 1) = k - 3$. That G is k-critical was observed by Dirac [13] and by Gallai [20]. Thus Dirac's bound is sharp for every $G \in \mathcal{D}_k$.

In 1974, Dirac [14] extended Theorem 3 as follows.

Theorem 4 [14]. *Let $k \geq 4$, and let G be a k-critical graph. If G is neither the K_k nor a member of $\mathcal{D}_k$, then*

$$\varepsilon(k, G) \geq \begin{cases} 2 & \text{if } k = 4, \\ k - 1 & \text{if } k \geq 5. \end{cases}$$

Shorter proofs of this result were found by Mitchem [40] and by Deuber et. al. [10].

For $k \geq 3$, let $\mathcal{F}_k$ denote the family of all graphs G whose vertex set consists of four non-empty pairwise disjoint sets A_1, A_2, B_1, B_2, where $|B_1| + |B_2| = |A_1| + |A_2| = k - 1$ and $|A_2| + |B_2| \leq k - 1$, and one additional vertex c such that $A = A_1 \cup A_2$ and $B = B_1 \cup B_2$ are cliques in G, $N_G(c) = A_1 \cup B_1$, and a vertex $a \in A$ is joined to a vertex $b \in B$ by an edge in G if and only if $a \in A_2$ and $b \in B_2$. Every such graph G has $2k - 1$ vertices and independence number 2. Consequently, G is not $(k - 1)$-colorable. Moreover, it is easy to check that the deletion of any edge results in a $(k - 1)$-colorable graph. Therefore, G is k-critical. Clearly, G is in $\mathcal{D}_k$ if and only if $|A_2| = 1$ or $|B_2| = 1$. Moreover, $\mathcal{D}_k \subseteq \mathcal{F}_k$.

Kostochka and Stiebitz [29] improved the bounds of Theorem 4 as follows.

Theorem 5 [29]. *Let $k \geq 4$ and G be a k-critical graph. If G is neither a K_k nor a member of $\mathcal{F}_k$, then $\varepsilon(k, G) \geq 2(k - 3)$.*

The bounds of this result are tight not only for graphs on $2k - 1$ vertices. There are examples of k-critical graphs G with $\varepsilon(k, G) = 2(k - 3)$ on $k + 2, 2k - 2, 2k - 1, 2k$ and $3k - 2$ vertices. However, for $k \geq 4$, it is possible to show that $\varepsilon(k, n) = 2(k - 3)$ if and only if $n \in \{k + 2, 2k - 2, 2k, 3k - 2\}$.

The *join* of vertex disjoint graphs G_1 and G_2, denoted by $G_1 \vee G_2$, is the graph obtained from their union by adding edges joining each vertex of G_1 to each vertex of G_2. It is evident that $\chi(G_1 \vee G_2) = \chi(G_1) + \chi(G_2)$. Moreover, $G_1 \vee G_2$ is critical if and only if both G_1 and G_2 are critical.

In one of his seminal papers from 1963, Gallai [21] proved that every k-critical graph with at most $2k - 2$ vertices is the join of two other critical graphs. This allowed him to find the minimum excess of k-critical graphs with at most $2k - 1$ vertices and to describe the extremal cases.

Theorem 6 [21]. *Let k, p be integers satisfying $k \geq 4$ and $2 \leq p \leq k - 1$. If G is a k-critical graph with $k + p$ vertices, then $\varepsilon(k, G) \geq p(k - p) - 2$, where equality holds if and only if G is the join of K_{k-p-1} and a graph in $\mathcal{D}_{p+1}$.*

Since Proposition 1 holds for list coloring, one might expect that for every $(k - 1)$-uniform list L, each L-critical graph G on $n > k$ vertices has $\varepsilon(k, G) \geq k - 3$. But Example 1 shows an L-critical graph $H(k)$ with $\varepsilon(k, H(k)) = 2$ for every k. On the other hand, if we forbid K_k as a subgraph, the situation changes.

Theorem 7 [31]. *Let $k \geq 4$. Let G be a hypergraph on n vertices not containing K_k, and let L be a list for G with $|L(v)| = k - 1$ for every $v \in V(G)$. If G is L-critical, then $2|E(G)| \geq (k-1)n + k - 3$. In particular, if G is a graph, then $\varepsilon(k, G) \geq k - 3$.*

The above results determine the values of $\varepsilon(k, n)$ (and hence $F(k, n)$) for $n \leq 2k$ and $n = 3k - 2$. Hajós construction with one of the graphs being K_k yields that

$$(4) \qquad \qquad \varepsilon(k, n + k - 1) \leq \varepsilon(k, n) + k - 3.$$

Ore [42] suggested that (4) holds with equality for every $n \geq k + 2$ (see also [23, p. 99]). In view of the above results, that would mean that

$$(5) \qquad \lim_{n \to \infty} \frac{2F(k, n)}{n} = k - \frac{2}{k - 1}.$$

The existing lower bounds are far from (5). The next section contains more discussion on the topic.

4. GALLAI–TYPE BOUNDS

The results of the previous section give bounds on $\varepsilon(k, n)$ that do not depend on n, while (5) (if true) would imply that $\varepsilon(k, n)$ grows asymptotically as $n(k - 3)/(k - 1)$. There is an attractive conjecture that for $n \geq 6$,

$$(6) \qquad F(4, n) \geq \left\lfloor \frac{5n}{3} \right\rfloor.$$

The first lower bound on $\varepsilon(k, n)$ depending on n was the abovementioned theorem of Gallai [20].

Theorem 8 [20]. *Let G be a k-critical graph. Then every block in the subgraph of G induced by vertices of degree $k - 1$ is a complete graph or an odd cycle. Furthermore, if $k \geq 4$ and $G \neq K_k$, then*

$$(7) \qquad 2|E(G)| \geq \left(k - 1 + \frac{k - 3}{k^2 - 3} \right) |V(G)|.$$

In particular, if $k \geq 4$ and $n \geq k + 2$, then

$$(8) \qquad \varepsilon(k, n) \geq \frac{k - 3}{k^2 - 3} n.$$

For $n = 2k$ this gives only $\varepsilon(k, n) \geq 2$ while Theorem 5 gives $\varepsilon(2n, n) \geq 2(k - 3)$; but in the long run the bound of Theorem 5 is much better.

Remark 1. The proof of Theorem 8 works for list coloring as well, so Inequality (7) holds also for every L-critical graph $G \neq K_k$ if L is a $(k - 1)$-uniform list for G and $k \geq 4$.

Theorem 8 yields $F(4, n) \geq \left(\frac{3}{2} + \frac{1}{26}\right)n$ while the conjecture (6) is that $F(4, n)$ is roughly $\left(\frac{3}{2} + \frac{1}{6}\right)n$.

Krivelevich [35, 36], using a result of Stiebitz [49] on the structure of critical graphs, improved this bound as follows.

Theorem 9 [35, 36]. *Suppose $k \geq 4$, and let G be a k-critical graph on more than k vertices. Then*

$$(9) \qquad 2|E(G)| \geq \left(k - 1 + \frac{k - 3}{k^2 - 2k - 1}\right)|V(G)|.$$

In particular, if $k \geq 4$ and $n \geq k + 2$, then

$$(10) \qquad \varepsilon(k, n) \geq \frac{k - 3}{k^2 - 2k - 1}n.$$

The improvement is better for small k. In particular, it gives $F(4, n) \geq \left(\frac{3}{2} + \frac{1}{14}\right)n$ for $n \geq 6$. Since Stiebitz's result [49] does not hold for list colorings, the proof of Theorem 9 does not generalize to list critical graphs.

Kostochka and Stiebitz [32] improved Krivelevich's bound for $k \geq 9$.

Theorem 10 [32]. *Suppose $k \geq 6$, and let G be a k-critical graph on more than k vertices. Then*

$$(11) \qquad 2|E(G)| \geq \left(k - 1 + \frac{2(k - 3)}{k^2 + 6k - 9 - \frac{6}{k-2}}\right)|V(G)|.$$

In particular, if $k \geq 4$ and $n \geq k + 2$, then

$$(12) \qquad \varepsilon(k, n) \geq \frac{2(k - 3)}{k^2 + 6k - 9 - \frac{6}{k-2}}n.$$

The technique of [32] generalizes to list colorings, with sufficiently weaker bounds.

Theorem 11 [32]. *Suppose $k \geq 9$, and let G be an L-critical graph, where L is a $(k - 1)$-uniform list for G. If $G \neq K_k$, then*

$$(13) \qquad 2|E(G)| \geq \left(k - 1 + \frac{1.2(k - 3)}{k^2 + k - 4 - \frac{4}{5k-10}}\right)|V(G)|.$$

There is still a gap of roughly $\frac{k-4}{k-1}n$ between (12) and the known upper bounds on $\varepsilon(k, n)$. And the conjecture (6) is an attractive challenge.

5. CRITICAL GRAPHS WITH NO LARGE CLIQUES

It is natural to ask whether the bound on the number of edges in a k-critical graph with n vertices can be improved when we have additional restrictions on the structure of the graph. A possible direction is to ask what is $F(k, n, s)$—the minimum number of edges in a k-critical graph on n vertices without cliques of size $s + 1$.

Together with Theorem 8, Dirac [13] proved the bound

$$2F(k, n, s) \geq (k - 1)n + (k - 3) + (k - s) \quad \text{if} \quad s \leq k \leq n - 2.$$

Weinstein [56] improved the bound to $2F(k, n, s) \geq (k - 1)n + (k - 3) + 2(k - s)$, but the surplus over $(k - 1)n$ still does not depend on n.

Krivelevich [36] improved the bound as follows.

Theorem 12 [36]. *Let k and s be integers satisfying $3 \leq s < k$. Let G be a k-critical graph not containing a clique of size $s + 1$. Then*

1. *if $s \leq 2k/3$, then* $\quad |E(G)| \geq \left(\dfrac{k}{2} - \dfrac{k - 2}{2(2k - s - 3)} \right) |V(G)|$;

2. *if $s \geq 2k/3$, then* $\quad |E(G)| \geq \left(\dfrac{k}{2} - \dfrac{(k - 2)s}{2(2ks - 2k - s^2)} \right) |V(G)|$;

Krivelevich [36] also gives a bit stronger bounds on the number of edges for critical graphs without short odd cycles, and shows nice applications of his bounds to other interesting problems.

The case of fixed s and large k was considered by Kostochka and Stiebitz [30].

Theorem 13 [30]. *For every fixed s and sufficiently large k, every L-critical graph G on n vertices without cliques of size $s + 1$ for any $(k - 1)$-uniform list L has at least $\bigl(k - o(k)\bigr) n$ edges. In particular, $F(k, n, s) \geq \bigl(k - o(k)\bigr) n$.*

This bound is almost twice larger than the previously mentioned bounds for large k. The bad side of the theorem is that it works only for really large k, when Johannson's theorem on coloring of sparse graphs with given maximum degree works. The good side of it is that the theorem is asymptotically (in k) tight even for graphs of arbitrary girth. A way to construct k-critical graphs G of arbitrary girth with $|E(G)|/|V(G)| < k - 1$ was shown in [24].

Abbott, Hare, and Zhou [3] constructed k-critical graphs G with density $|E(G)|/|V(G)| < k - 7/3$ for girth 4 and density $|E(G)|/|V(G)| < k - 2$ for girth 5. Kostochka and Nešetřil [28] proved that there exist k-critical graphs G with $|E(G)|/|V(G)| < k - 2$ and arbitrarily large girth.

But for small and moderate k, finding least possible average degree of a triangle-free k-critical graph is an interesting open problem.

6. CRITICAL HYPERGRAPHS WITH FEW EDGES

Famous Local Lemma [18] implies that every k-critical r-uniform hypergraph has maximum degree at least $(k - 1)^{r-1}/4r$. One might expect that the average degree of k-critical r-uniform hypergraphs is also always superlinear in k for fixed r. In fact, Erdős and Lovász [18, p. 612] conjectured this for simple hypergraphs. But this is not the case.

Lovász [38, 39], Woodall [53], Seymour [47], and Burstein [9] proved that $|E(H)| \geq |V(H)|$ for every 3-critical hypergraph H. Kostochka and Nešetřil [28] extended results of Burstein [9] and of Abbott, Hare, and Zhou [1, 3] by proving the following upper bound on the minimum of $|E(H)|$ in terms of $|V(H)|$.

Theorem 14 [28]. *For each $r \geq 3$, $k \geq 4$, $g \geq 3$ and $\varepsilon > 0$, there exists an r-uniform k-critical hypergraph H with girth at least g and $|E(H)|/|V(H)| < k - 2 + \varepsilon$.*

And for large k, this is almost matched by the following lower bound due to Kostochka and Stiebitz [30].

Theorem 15 [30]. *Let H be a hypergraph with no edges of size 2. If H is L-critical for a k-uniform list assignment L, then $|E(H)|/|V(H)| \geq k(1 - 3/\sqrt[3]{k})$.*

The advantage of Theorem 15 is that it works for list coloring, and not only for uniform hypergraphs. The girth is also not an issue. The disadvantage is that it provides no information when $k < 27$.

Note that all known examples of r-uniform k-critical hypergraph with small average degree have many vertices. Thus it makes sense to ask about $m(r, k)$—*the minimum number of edges in an r-uniform not k-colorable hypergraph (with no restriction on the number of vertices)*. A first thought here would be that the complete r-uniform hypergraph on $1 + k(r - 1)$ vertices gives the answer. And for $r = 2$ (ordinary graphs) this is the case. But already for $r = 3$ the Fano plane with 7 edges beats K_5^3 with 10 edges. Erdős and Hajnal [17] suggested that if k is very large in comparison with r, then the complete hypergraph still is the best construction, but Alon [4] disproved this conjecture.

Finding good estimates on $m(r, k)$, and especially on $m(r, 2)$, was one of the favorite topics of Paul Erdős for a long time. He proved in [15, 16] the first nontrivial bounds on $m(r, 2)$:

$$2^{r-1} \le m(r, 2) \le r^2 2^r.$$

The proofs of both upper and lower bounds are simple, so we present them here.

Lemma 16. *For every $r \ge 2$ and $k \ge 2$,*

(14) $$k^{r-1} \le m(r, k) \le 20r^2 k^r \ln k.$$

Proof. Suppose that an r-uniform hypergraph $H = (V, E)$ has less than k^{r-1} edges. Consider a random coloring f of V with k colors such that every vertex gets colored with color i with probability $1/k$ for every $1 \le i \le k$ independently of all other vertices. Then for every edge $e \in E$, the probability that e is monochromatic is k^{1-r} and the expected number of monochromatic edges is $|E|k^{1-r} < 1$. Thus there exists a k-coloring of V with no monochromatic edges.

To prove the upper bound, let $m = \lfloor 20r^2 k^r \ln k \rfloor$. If $m \ge \binom{1+kr}{r}$, then the complete hypergraph K_{1+kr}^r witnesses the bound, so we assume the opposite. Consider a random hypergraph $G(r, k, m)$ on a set V of kr^2 labelled vertices, where every of $\binom{kr^2}{r}$ r-subsets of V belongs to $E\big(G(r, k, m)\big)$ with probability $p = 0.5m\binom{kr^2}{r}^{-1}$ independently of all other r-subsets. Note that

$$\mathbf{Pr}\Big\{\big|E\big(G(r, k, m)\big)\big| > m\Big\} < 1/2.$$

For a given $W \subset V$ with $|W| = r^2$, the probability that W is independent is at most

$$(1-p)^{\binom{r^2}{r}} \le \exp\left\{-\binom{r^2}{r} \cdot 0.5m\binom{kr^2}{r}^{-1}\right\} \le \exp\left\{-\frac{0.5m}{ek^r}\right\}.$$

Therefore, since $m > 19.5r^2k^r \ln k$, the probability that there is some independent $W \subset V$ with $|W| = r^2$ is at most

$$\binom{kr^2}{r^2} \exp\left\{-\frac{m}{2ek^r}\right\} \leq (ek)^{r^2} \exp\left\{-\frac{19.5r^2 \ln k}{2e}\right\}$$

$$\leq \exp\left\{r^2(1 + \ln k) - 3.5r^2 \ln k\right\}.$$

The last expression is at most $\exp\{-0.5r^2\} < 1/4$, and with positive probability $G(r, k, m)$ has at most m edges and has no independent set of size r^2, which means that it is not k-colorable. This proves the lemma. ∎

Remark 2. The proof of the lower bound works for list colorings as well.

Beck [8] improved the lower bound for $m(r, 2)$ to $2^r r^{1/3-\varepsilon}$ and Spencer [48] presented a simpler proof of the Beck's bound based on random recoloring. Recently, Radhakrishnan and Srinivasan [44] improved the lower bound further.

Theorem 17 [44]. *For every $c < 1/\sqrt{2}$, there exists an $r_0 = r_0(c)$ such that*

$$m(r, 2) \geq c2^r \sqrt{r/\ln r}.$$

for every $r > r_0$.

Remark 3. In fact, the proof of Theorem 17 also can be adapted for list coloring. So, the result holds for L-critical r-uniform hypergraphs for every 2-uniform list L.

Erdős [16] and Erdős and Lovász [18] said that "perhaps, the order of magnitude of $m(r, 2)$ is $r2^r$". The following result supports the insight of Erdős.

Theorem 18 [25]. *For every positive integer k, let $c = c(k) = \exp\{-4k^2\}$ and $r_k = \exp\{2c_k^{-2}\}$. Let n be a positive integer such that $k \geq 2^n$. Then for every $r > r_k$,*

$$m(r, k) \geq ck^r \left(\frac{r}{\ln r}\right)^{\frac{n}{n+1}}.$$

Note that the proof of Theorem 18 does not work for list coloring. Mubayi and Tetali [41] have some other results for fixed k and large r.

Recall that the ratio of the RHS of (14) to the LHS is $20r^2k \ln k$. When k is larger than r, then the factor $k \ln k$ becomes more important than r^2. Alon [4] improved both bounds in (14) for k large in comparison with r.

Theorem 19 [4]. *For every positive integers $r \geq 3$ and $k \geq 2$,*

$$m(r,k) \leq \left(\frac{k(r-1)+1}{r}\right)\frac{\ln r}{\ln r - 1}\left\lfloor\frac{r}{\ln r}\right\rfloor^{-1} < \left(\frac{k(r-1)+1}{r}\right)$$

and

$$m(r,k) > (r-1)\left\lceil\frac{k}{r}\right\rceil\left\lfloor\frac{r-1}{r}\right\rfloor^{r-1} \sim \frac{1}{e}k^r.$$

Furthermore, if $r \to \infty$ and $k/r \to \infty$, then

$$m(r,k) = O\left(\left(\frac{k(r-1)+1}{r}\right)r^{1.5}\ln r\left(\frac{3}{4}\right)^r\right)$$

Note that when k is much larger than r, the complete hypergraph $K^r_{1+k(r-1)}$ gives a better upper bound than (14), but Alon's bound is even better. The proof of the lower bound is amazingly simple: he first colors vertices of a hypergraph at random using most of the colors, but not all. Then he uncolors a vertex in every monochromatic edge and spends a new color for every $r - 1$ uncolored vertices. This proof does not work for list coloring; thus it would be interesting to find a reasonable lower bound for the number of edges in L-critical r-uniform hypergraphs for arbitrary k-uniform lists L. Also, with respect to k, the upper and lower bounds are of the same order, but with respect to r, the gap probably could be narrowed.

If a hypergraph $H = (V, E)$ is not uniform but $\sum_{e \in E} 2^{-|e|} \leq 1/2$, then a random 2-coloring (as in the proof of Lemma 16) with positive probability is proper. Erdős and Lovász [18] conjectured that the minimum value $\phi(n)$ of $\sum_{e \in E} 2^{-|e|}$ over non-2-colorable hypergraphs with the minimum size of an edge equal to n tends to infinity as n tends to infinity. Beck [8] proved this conjecture. The lower bound on $\phi(n)$ in his proof tends to infinity rather slowly. It would be interesting to estimate the rate of growth of $\phi(n)$.

7. ON CRITICAL SIMPLE HYPERGRAPHS

A hypergraph is called *simple* (sometimes, *linear*) if no two distinct edges share more than one vertex. Let $m^*(r,k)$ denote the minimum number of edges in an r-uniform not k-colorable *simple* hypergraph. Since Fano plane is a simple hypergraph, $m^*(3,2) = m(3,2) = 7$. But in general, $m^*(r,k)$ grows much faster than $m(r,k)$. In their seminal paper [18], Erdős and

Lovász thoroughly studied $m^*(r, k)$. In fact, the celebrated Local Lemma appeared in this paper and its first application was to give lower bounds on $m^*(r, k)$.

Theorem 20 [18]. *Let $s \geq 2$, $r \geq 2$, $k \geq 2$, $n = 4 \cdot 20^{s-1}r^{3s-2}k^{(s-1)(r+1)}$, $m = 4 \cdot 20^s r^{3s-2}k^{s(r+1)}$, $d = 20r^2 k^{r-1}$. Then there exists an r-uniform hypergraph H on kn vertices with at most m edges and with degrees at most d which does not contain any circuit of length $\leq s$ and in which each set of n vertices contains an edge. In particular, H is not k-colorable.*

Since for a hypergraph being simple is the same as to have no 2-circuits, pluging in $s = 2$ yields

$$(15) \qquad m^*(r, k) \leq 1600r^2 k^{2(r+1)},$$

and this is still the best known bound for r large in comparison with k.

Theorem 21 [18]. *Let $r \geq 2$, $k \geq 2$. Then*

$$m^*(r, k) \geq \frac{k^{2(r-2)}}{16r(r-1)^2}.$$

This bound can be improved by a factor of $r/2$ as follows. Theorem 5 in [18] says that *every simple $(k + 1)$-chromatic r-uniform hypergraph contains at least $k^{r-2}/4(r-1)$ vertices with degree at least $k^{r-2}/4(r-1)$.* Then simply the sum of degrees of vertices is used. But one can be less generous. Let G be a $(k+1)$-chromatic r-uniform hypergraph. Order the vertices $v_1, v_2, \ldots$ of G so that $\deg_G(v_1) \geq \deg_G(v_2) \geq \ldots$ and delete one by one vertices in this order together with the incident edges. The degree of a vertex v_i at the moment of deletion is at least $\deg_G(v_i) - (i - 1)$, because G is simple. Thus by the cited above Theorem 5 in [18], after deleting vertex $v_{k^{r-2}/4(r-1)}$ we have deleted at least

$$\frac{k^{r-2}}{4(r-1)} + \left(\frac{k^{r-2}}{4(r-1)} - 1\right) + \left(\frac{k^{r-2}}{4(r-1)} - 2\right) + \cdots + 1 \geq \frac{k^{2r-4}}{32(r-1)^2}$$

edges. This proves the bound.

For $k = 2$, the lower bound can be improved further. Szabó [50] proved that *for every $\varepsilon > 0$ there exists $r_0(\varepsilon)$ such that for $r \geq r_0(\varepsilon)$ every 3-chromatic r-uniform simple hypergraph has a vertex of degree at least $2^r r^{-\varepsilon}$.* Using this result one gets along the lines of the proof of Theorem 5 in [18] and of the previous paragraph that for every $\varepsilon > 0$ there exists $r_0(\varepsilon)$ such that

$$m^*(r, 2) \geq \frac{k^{2r-2}}{2(r-1)^{2\varepsilon}}$$

for $r \geq r_0(\varepsilon)$.

For k very large in comparison with r, the bounds on $m^*(r, k)$ were improved and generalized to partial (r, l)-systems. A *partial (r, l)-system* is an r-uniform hypergraph in which every set of l vertices is contained in at most one edge. Let $m(r, k, l)$ be the minimum number of edges in an (r, l)-system that is not k-colorable. Thus, a simple r-uniform hypergraph is a partial $(r, 2)$-system and $m^*(r, k) = m(r, k, 2)$.

The works [43, 45, 22] on Steiner systems with small independence number yield results for partial (r, l)-systems, and imply upper bounds on $m^*(r, k)$ that improve (15) for k very large in comparison with r. In particular, Grable, Phelps and Rödl [22] constructed simple hypergraphs (in fact, Steiner systems) with chromatic number at least $k + 1$ and at most $c4^r r^2 k^{2r-2} \ln^2 k$ edges for every r and infinitely many k. Thus, for such r and k,

(16) $$m^*(r, k) \leq c4^r r^2 k^{2r-2} \ln^2 k.$$

Kostochka, Mubayi, Rödl, and Tetali [27] proved that for every $r \geq 3$, $l \geq 2$,

$$m(r, k, l) \leq \frac{(2r^l)^{\frac{3l}{l-1}}}{r(r-1)\ldots(r-l+1)} (k^{r-1} \ln 3k)^{\frac{l}{l-1}}.$$

For fixed r and huge k, this bound was matched by the following lower bound.

Theorem 22 [27]. *Let $r > l \geq 2$ be fixed. Then there exists C depending only on r and l such that*

$$m(r, k, l) \geq C \left(k^{r-1} \ln k\right)^{l/(l-1)}.$$

The proof of Theorem 22 does not work for list coloring.

8. VARIATIONS: PANCHROMATIC AND STRONG COLORINGS

One of reasonable generalizations of hypergraph coloring is the *panchromatic k-coloring*—a k-coloring such that every edge meets every of k colors. Then the ordinary 2-coloring is a panchromatic 2-coloring. Let $p(r, k)$ denote the minimum number of edges in an r-uniform hypergraph not admitting any panchromatic k-coloring. By above, $p(r, 2) = m(r, 2)$.

Theorem 3 in the already mentioned paper [18] by Erdős and Lovász speaks on panchromatic colorings.

Theorem 23 [18]. *If each edge of an r-uniform hypergraph H meets at most $k^{r-1}/4(k-1)^r$ other edges, then H is panchromatically k-colorable.*

This implies that

$$p(r, k) > 1 + \frac{k^{r-1}}{4(k-1)^r} > 1 + \frac{1}{4k}e^{r/k}.$$

Let $N(k, r)$ denote the minimum number of vertices in a k-partite graph with list chromatic number greater than r. Among other results, Erdős, Rubin, and Taylor [19] proved that $N(2, r)$ is closely connected with $m(r, 2)$:

$$m(r, 2) \leq N(2, r) \leq 2m(r, 2).$$

An interesting feature of this results is that *ordinary* coloring of r-uniform hypergraphs relates to *list* coloring of bipartite graphs. This relation can be easily extended to panchromatic colorings with more colors:

Theorem 24 [26]. *For every $r \geq 2$ and $k \geq 2$, $p(r, k) \leq N(k, r) \leq k\,p(r, k)$.*

It follows from Alon's results in [5] that for some $0 < c_1 < c_2$ and every $r \geq 2$ and $k \geq 2$,

$$\exp\{c_1 r/k\} \leq N(k, r) \leq k \exp\{c_2 r/k\}.$$

Therefore, by Theorem 4 we get reasonable bounds on $p(r, k)$ for fixed k and large r:

$$\exp\{c_1 r/k\}/k \leq p(r, k) \leq k \exp\{c_2 r/k\}.$$

Recall that Theorem 23 also yields the lower bound on $p(r, k)$ with $c_1 = 1/4$ and thus itself implies the lower bound $1 + \frac{1}{4k}e^{r/k}$ on $N(k, r)$.

One can also define panchromatic list colorings: If each vertex v of H is assigned a list $L(v)$ of k colors, then a *panchromatic L-coloring* of H is a coloring in which each vertex is given a color from its own list and each edge contains vertices with at least k different colours.

Kostochka and Woodall [33] obtained bounds on the minimum number of edges in hypergraphs being edge critical with respect to panchromatic colorings.

Theorem 25 [33]. *Let $k \geq 2$ and let $H = (V, E)$ be a hypergraph in which every edge has at least k vertices, and every vertex is given a list $L(v)$ of k colors. If H is not panchromatically L-colorable, but after deleting any edge becomes panchromatically L-colorable, then $|E| \geq (|V| + k - 2)/(k - 1)$. This bound is attained for every $k \geq 2$ for ordinary panchromatic colorings.*

If the condition 'every edge has at least k vertices' is replaced with 'k-uniform', then the inequality can be strengthened.

Theorem 26 [33]. *Let $k \geq 2$ and let $H = (V, E)$ be a k-uniform hypergraph and every vertex is given a list $L(v)$ of k colors. If H is not panchromatically L-colorable, but after deleting any edge becomes panchromatically L-colorable, then $|E| \geq |V|(k + 2)/k^2$.*

This bound is unlikely to be sharp. The following result says that even if panchromatically critical hypergraphs are not dense themselves, they must contain dense subgraphs.

Theorem 27 [33]. *Let $k \geq 4$, $k \neq 5$, and let $H = (V, E)$ be a k-uniform hypergraph such that*

$$\left| E\big(H(V')\big) \right| \leq \frac{k|V'| - k + 1}{k^2 - 2k + 2}, \quad \text{whenever} \quad V' \subseteq V, \quad E\big(H(V')\big) \neq \emptyset.$$

Then H is panchromatically k-colorable. For $k \in \{3, 5\}$, the same conclusion follows if the final $+1$ in the numerator is omitted.

This bound is sharp if $k \notin \{3, 5\}$. Note that a panchromatic k-coloring of a k-uniform hypergraph is a *strong coloring*, i.e. the coloring in which every two distinct vertices sharing an edge must have different colors. Every strong coloring of a hypergraph H corresponds to a proper edge coloring of the hypergraph H^* dual to H. The problem of estimating the edge chromatic number of uniform hypergraphs with a given maximum degree and moderate codegree attracted a lot of attention after Rödl's solution of the Erdős–Hanani Problem. A remarkable sequence of significant papers due to Rödl, Frankl, Pippenger, Spencer, Kahn, Grable, Alon, Kim, Molloy, Reed, and Vu was devoted to this topic. Theorem 27 can be interpreted as a (somewhat unusual) sufficient condition for the edge-chromatic number of a hypergraph H to equal its trivial lower bound, the maximum vertex degree $\Delta(H)$.

Theorem 28 [33]. *Let H be a hypergraph with maximum degree r, where $r = 4$ or $r \geq 6$. If every vertex subset S is incident with at least $\left((r^2 - 2r + 2)|S| + r - 1\right)/r$ edges, then H is r-edge-colorable.*

By the definition, the strong chromatic number of a hypergraph $H = (V, E)$ equals the chromatic number of its *skeleton, $S(H)$*—the graph on V whose vertices are adjacent if and only if they share some edge in H. Deletion of an edge from a hypergraph is a rather rough action with respect to strong coloring: deletion of an edge of size r may reduce the strong chromatic number by $r - 1$. A subtler operation is *splitting:* if $H = (V, E)$ is a hypergraph, $v \in e \in E$, and $\deg_H(v) \geq 2$, then the (v, e)-*splitting* of H is obtained by replacing the edge e by the edge $e - v + v'$, where v' is a new vertex. Then deleting an edge e can be performed as a sequence of (v, e)-splittings over the vertices $v \in e$ of degree at least two. The (v, e)-splitting corresponds to cutting edge v in the dual hypergraph H^* into two pieces, one of which has size one.

Kostochka and Woodall [34] considered *splitting-critical* hypergraphs with respect to strong coloring and strong list-coloring. It appears that for $k \geq r + 2$, the sparsest k-splitting-critical r-uniform hypergraphs are obtained from sparse k-critical graphs by adding to every edge $r - 2$ new vertices (of degree one in the resulting hypergraph). On the other hand, the sparsest $(r + 1)$-splitting-critical r-uniform hypergraphs cannot be obtained this way. If k is large in comparison with r and the skeleton $S(H)$ of a k-splitting-critical r-uniform hypergraph H has no large cliques, then the lower bound on the number of edges in H can be improved.

Theorem 29 [34]. *Let $s \geq r$ be positive integers and let k be sufficiently large with respect to s. Let H be a list-k-splitting-critical r-uniform hypergraph with respect to strong coloring whose skeleton $S(H)$ does not contain a complete subgraph on $s + 1$ vertices. Then*

$$|E(H)| \geq k\left(1 - 6(\ln k)^{-1/3}\right)\left(|V(H)| - (r - 2)|E(H)|\right).$$

As with Theorem 13, the bad side of the last theorem is that the proof works only for really large k, and the good side of it is that the bound is asymptotically (in k) sharp even for hypergraphs of large girth.

9. CONCLUDING REMARKS

Certainly, the survey is not full. Essentially, it describes problems I am interested in. The reader might look into [23, Chapter 5] and [46] for more problems on color-critical graphs and hypergraphs. Maybe some proofs of the results above can be simplified using recent impressive results of Vu (see, e.g., [55]).

I thank Michael Stiebitz and Douglas Woodall for our discussions on the topic and their helpful comments on a earlier version of this survey. Some pieces of our joint works were used in this text. Thanks for helpful comments are also due to Oleg Borodin and Bjarne Toft.

REFERENCES

[1] H. L. Abbott and D. R. Hare, Sparse color-critical hypergraphs, *Combinatorica,* **9** (1989), 233–243.

[2] H. L. Abbott, D. R. Hare and B. Zhou, Sparse color-critical graphs and hypergraphs with no short cycles, *J. Graph Theory,* **18** (1994), 373–388.

[3] H. L. Abbott, D. R. Hare and B. Zhou, Color-critical graphs and hypergraphs with few edges and no short cycles, *Discrete Math.,* **182** (1998), 3–11.

[4] N. Alon, Hypergraphs with high chromatic number, *Graphs and Combinatorics,* **1** (1985), 387–389.

[5] N. Alon, Choice number of graphs: a probabilistic approach, *Combinatorics, Probability and Computing,* **1** (1992), 107–114.

[6] N. Alon, Restricted colorings of graphs, in: K. Walker, ed., "Surveys in Combinatorics, 1993", London Math. Soc. Lecture Note Series, **187** (Cambridge Univ. Press, Cambridge, UK, 1993), 1–33.

[7] D. Archdeacon, J. Hutchinson, A. Nakamoto, S. Negami and K. Ota, Chromatic numbers of quadrangulations on closed surfaces, *J. Graph Theory* **37** (2001), 100–114.

[8] J. Beck, On 3-chromatic hypergraphs, *Discrete Math.,* **24** (1978), 127–137.

[9] M. I. Burstein, Critical hypergraphs with minimal number of edges (Russian), *Bull. Acad. Sci. Georgian SSR*, **83** (1976), 285–288.

[10] W. A. Deuber, A. V. Kostochka and H. Sachs, A shorter proof of Dirac's theorem on the number of edges in chromatically critical graphs, *Diskretnyi Analiz i Issledovanie Operacii*, **3** (1996), No. 4, 28–34 (in Russian).

[11] G. A. Dirac, Note on the colouring of graphs, *Math. Z.*, **54** (1951), 347–353.

[12] G. A. Dirac, The structure of *k*-chromatic graphs, *Fund. Math.*, **40** (1953), 42–55.

[13] G. A. Dirac, A theorem of R. L. Brooks and a conjecture of H. Hadwiger, *Proc. London Math. Soc.*, (3) **7** (1957), 161–195.

[14] G. A. Dirac, The number of edges in critical graphs, *J. Reine u. Angew. Math.*, **268/269** (1974), 150–164.

[15] P. Erdős, On a combinatorial problem, I, *Nordisk Mat. Tidskrift,* **11** (1963), 5–10.

[16] P. Erdős, On a combinatorial problem, II, *Acta Mathematica of the Academy of Sciences,* Hungary, **15** (1964), 445–447.

[17] P. Erdős and A. Hajnal, On a property of families of sets, *Acta Mathematica of the Academy of Sciences,* Hungary, **12** (1961), 87–123.

[18] P. Erdős and L. Lovász, Problems and Results on 3-chromatic hypergraphs and some related questions, in: *Infinite and Finite Sets,* A. Hajnal et. al., editors, Colloq. Math. Soc. J. Bolyai, **11**, North Holland, Amsterdam, 609–627, 1975.

[19] P. Erdős, A. L. Rubin and H. Taylor, Choosability in graphs, in: *Proc. West Coast Conference on Combinatorics, Graph Theory and Computing, Arcata, 1979, Congr. Numer.,* 26 (1980), 125–157.

[20] T. Gallai, Kritische Graphen I, *Publ. Math. Inst. Hungar. Acad. Sci.,* **8** (1963), 165–192.

[21] T. Gallai, Kritische Graphen II, *Publ. Math. Inst. Hungar. Acad. Sci.,* **8** (1963), 373–395.

[22] D. Grable, K. Phelps and V Rödl, The minimum independence number for designs, *Combinatorica,* **15** (1995), 175–185.

[23] T. R. Jensen and B. Toft, Graph coloring problems, Wiley–Interscience, 1995.

[24] A. V. Kostochka, Constructing strictly *k*-degenerate *k*-chromatic graphs of arbitrary girth, *Abstracts of the V All-Union Conference on the Problems of Theoretical Cybernetics,* Novosibirsk, 1980, 130–131 (in Russian).

[25] A. V. Kostochka, Coloring uniform hypergraphs with few colors, submitted.

[26] A. V. Kostochka, On a theorem by Erdős, Rubin and Taylor, submitted.

[27] A. V. Kostochka, D. Mubayi, V. Rödl and P. Tetali, On the chromatic number of set-systems, *Random Structures and Algorithms,* **19** (2001), 87–98.

[28] A. V. Kostochka and J. Nešetřil, Properties of Descartes' construction of triangle-free graphs with high chromatic number, *Combinatorics, Probability and Computing,* **8** (1999), 467–472.

[29] A. V. Kostochka and M. Stiebitz, Excess in colour-critical graphs, *Bolyai Society Mathematical Studies,* **7** (1999), 87–99.

[30] A. V. Kostochka and M. Stiebitz, On the number of edges in colour-critical graphs and hypergraphs, *Combinatorica,* **20** (2000), 521–530.

[31] A. V. Kostochka and M. Stiebitz, A list version of Dirac's theorem on the number of edges in colour-critical graphs, *Journal of Graph Theory,* **39** (2002), 165–167.

[32] A. V. Kostochka and M. Stiebitz, A new lower bound on the number of edges in colour-critical graphs, to appear in *J. Combinatorial Theory B.*

[33] A. V. Kostochka and D. R. Woodall, Density conditions for panchromatic colourings of hypergraphs, *Combinatorica,* **21** (2001), 515–541.

[34] A. V. Kostochka and D. R. Woodall, On the number of edges in hypergraphs critical with respect to strong colourings, *European Journal of Combinatorics,* **21** (2000), 249–255.

[35] M. Krivelevich, An improved bound on the minimal number of edges in color-critical graphs, *Electron J. Combin.,* **5** (1998), no. 1, Research Paper 4, 4 pp.

[36] M. Krivelevich, On the minimal number of edges in color-critical graphs, *Combinatorica,* **17** (1997), 401–426.

[37] H. V. Kronk and J. Mitchem, On Dirac's generalization of Brooks' theorem, *Canad. J. Math.,* **24** (1972), 805–807.

[38] L. Lovász, A generalization of Konig's theorem, *Acta Math. Acad. Sci. Hungar.,* **21** (1970), 443–446.

[39] L. Lovász, Coverings and colorings of hypergraphs, in: *Congressus Numer.,* **8** (1973), 3–12.

[40] J. Mitchem, A new proof of a theorem of Dirac on the number of edges in critical graphs, *J. Reine u. Angew. Math.,* **299/300** (1978), 84–91.

[41] D. Mubayi and P. Tetali, Generalizing Property B to many colors, manuscript.

[42] O. Ore, The Four Colour Problem, Academic Press, New York, 1967.

[43] K. Phelps, V Rödl, Steiner Triple Systems with Minimum Independence Number, *Ars combinatoria,* **21** (1986), 167–172.

[44] J. Radhakrishnan and A. Srinivasan, Improved bounds and algorithms for hypergraph two-coloring, *Random Structures and Algorithms,* **16** (2000), 4–32.

[45] V. Rödl, E. Šinajová, Note on Independent Sets in Steiner Systems, *Random Structures and Algorithms,* **5** (1994), 183–190.

[46] H. Sachs and M. Stiebitz, On constructive methods in the theory of colour-critical graphs, *Discrete Math.,* **74** (1989), 201–226.

[47] P. D. Seymour, On the two-coloring of hypergraphs, *Quart. J. Math. Oxford,* **25** (1974), 303–312.

[48] J. Spencer, Coloring n-sets red and blue, *J. Comb. Theory Ser. A,* **30** (1981), 112–113.

[49]   M. Stiebitz, Proof of a conjecture of T. Gallai concerning connectivity properties of colour-critical graphs, *Combinatorica,* **2** (1982), 315–323.

[50]   Szabó, An application of Lovász' Local Lemma—a new lower bound for the van der Waerden number, *Random Structures and Algorithms,* **1** (1990), 344–360.

[51]   B. Toft, Colour-critical graphs and hypergraphs, *J. Combin. Th. Ser. B,* **16** (1974), 145–161.

[52]   B. Toft, Colouring, stable sets and perfect graphs, Graham, R. L. (ed.) et al., Handbook of combinatorics. Vol. 1–2. Amsterdam: Elsevier (North-Holland), 1995, 233–288.

[53]   D. R. Woodall, Property B and the four-color problem. *Combinatorics.* Institute of Mathematics and its Applications, Southend-on-sea, England (1972), 322–340.

[54]   V. G. Vizing, Colouring the vertices of a graph with prescribed colours, *Metody Diskretnogo Analiza v Teorii Kodov i Skhem,* No. **29** (1976), 3–10 (in Russian).

[55]   V. H. Vu, A general upper bound on the list chromatic number of locally sparse graphs, *Combinatorics, Probability and Computing,* **11** (2002), 103–111.

[56]   J. Weinstein, Excess in critical graphs, *J. Combin. Th.(B),* **18** (1975), 24–31.

Alexandr Kostochka

University of Illinois at
Urbana–Champaign
Urbana
IL 61801

and

Institute of Mathematics
Novosibirsk 630090
Russia

kostochk@math.uiuc.edu

BOLYAI SOCIETY
MATHEMATICAL STUDIES, 15

Conference on Finite
and Infinite Sets
Budapest, pp. 199–262.

Pseudo-random Graphs

M. KRIVELEVICH* and B. SUDAKOV†

1. Introduction

Random graphs have proven to be one of the most important and fruitful concepts in modern Combinatorics and Theoretical Computer Science. Besides being a fascinating study subject for their own sake, they serve as essential instruments in proving an enormous number of combinatorial statements, making their role quite hard to overestimate. Their tremendous success serves as a natural motivation for the following very general and deep informal questions: what are the essential properties of random graphs? How can one tell when a given graph behaves like a random graph? How to create deterministically graphs that look random-like? This leads us to a concept of *pseudo-random graphs*.

Speaking very informally, a pseudo-random graph $G = (V, E)$ is a graph that behaves like a truly random graph $G(|V|, p)$ of the same edge density $p = |E|/\binom{|V|}{2}$. Although the last sentence gives some initial idea about this concept, it is not very informative, as first of all it does not say in which aspect the pseudo-random graph behavior is similar to that of the corresponding random graph, and secondly it does not supply any quantitative measure of this similarity. There are quite a few possible graph parameters that can potentially serve for comparing pseudo-random and

*Research supported in part by a USA-Israel BSF Grant, by a grant from the Israel Science Foundation and by a Bergmann Memorial Grant.

†Research supported in part by NSF grants DMS-0355497, DMS-0106589, and by an Alfred P. Sloan fellowship. Part of this research was done while visiting Microsoft Research.

random graphs (and in fact quite a few of them are equivalent in certain, very natural sense, as we will see later), but probably the most important characteristics of a truly random graph is its *edge distribution*. We can thus make a significant step forward and say that a pseudo-random graph is a graph with edge distribution resembling the one of a truly random graph with the same edge density. Still, the quantitative measure of this resemblance remains to be introduced.

Although first examples and applications of pseudo-random graphs appeared very long time ago, it was Andrew Thomason who launched systematic research on this subject with his two papers [79], [80] in the mid-eighties. Thomason introduced the notion of jumbled graphs, enabling to measure in quantitative terms the similarity between the edge distributions of pseudo-random and truly random graphs. He also supplied several examples of pseudo-random graphs and discussed many of their properties. Thomason's papers undoubtedly defined directions of future research for many years.

Another cornerstone contribution belongs to Chung, Graham and Wilson [26] who in 1989 showed that many properties of different nature are in certain sense equivalent to the notion of pseudo-randomness, defined using the edge distribution. This fundamental result opened many new horizons by showing additional facets of pseudo-randomness.

Last years brought many new and striking results on pseudo-randomness by various researchers. There are two clear trends in recent research on pseudo-random graphs. The first is to apply very diverse methods from different fields (algebraic, linear algebraic, combinatorial, probabilistic etc.) to construct and study pseudo-random graphs. The second and equally encouraging is to find applications, in many cases quite surprising, of pseudo-random graphs to problems in Graph Theory, Computer Science and other disciplines. This mutually enriching interplay has greatly contributed to significant progress in research on pseudo-randomness achieved lately.

The aim of this survey is to provide a systematic treatment of the concept of pseudo-random graphs, probably the first since the two seminal contributions of Thomason [79], [80]. Research in pseudo-random graphs has developed tremendously since then, making it impossible to provide full coverage of this subject in a single paper. We are thus forced to omit quite a few directions, approaches, theorem proofs from our discussion. Nevertheless we will attempt to provide the reader with a rather detailed and illustrative account of the current state of research in pseudo-random graphs.

Although, as we will discuss later, there are several possible formal approaches to pseudo-randomness, we will mostly emphasize the approach based on graph eigenvalues. We find this approach, combining linear algebraic and combinatorial tools in a very elegant way, probably the most appealing, convenient and yet quite powerful.

This survey is structured as follows. In the next section we will discuss various formal definitions of the notion of pseudo-randomness, from the so called jumbled graphs of Thomason to the (n, d, λ)-graphs defined by Alon, where pseudo-randomness is connected to the eigenvalue gap. We then describe several known constructions of pseudo-random graphs, serving both as illustrative examples for the notion of pseudo-randomness, and also as test cases for many of the theorems to be presented afterwards. The strength of every abstract concept is best tested by properties it enables to derive. Pseudo-random graphs are certainly not an exception here, so in Section 4 we discuss various properties of pseudo-random graphs. Section 5, the final section of the paper, is devoted to concluding remarks.

2. Definitions of pseudo-random graphs

Pseudo-random graphs are much more of a general concept describing some graph theoretic phenomenon than of a rigid well defined notion – the fact reflected already in the plural form of the title of this section! Here we describe various formal approaches to the concept of pseudo-randomness. We start with stating known facts on the edge distribution of random graphs, that will serve later as a benchmark for all other definitions. Then we discuss the notion of jumbled graphs introduced by Thomason in the mid-eighties. Then we pass on to the discussion of graph properties, equivalent in a weak (qualitative) sense to the pseudo-random edge distribution, as revealed by Chung, Graham and Wilson in [26]. Our next item in this section is the definition of pseudo-randomness based on graph eigenvalues – the approach most frequently used in this survey. Finally, we discuss the related notion of strongly regular graphs, their eigenvalues and their relation to pseudo-randomness.

2.1. Random graphs

As we have already indicated in the Introduction, pseudo-random graphs are modeled after truly random graphs, and therefore mastering the edge distribution in random graphs can provide the most useful insight on what can be expected from pseudo-random graphs. The aim of this subsection is to state all necessary definitions and results on random graphs. We certainly do not intend to be comprehensive here, instead referring the reader to two monographs on random graphs [20], [49], devoted entirely to the subject and presenting a very detailed picture of the current research in this area.

A *random graph* $G(n, p)$ is a probability space of all labeled graphs on n vertices $\{1, \ldots, n\}$, where for each pair $1 \leq i < j \leq n$, (i, j) is an edge of $G(n, p)$ with probability $p = p(n)$, independently of any other edges. Equivalently, the probability of a graph $G = (V, E)$ with $V = \{1, \ldots, n\}$ in $G(n, p)$ is $Pr[G] = p^{|E(G)|}(1 - p)^{\binom{n}{2} - |E(G)|}$. We will occasionally mention also the probability space $G_{n,d}$, this is the probability space of all d-regular graphs on n vertices endowed with the uniform measure, see the survey of Wormald [83] for more background. We also say that a graph property $\mathcal{A}$ holds *almost surely*, or a.s. for brevity, in $G(n, p)$ $(G_{n,d})$ if the probability that $G(n, p)$ $(G_{n,d})$ has $\mathcal{A}$ tends to one as the number of vertices n tends to infinity.

From our point of view the most important parameter of random graph $G(n, p)$ is its edge distribution. This characteristics can be easily handled due to the fact that $G(n, p)$ is a product probability space with independent appearances of different edges. Below we cite known results on the edge distribution in $G(n, p)$.

Theorem 2.1. *Let* $p = p(n) \leq 0.99$. *Then almost surely* $G \in G(n, p)$ *is such that if* U *is any set of* u *vertices, then*

$$\left| e(U) - p\binom{u}{2} \right| = O\left(u^{3/2} p^{1/2} \log^{1/2}(2n/u) \right).$$

Theorem 2.2. *Let* $p = p(n) \leq 0.99$. *Then almost surely* $G \in G(n, p)$ *is such that if* U, W *are disjoint sets of vertices satisfying* $u = |U| \leq w = |W|$, *then*

$$\left| e(U, W) - puw \right| = O\left(u^{1/2} w p^{1/2} \log^{1/2}(2n/w) \right).$$

The proof of the above two statements is rather straightforward. Notice that both quantities $e(U)$ and $e(U, W)$ are binomially distributed random variables with parameters $\binom{u}{2}$ and p, and uw and p, respectively. Applying standard Chernoff-type estimates on the tails of the binomial distribution (see, e.g., Appendix A of [18]) and then the union bound, one gets the desired inequalities.

It is very instructive to notice that we get less and less control over the edge distribution as the set size becomes smaller. For example, in the probability space $G(n, 1/2)$ every subset is expected to contain half of its potential edges. While this is what happens almost surely for large enough sets due to Theorem 2.1, there will be almost surely sets of size about $2 \log_2 n$ containing all possible edges (i.e. cliques), and there will be almost surely sets of about the same size, containing no edges at all (i.e. independent sets).

For future comparison we formulate the above two theorems in the following unified form:

Corollary 2.3. *Let* $p = p(n) \leq 0.99$. *Then almost surely in* $G(n,p)$ *for every two (not necessarily) disjoint subsets of vertices* $U, W \subset V$ *of cardinalities* $|U| = u$, $|W| = w$, *the number* $e(U, W)$ *of edges of* G *with one endpoint in* U *and the other one in* W *satisfies:*

$$(1) \qquad \left| e(U, W) - puw \right| = O\left(\sqrt{uwnp} \right).$$

(A notational agreement here and later in the paper: if an edge e belongs to the intersection $U \cap W$, then e is counted twice in $e(U, W)$.)

Similar bounds for edge distribution hold also in the space $G_{n,d}$ of d-regular graphs, although they are significantly harder to derive there.

Inequality (1) provides us with a quantitative benchmark, according to which we will later measure the uniformity of edge distribution in pseudo-random graphs on n vertices with edge density $p = \left| E(G) \right| / \binom{n}{2}$.

It is interesting to draw comparisons between research in random graphs and in pseudo-random graphs. In general, many properties of random graphs are much easier to study than the corresponding properties of pseudo-random graphs, mainly due to the fact that along with the almost uniform edge distribution described in Corollary 2.3, random graphs possess as well many other nice features, first and foremost of them being that they are in fact very simply defined product probability spaces. Certain graph properties can be easily shown to hold almost surely in $G(n, p)$ while they are not necessarily valid in pseudo-random graphs of the same edge density. We will see quite a few such examples in the next section. A general line of research appears to be not to use pseudo-random methods to get new results for random graphs, but rather to try to adapt techniques developed for random graphs to the case of pseudo-random graphs, or alternatively to develop original techniques and methods.

2.2. Thomason's jumbled graphs

In two fundamental papers [79], [80] published in 1987 Andrew Thomason introduced the first formal quantitative definition of pseudo-random graphs. It appears quite safe to attribute the launch of the systematic study of pseudo-randomness to Thomason's papers.

Thomason used the term "jumbled" graphs in his papers. A graph $G = (V, E)$ is said to be (p, α)-jumbled if p, α are real numbers satisfying $0 < p < 1 \leq \alpha$ if every subset of vertices $U \subset V$ satisfies:

$$(2) \qquad \left| e(U) - p \binom{|U|}{2} \right| \leq \alpha |U|.$$

The parameter p can be thought of as the density of G, while α controls the deviation from the ideal distribution. According to Thomason, the word "jumbled" is intended to convey the fact that the edges are evenly spread throughout the graph.

The motivation for the above definition can be clearly traced to the attempt to compare the edge distribution in a graph G to that of a truly random graph $G(n, p)$. Applying it indeed to $G(n, p)$ and recalling (1) we conclude that the random graph $G(n, p)$ is almost surely $O(\sqrt{np})$-jumbled.

Thomason's definition has several trivial yet very nice features. Observe for example that if G is (p, α)-jumbled then the complement $\bar{G}$ is $(1 - p, \alpha)$-jumbled. Also, the definition is hereditary – if G is (p, α)-jumbled, then so is every induced subgraph H of G.

Note that being $(p, \Theta(np))$-jumbled for a graph G on n vertices and $\binom{n}{2} p$ edges does not say too much about the edge distribution of G as the number of edges in linear sized sets can deviate by a percentage from their expected value. However as we shall see very soon if G is known to be $(p, o(np))$-jumbled, quite a lot can be said about its properties. Of course, the smaller is the value of α, the more uniform or jumbled is the edge distribution of G. A natural question is then how small can be the parameter $\alpha = \alpha(n, p)$ for a graph $G = (V, E)$ on $|V| = n$ vertices with edge density $p = |E| / \binom{n}{2}$? Erdős and Spencer proved in [35] that α satisfies $\alpha = \Omega(\sqrt{n})$ for a constant p; their method can be extended to show $\alpha = \Omega(\sqrt{np})$ for all values of

$p = p(n)$. We thus may think about $(p, O(\sqrt{np}))$-jumbled graphs on n vertices as in a sense best possible pseudo-random graphs.

Although the fact that G is (p, α)-jumbled carries in it a lot of diverse information on the graph, it says almost nothing (directly at least) about small subgraphs, i.e. those spanned by subsets U of size $|U| = o(\alpha/p)$. Therefore in principle a (p, α)-jumbled graph can have subsets of size $|U| = O(\alpha/p)$ spanning by a constant factor less or more edges then predicted by the uniform distribution. In many cases however quite a meaningful local information (such as the presence of subgraphs of fixed size) can still be salvaged from global considerations as we will see later.

Condition (2) has obviously a global nature as it applies to *all* subsets of G, and there are exponentially many of them. Therefore the following result of Thomason, providing a sufficient condition for pseudo-randomness based on degrees and co-degrees only, carries a certain element of surprise in it.

Theorem 2.4 [79]. *Let G be a graph on n vertices with minimum degree np. If no pair of vertices of G has more than $np^2 + l$ common neighbors, then G is $(p, \sqrt{(p+l)n})$-jumbled.*

The above theorem shows how the pseudo-randomness condition of (2) can be ensured/checked by testing only a polynomial number of easily accessible conditions. It is very useful for showing that specific constructions are jumbled. Also, it can find algorithmic applications, for example, a very similar approach has been used by Alon, Duke, Lefmann, Rödl and Yuster in their Algorithmic Regularity Lemma [9].

As observed by Thomason, the minimum degree condition of Theorem 2.4 can be dropped if we require that every pair of vertices has $(1+o(1)) np^2$ common neighbors. One cannot however weaken the conditions of the theorem so as to only require that every *edge* is in at most $np^2 + l$ triangles.

Another sufficient condition for pseudo-randomness, this time of global nature, has also been provided in [79], [80]:

Theorem 2.5 [79]. *Let G be a graph of order n, let ηn be an integer between 2 and $n - 2$, and let $\omega > 1$ be a real number. Suppose that each induced subgraph H of order ηn satisfies $\left| e(H) - p\binom{\eta n}{2} \right| \le \eta n \alpha$. Then G is $\left(p, 7\sqrt{n\alpha/\eta}/(1-\eta) \right)$-jumbled. Moreover G contains a subset $U \subseteq V(G)$ of size $|U| \ge \left(1 - \frac{380}{n(1-\eta)^2 w} \right) n$ such that the induced subgraph $G[U]$ is $(p, \omega\alpha)$-jumbled.*

Thomason also describes in [79], [80] several properties of jumbled graphs. We will not discuss these results in details here as we will mostly adopt a different approach to pseudo-randomness. Occasionally however we will compare some of later results to those obtained by Thomason.

2.3. Equivalent definitions of weak pseudo-randomness

Let us go back to the jumbledness condition (2) of Thomason. As we have already noted it becomes non-trivial only when the error term in (2) is $o(n^2p)$. Thus the latter condition can be considered as the weakest possible condition for pseudo-randomness.

Guided by the above observation we now define the notion of weak pseudo-randomness as follows. Let (G_n) be a sequence of graphs, where G_n has n vertices. Let also $p = p(n)$ is a parameter ($p(n)$ is a typical density of graphs in the sequence). We say that the sequence (G_n) is *weakly pseudo-random* if the following condition holds:

$$(3) \qquad \text{For all subsets } U \subseteq V(G_n), \qquad \left| e(U) - p\binom{|U|}{2} \right| = o(n^2p).$$

For notational convenience we will frequently write $G = G_n$, tacitly assuming that (G) is in fact a sequence of graphs.

Notice that the error term in the above condition of weak pseudo-randomness does not depend on the size of the subset U. Therefore it applies essentially only to subsets U of linear size, ignoring subsets U of size $o(n)$. Hence (3) is potentially much weaker than Thomason's jumbledness condition (2).

Corollary 2.3 supplies us with the first example of weakly pseudo-random graphs – a random graph $G(n, p)$ is weakly pseudo-random as long as $p(n)$ satisfies $np \to \infty$. We can thus say that if a graph G on n vertices is weakly pseudo-random for a parameter p, then the edge distribution of G is close to that of $G(n, p)$.

In the previous subsection we have already seen examples of conditions implying pseudo-randomness. In general one can expect that conditions of various kinds that hold almost surely in $G(n, p)$ may imply or be equivalent to weak pseudo-randomness of graphs with edge density p.

Let us first consider the case of the constant edge density p. This case has been treated extensively in the celebrated paper of Chung, Graham and

Wilson from 1989 [26], where they formulated several equivalent conditions for weak pseudo-randomness. In order to state their important result we need to introduce some notation.

Let $G = (V, E)$ be a graph on n vertices. For a graph L we denote by $N_G^*(L)$ the number of labeled induced copies of L in G, and by $N_G(L)$ the number of labeled not necessarily induced copies of L in G. For a pair of vertices $x, y \in V(G)$, we set $s(x, y)$ to be the number of vertices of G joined to x and y the same way: either to both or to none. Also, $codeg(x, y)$ is the number of common neighbors of x and y in G. Finally, we order the eigenvalues λ_i of the adjacency matrix $A(G)$ so that $|\lambda_1| \geq |\lambda_2| \geq \ldots \geq |\lambda_n|$.

Theorem 2.6 [26]. *Let $p \in (0, 1)$ be fixed. For any graph sequence (G_n) the following properties are equivalent:*

$P_1(l)$: *For a fixed $l \geq 4$ for all graphs L on l vertices,*

$$N_G^*(L) = \left(1 + o(1)\right) n^l p^{|E(L)|} (1 - p)^{\binom{l}{2} - |E(L)|}.$$

$P_2(t)$: *Let C_t denote the cycle of length t. Let $t \geq 4$ be even,*

$$e(G_n) = \frac{n^2 p}{2} + o(n^2) \quad \text{and} \quad N_G(C_t) \leq (np)^t + o(n^t).$$

P_3: $\quad e(G_n) \geq \frac{n^2 p}{2} + o(n^2) \quad$ *and* $\quad \lambda_1 = \left(1 + o(1)\right) np, \quad \lambda_2 = o(n)$.

P_4: *For each subset $U \subset V(G)$, $\quad e(U) = \frac{p}{2}|U|^2 + o(n^2)$.*

P_5: *For each subset $U \subset V(G)$ with $|U| = \lfloor \frac{n}{2} \rfloor$, we have*
$\quad e(U) = \left(\frac{p}{8} + o(1)\right) n^2$.

P_6: $\quad \sum_{x,y \in V} \left| s(x, y) - \left(p^2 + (1 - p)^2\right) n \right| = o(n^3)$.

P_7: $\quad \sum_{x,y \in V} \left| codeg(x, y) - p^2 n \right| = o(n^3)$.

Note that condition P_4 of this remarkable theorem is in fact identical to our condition (3) of weak pseudo-randomness. Thus according to the theorem all conditions P_1–P_3, P_5–P_7 are in fact equivalent to weak pseudo-randomness!

As noted by Chung et al. probably the most surprising fact (although possibly less surprising for the reader in view of Theorem 2.4) is that apparently the weak condition $P_2(4)$ is strong enough to imply weak pseudo-randomness.

It is quite easy to add another condition to the equivalence list of the above theorem: for all $U, W \subset V$, $e(U, W) = p|U||W| + o(n^2)$.

A condition of a very different type, related to the celebrated Szemerédi Regularity Lemma has been added to the above list by Simonovits and Sós in [73]. They showed that if a graph G possesses a Szemerédi partition in which almost all pairs have density p, then G is weakly pseudo-random, and conversely if G is weakly pseudo-random then in every Szemerédi partition all pairs are regular with density p. An extensive background on the Szemerédi Regularity Lemma, containing in particular the definitions of the above used notions, can be found in a survey paper of Komlós and Simonovits [55].

The reader may have gotten the feeling that basically every property of random graphs $G(n, p)$ ensures weak pseudo-randomness. This feeling is quite misleading, and one should be careful while formulating properties equivalent to pseudo-randomness. Here is an example provided by Chung et al. Let G be a graph with vertex set $\{1, \ldots, 4n\}$ defined as follows: the subgraph of G spanned by the first $2n$ vertices is a complete bipartite graph $K_{n,n}$, the subgraph spanned by the last $2n$ vertices is the complement of $K_{n,n}$, and for every pair $(i, j), 1 \leq i \leq 2n, 2n + 1 \leq j \leq 4n$, the edge (i, j) is present in G independently with probability 0.5. Then G is almost surely a graph on $4n$ vertices with edge density 0.5. One can verify that G has properties $P_1(3)$ and $P_2(2t + 1)$ for every $t \geq 1$, but is obviously very far from being pseudo-random (contains a clique and an independent set of one quarter of its size). Hence $P_1(3)$ and $P_2(2t + 1)$ are not pseudo-random properties. This example shows also the real difference between even and odd cycles in this context – recall that Property $P_2(2t)$ does imply pseudo-randomness.

A possible explanation to the above described somewhat disturbing phenomenon has been suggested by Simonovits and Sós in [74]. They noticed that the above discussed properties are not hereditary in the sense that the fact that the whole graph G possesses one of these properties does not imply that large induced subgraphs of G also have it. A property is called *hereditary* in this context if it is assumed to hold for all sufficiently large subgraphs F of our graph G with the same error term as for G. Simonovits and Sós proved that adding this hereditary condition gives significant extra strength to many properties making them pseudo-random.

Theorem 2.7 [74]. *Let L be a fixed graph on l vertices, and let $p \in (0,1)$ be fixed. Let (G_n) be a sequence of graphs. If for every induced subgraph $H \subseteq G$ on h vertices,*

$$N_H(L) = p^{|E(L)|} h^l + o(n^l),$$

then (G_n) is weakly pseudo-random, i.e. property P_4 holds.

Two main distinctive features of the last result compared to Theorem 2.6 are: (a) $P_1(3)$ assumed hereditarily implies pseudo-randomness; and (b) requiring the right number of copies of a *single* graph L on l vertices is enough, compared to Condition $P_1(l)$ required to hold for *all* graphs on l vertices simultaneously.

Let us switch now to the case of vanishing edge density $p(n) = o(1)$. This case has been treated in two very recent papers of Chung and Graham [25] and of Kohayakawa, Rödl and Sissokho [50]. Here the picture becomes significantly more complicated compared to the dense case. In particular, there exist graphs with very balanced edge distribution not containing a single copy of some fixed subgraphs (see the Erdős–Rényi graph and the Alon graph in the next section (Examples 6, 9, resp.)).

In an attempt to find properties equivalent to weak pseudo-randomness in the sparse case, Chung and Graham define the following properties in [25]:

CIRCUIT(t): The number of closed walks $w_0, w_1, \ldots, w_t = w_0$ of length t in G is $\left(1 + o(1)\right) (np)^t$;

CYCLE(t): The number of labeled t-cycles in G is $\left(1 + o(1)\right) (np)^t$;

EIG: The eigenvalues λ_i, $|\lambda_1| \geq |\lambda_2| \geq \ldots |\lambda_n|$, of the adjacency matrix of G satisfy:

$$\lambda_1 = \left(1 + o(1)\right) np,$$

$$|\lambda_i| = o(np), i > 1.$$

DISC: For all $X, Y \subset V(G)$,

$$\left| e(X, Y) - p|X||Y| \right| = o(pn^2).$$

(DISC here is in fact DICS(1) in [25]).

Theorem 2.8 [25]. *Let* $(G = G_n : n \to \infty)$ *be a sequence of graphs with* $e(G_n) = \left(1 + o(1)\right) p\binom{n}{2}$. *Then the following implications hold for all* $t \geq 1$:

$$CIRCUIT(2t) \Rightarrow EIG \Rightarrow DISC.$$

Proof. To prove the first implication, let A be the adjacency matrix of G, and consider the trace $Tr(A^{2t})$. The (i,i)-entry of A^{2t} is equal to the number of closed walks of length $2t$ starting and ending at i, and hence $Tr(A^{2t}) = \left(1 + o(1)\right) (np)^{2t}$. On the other hand, since A is symmetric it is similar to the diagonal matrix $D = diag(\lambda_1, \lambda_2, \ldots, \lambda_n)$, and therefore $Tr(A^{2t}) = \sum_{i=1}^{2t} \lambda_i^{2t}$. We obtain:

$$\sum_{i=1}^{n} \lambda_i^{2t} = \left(1 + o(1)\right) (np)^{2t}.$$

Since the first eigenvalue of G is easily shown to be as large as its average degree, it follows that $\lambda_1 \geq 2|E(G)|/|V(G)| = \left(1 + o(1)\right) np$. Combining these two facts we derive that $\lambda_1 = \left(1 + o(1)\right) np$ and $|\lambda_i| = o(np)$ as required.

The second implication will be proven in the next subsection. ∎

Both reverse implications are false in general. To see why $DISC \not\Rightarrow EIG$ take a graph G_0 on $n - 1$ vertices with all degrees equal to $\left(1 + o(1)\right) n^{0.1}$ and having property $DISC$ (see next section for examples of such graphs). Now add to G_0 a vertex v^* and connect it to any set of size $n^{0.8}$ in G_0, let G be the obtained graph. Since G is obtained from G_0 by adding $o\left(|E(G_0)|\right)$ edges, G still satisfies $DISC$. On the other hand, G contains a star S of size $n^{0.8}$ with a center at v^*, and hence $\lambda_1(G) \geq \lambda_1(S) = \sqrt{n^{0.8} - 1} \gg |E(G)|/n$ (see, e.g. Chapter 11 of [64] for the relevant proofs). This solves an open question from [25].

The Erdős–Rényi graph from the next section is easily seen to satisfy EIG, but fails to satisfy $CIRCUIT(4)$. Chung and Graham provide an alternative example in [25] (Example 1).

The above discussion indicates that one probably needs to impose some additional condition on the graph G to glue all these pieces together and to make the above stated properties equivalent. One such condition has been suggested by Chung and Graham who defined:

U(t): For some absolute constant c, all degrees in G satisfy: $d(v) < cnp$, and for every pair of vertices $x, y \in G$ the number $e_{t-1}(x,y)$ of walks of length $t - 1$ from x to y satisfies: $e_{t-1}(x,y) \leq cn^{t-2}p^{t-1}$.

Notice that $U(t)$ can only hold for $p > c'n^{-1+1/(t-1)}$, where c' depends on c. Also, every dense graph ($p = \Theta(1)$) satisfies $U(t)$.

As it turns out adding property $U(t)$ makes all the above defined properties equivalent and thus equivalent to the notion of weak pseudo-randomness (that can be identified with property $DISC$):

Theorem 2.9 [25]. *Suppose for some constant $c > 0$, $p(n) > cn^{-1+1/(t-1)}$, where $t \geq 2$. For any family of graphs G_n, $\left| E(G_n) \right| = \left(1 + o(1)\right) p\binom{n}{2}$, satisfying $U(t)$, the following properties are all equivalent: $CIRCUIT(2t)$, $CYCLE(2t)$, EIG and $DISC$.*

Theorem 2.9 can be viewed as a sparse analog of Theorem 2.6 as it also provides a list of conditions equivalent to weak pseudo-randomness.

Further properties implying or equivalent to pseudo-randomness, including local statistics conditions, are given in [50].

2.4. Eigenvalues and pseudo-random graphs

In this subsection we describe an approach to pseudo-randomness based on graph eigenvalues – the approach most frequently used in this survey. Although the eigenvalue-based condition is not as general as the jumbledness condition of Thomason or some other properties described in the previous subsection, its power and convenience are so appealing that they certainly constitute a good enough reason to prefer this approach. Below we first provide a necessary background on graph spectra and then derive quantitative estimates connecting the eigenvalue gap and edge distribution.

Recall that the *adjacency matrix* of a graph $G = (V, E)$ with vertex set $V = \{1, \ldots, n\}$ is an n-by-n matrix whose entry a_{ij} is 1 if $(i, j) \in E(G)$, and is 0 otherwise. Thus A is a $0, 1$ symmetric matrix with zeroes along the main diagonal, and we can apply the standard machinery of eigenvalues and eigenvectors of real symmetric matrices. It follows that all eigenvalues of A (usually also called the eigenvalues of the graph G itself) are real, and we denote them by $\lambda_1 \geq \lambda_2 \geq \ldots \geq \lambda_n$. Also, there is an orthonormal basis $B = \{x_1, \ldots, x_n\}$ of the euclidean space R^n composed of eigenvectors of A: $Ax_i = \lambda_i x_i$, $x_i^t x_i = 1$, $i = 1, \ldots, n$. The matrix A can be decomposed then as: $A = \sum_{i=1}^{n} \lambda_i x_i x_i^t$ – the so called spectral decomposition of A. (Notice that the product xx^t, $x \in R^n$, is an n-by-n matrix of rank 1; if $x, y, z \in R^n$ then $y^t(xx^t)z = (y^t x)(x^t z)$). Every vector $y \in R^n$ can be

easily represented in basis B: $y = \sum_{i=1}^{n} (y^t x_i) x_i$. Therefore, for $y, z \in R^n$,
$y^t z = \sum_{i=1}^{n} (y^t x_i)(z^t x_i)$ and $\|y\|^2 = y^t y = \sum_{i=1}^{n} (y^t x_i)^2$.

All the above applies in fact to all real symmetric matrices. Since the adjacency matrix A of a graph G is a matrix with non-negative entries, one can derive some important extra features of A, most notably the Perron–Frobenius Theorem, that reads in the graph context as follows: if G is connected then the multiplicity of λ_1 is one, all coordinates of the first eigenvector x_1 can be assumed to be strictly positive, and $|\lambda_i| \leq \lambda_1$ for all $i \geq 2$. Thus, graph spectrum lies entirely in the interval $[-\lambda_1, \lambda_1]$.

For the most important special case of regular graphs Perron–Frobenius implies the following corollary:

Proposition 2.10. *Let G be a d-regular graph on n vertices. Let $\lambda_1 \geq \lambda_2 \geq \ldots \geq \lambda_n$ be the eigenvalues of G. Then $\lambda_1 = d$ and $-d \leq \lambda_i \leq d$ for all $1 \leq i \leq n$. Moreover, if G is connected then the first eigenvector x_1 is proportional to the all one vector $(1, \ldots, 1)^t \in R^n$, and $\lambda_i < d$ for all $i \geq 2$.*

To derive the above claim from the Perron–Frobenius Theorem observe that $e = (1, \ldots, 1)$ is immediately seen to be an eigenvector of $A(G)$ corresponding to the eigenvalue d: $Ae = de$. The positivity of the coordinates of e implies then that e is not orthogonal to the first eigenvector, and hence is in fact proportional to x_1 of $A(G)$. Proposition 2.10 can be also proved directly without relying on the Perron–Frobenius Theorem.

We remark that $\lambda_n = -d$ is possible, in fact it holds if and only if the graph G is bipartite.

All this background information, presented above in a somewhat condensed form, can be found in many textbooks in Linear Algebra. Readers more inclined to consult combinatorial books can find it for example in a recent monograph of Godsil and Royle on Algebraic Graph Theory [46].

We now prove a well known theorem (see its variant, e.g., in Chapter 9, [18]) bridging between graph spectra and edge distribution.

Theorem 2.11. *Let G be a d-regular graph on n vertices. Let $d = \lambda_1 \geq \lambda_2 \geq \ldots \lambda_n$ be the eigenvalues of G. Denote*

$$\lambda = \max_{2 \leq i \leq n} |\lambda_i|.$$

Then for every two subsets $U, W \subset V$,

$$(4) \qquad \left| e(U, W) - \frac{d|U||W|}{n} \right| \leq \lambda \sqrt{|U||W| \left(1 - \frac{|U|}{n}\right)\left(1 - \frac{|W|}{n}\right)}.$$

Proof. Let $B = \{x_1, \ldots, x_n\}$ be an orthonormal basis of R^n composed from eigenvectors of A: $Ax_i = \lambda_i x_i$, $1 \le i \le n$. We represent $A = \sum_{i=1}^{n} \lambda_i x_i x_i^t$. Denote

$$A_1 = \lambda_1 x_1 x_1^t,$$

$$\mathcal{E} = \sum_{i=2}^{n} \lambda_i x_i x_i^t,$$

then $A = A_1 + \mathcal{E}$.

Let $u = |U|$, $w = |W|$ be the cardinalities of U, W, respectively. We denote the characteristic vector of U by $\chi_U \in R^n$, i.e. $\chi_U(i) = 1$ if $i \in U$, and $\chi_U(i) = 0$ otherwise. Similarly, let $\chi_W \in R^n$ be the characteristic vector of W. We represent χ_U, χ_W according to B:

$$\chi_U = \sum_{i=1}^{n} \alpha_i x_i, \quad \alpha_i = \chi_U^t x_i, \quad \sum_{i=1}^{n} \alpha_i^2 = \|\chi_U\|^2 = u,$$

$$\chi_W = \sum_{i=1}^{n} \beta_i x_i, \quad \beta_i = \chi_W^t x_i, \quad \sum_{i=1}^{n} \beta_i^2 = \|\chi_W\|^2 = w.$$

It follows easily from the definitions of A, χ_U and χ_W that the product $\chi_U^t A \chi_W$ counts exactly the number of edges of G with one endpoint in U and the other one in W, i.e.

$$e(U, W) = \chi_U^t A \chi_W = \chi_U^t A_1 \chi_W + \chi_U^t \mathcal{E} \chi_W.$$

Now we estimate the last two summands separately, the first of them will be the main term for $e(U, W)$, the second one will be the error term. Substituting the expressions for χ_U, χ_W and recalling the orthonormality of B, we get:

$$(5) \qquad \chi_U^t A_1 \chi_W = \left(\sum_{i=1}^{n} \alpha_i x_i \right)^t (\lambda_1 x_1 x_1^t) \left(\sum_{j=1}^{n} \beta_j x_j \right)$$

$$= \sum_{i=1}^{n} \sum_{j=1}^{n} \alpha_i \lambda_1 \beta_j (x_i^t x_1)(x_1^t x_j) = \alpha_1 \beta_1 \lambda_1.$$

Similarly,

$$(6) \quad \chi_U^t \mathcal{E} \chi_W = \left(\sum_{i=1}^{n} \alpha_i x_i \right)^t \left(\sum_{j=2}^{n} \lambda_j x_j x_j^t \right) \left(\sum_{k=1}^{n} \beta_k x_k \right) = \sum_{i=2}^{n} \alpha_i \beta_i \lambda_i.$$

Recall now that G is d-regular. Then according to Proposition 2.10, $\lambda_1 = d$ and $x_1 = \frac{1}{\sqrt{n}}(1,\ldots,1)^t$. We thus get: $\alpha_1 = \chi_U^t x_1 = u/\sqrt{n}$ and $\beta_1 = \chi_W^t x_1 = w/\sqrt{n}$. Hence it follows from (5) that $\chi_U^t A_1 \chi_W = duw/n$.

Now we estimate the absolute value of the error term $\chi_U^t \mathcal{E} \chi_W$. Recalling (6), the definition of λ and the obtained values of α_1, β_1, we derive, applying Cauchy–Schwartz:

$$|\chi_U^t \mathcal{E} \chi_W| = \left| \sum_{i=2}^n \alpha_i \beta_i \lambda_i \right| \leq \lambda \left| \sum_{i=2}^n \alpha_i \beta_i \right| \leq \lambda \sqrt{\sum_{i=2}^n \alpha_i^2 \sum_{i=2}^n \beta_i^2}$$

$$= \lambda \sqrt{\left(\|\chi_U\|^2 - \alpha_1^2 \right) \left(\|\chi_W\|^2 - \beta_1^2 \right)} = \lambda \sqrt{\left(u - \frac{u^2}{n} \right) \left(w - \frac{w^2}{n} \right)}.$$

The theorem follows. ∎

The above proof can be extended to the irregular (general) case. Since the obtained quantitative bounds on edge distribution turn out to be somewhat cumbersome, we will just indicate how they can be obtained. Let $G = (V, E)$ be a graph on n vertices with *average* degree d. Assume that the eigenvalues of G satisfy $\lambda < d$, with λ as defined in the theorem. Denote

$$K = \sum_{v \in V} \left(d(v) - d \right)^2.$$

The parameter K is a measure of irregularity of G. Clearly $K = 0$ if and only if G is d-regular. Let $e = \frac{1}{\sqrt{n}}(1,\ldots,1)^t$. We represent e in the basis $B = \{x_1,\ldots,x_n\}$ of the eigenvectors of $A(G)$:

$$e = \sum_{i=1}^n \gamma_i x_i, \quad \gamma_i = e^t x_i, \quad \sum_{i=1}^n \gamma_i^2 = \|e\|^2 = 1.$$

Denote $z = \frac{1}{\sqrt{n}}\left(d(v_1) - d,\ldots,d(v_n) - d\right)^t$, then $\|z\|^2 = K/n$. Notice that $Ae = \frac{1}{\sqrt{n}}\left(d(v_1),\ldots,d(v_n)\right)^t = de + z$, and therefore $z = Ae - de = \sum_{i=1}^n \gamma_i(\lambda_i - d)x_i$. This implies:

$$\frac{K}{n} = \|z\|^2 = \sum_{i=1}^n \gamma_i^2(\lambda_i - d)^2 \geq \sum_{i=2}^n \gamma_i^2(\lambda_i - d)^2$$

$$\geq (d - \lambda)^2 \sum_{i=2}^n \gamma_i^2.$$

Hence $\sum_{i=2}^{n} \gamma_i^2 \leq \frac{K}{n(d-\lambda)^2}$. It follows that $\gamma_1^2 = 1 - \sum_{i=2}^{n} \gamma_i^2 \geq 1 - \frac{K}{n(d-\lambda)^2}$ and

$$\gamma_1 \geq \gamma_1^2 \geq 1 - \frac{K}{n(d-\lambda)^2}.$$

Now we estimate the distance between the vectors e and x_1 and show that they are close given that the parameter K is small.

$$\|e - x_1\|^2 = (e - x_1)^t(e - x_1) = e^t e + x_1^t x_1 - 2e^t x_1 = 1 + 1 - 2\gamma_1 = 2 - 2\gamma_1$$

$$\leq \frac{2K}{n(d-\lambda)^2}.$$

We now return to expressions (5) and (6) from the proof of Theorem 2.11. In order to estimate the main term $\chi_U^t A_1 \chi_W$, we bound the coefficients α_1, β_1 and λ_1 as follows:

$$\alpha_1 = \chi_U^t x_1 = \chi_U^t e + \chi_U^t(x_1 - e) = \frac{u}{\sqrt{n}} + \chi_U^t(x_1 - e),$$

and therefore

$$(7) \qquad \left| \alpha_1 - \frac{u}{\sqrt{n}} \right| = |\chi_U^t(x_1 - e)| \leq \|\chi_U\| \cdot \|x_1 - e\| \leq \frac{\sqrt{\frac{2Ku}{n}}}{d - \lambda}.$$

In a similar way one gets:

$$(8) \qquad \left| \beta_1 - \frac{w}{\sqrt{n}} \right| \leq \frac{\sqrt{\frac{2Kw}{n}}}{d - \lambda}.$$

Finally, to estimate from above the absolute value of the difference between λ_1 and d we argue as follows:

$$\frac{K}{n} = \|z\|^2 = \sum_{i=1}^{n} \gamma_i^2 (\lambda_i - d)^2 \geq \gamma_1^2 (\lambda_1 - d)^2,$$

and therefore

$$(9) \qquad |\lambda_1 - d| \leq \frac{1}{\gamma_1} \sqrt{\frac{K}{n}} \leq \frac{n(d-\lambda)^2}{n(d-\lambda)^2 - K} \sqrt{\frac{K}{n}}.$$

Summarizing, we see from (7), (8) and (9) that the main term in the product $\chi_U^t A_1 \chi_W$ is equal to $\frac{duw}{n}$, just as in the regular case, and the error term is governed by the parameter K.

In order to estimate the error term $\chi_U^t \mathcal{E} \chi_W$ we use (6) to get:

$$\left| \chi_U^t \mathcal{E} \chi_W \right| = \left| \sum_{i=2}^{n} \alpha_i \beta_i \lambda_i \right| \leq \lambda \left| \sum_{i=2}^{n} \alpha_i \beta_i \right| \leq \lambda \sqrt{\sum_{i=2}^{n} \alpha_i^2 \sum_{i=2}^{n} \beta_i^2}$$

$$\leq \lambda \sqrt{\sum_{i=1}^{n} \alpha_i^2 \sum_{i=1}^{n} \beta_i^2} = \lambda \|\chi_U\| \, \|\chi_W\| = \lambda \sqrt{uw}. \qquad \blacksquare$$

Applying the above developed techniques we can prove now the second implication of Theorem 2.8. Let us prove first that EIG implies $K = o(nd^2)$, where $d = (1 + o(1)) np$ is as before the average degree of G. Indeed, for every vector $v \in R^n$ we have $\|Av\| \leq \lambda_1 \|v\|$, and therefore

$$\lambda_1^2 n = \lambda_1^2 e^t e \geq (Ae)^t (Ae) = \sum_{v \in V} d^2(v).$$

Hence from EIG we get: $\sum_{v \in V} d^2(v) \leq (1 + o(1)) nd^2$. As $\sum_v d(v) = nd$, it follows that:

$$K = \sum_{v \in V} \left(d(v) - d \right)^2 = \sum_{v \in V} d^2(v) - 2d \sum_{v \in V} d(v) + nd^2$$

$$= (1 + o(1)) nd^2 - 2nd^2 + nd^2 = o(nd^2),$$

as promised. Substituting this into estimates (7), (8), (9) and using $\lambda = o(d)$ of EIG we get:

$$\alpha_1 = \frac{u}{\sqrt{n}} + o(\sqrt{u}),$$

$$\beta_1 = \frac{w}{\sqrt{n}} + o(\sqrt{w}),$$

$$\lambda_1 = (1 + o(1)) d,$$

and therefore

$$\chi_U^t A_1 \chi_W = \frac{duw}{n} + o(dn).$$

Also, according to EIG, $\lambda = o(d)$, which implies:

$$\chi_U^t \mathcal{E} \chi_w = o(d\sqrt{uw}) = o(dn),$$

and the claim follows. $\blacksquare$

Theorem 2.11 is a truly remarkable result. Not only it connects between two seemingly unrelated graph characteristics – edge distribution and spectrum, it also provides a very good quantitative handle for the uniformity of edge distribution, based on easily computable, both theoretically and practically, graph parameters – graph eigenvalues. According to the bound (4), a polynomial number of parameters can control quite well the number of edges in exponentially many subsets of vertices.

The parameter λ in the formulation of Theorem 2.11 is usually called the *second eigenvalue* of the d-regular graph G (the first and the trivial one being $\lambda_1 = d$). There is certain inaccuracy though in this term, as in fact $\lambda = \max\{\lambda_2, -\lambda_n\}$. Later we will call, following Alon, a d-regular graph G on n vertices in which all eigenvalues, but the first one, are at most λ in their absolute values, an (n, d, λ)-*graph*.

Comparing (4) with the definition of jumbled graphs by Thomason we see that an (n, d, λ)-graph G is $(d/n, \lambda)$-jumbled. Hence the parameter λ (or in other words, the so called *spectral gap* – the difference between d and λ) is responsible for pseudo-random properties of such a graph. The smaller the value of λ compared to d, the more close is the edge distribution of G to the ideal uniform distribution. A natural question is then: how small can be λ? It is easy to see that as long as $d \leq (1 - \varepsilon)n$, $\lambda = \Omega(\sqrt{d})$. Indeed, the trace of A^2 satisfies:

$$nd = 2|E(G)| = Tr(A^2) = \sum_{i=1}^{n} \lambda_i^2 \leq d^2 + (n-1)\lambda_2 \leq (1-\varepsilon)nd + (n-1)\lambda^2,$$

and $\lambda = \Omega(\sqrt{d})$ as claimed. More accurate bounds are known for smaller values of d (see, e.g. [69]). Based on these estimates we can say that an (n, d, λ)-graph G, for which $\lambda = \Theta(\sqrt{d})$, is a very good pseudo-random graph. We will see several examples of such graphs in the next section.

2.5. Strongly regular graphs

A *strongly regular graph* $srg(n, d, \eta, \mu)$ is a d-regular graph on n vertices in which every pair of adjacent vertices has exactly η common neighbors and every pair of non-adjacent vertices has exactly μ common neighbors. (We changed the very standard notation in the above definition so as to avoid interference with other notational conventions throughout this paper and to make it more coherent, usually the parameters are denoted (v, k, λ, μ)).

Two simple examples of strongly regular graph are the pentagon C_5 that has parameters $(5, 2, 0, 1)$, and the Petersen graph whose parameters are $(10, 3, 0, 1)$. Strongly regular graphs were introduced by Bose in 1963 [21] who also pointed out their tight connections with finite geometries. As follows from the definition, strongly regular graphs are highly regular structures, and one can safely predict that algebraic methods are extremely useful in their study. We do not intend to provide any systematic coverage of this fascinating concept here, addressing the reader to the vast literature on the subject instead (see, e.g., [24]). Our aim here is to calculate the eigenvalues of strongly regular graphs and then to connect them with pseudo-randomness, relying on results from the previous subsection.

Proposition 2.12. *Let G be a connected strongly regular graph with parameters (n, d, η, μ). Then the eigenvalues of G are: $\lambda_1 = d$ with multiplicity $s_1 = 1$,*

$$\lambda_2 = \frac{1}{2}\left(\eta - \mu + \sqrt{(\eta - \mu)^2 + 4(d - \mu)}\right)$$

and

$$\lambda_3 = \frac{1}{2}\left(\eta - \mu - \sqrt{(\eta - \mu)^2 + 4(d - \mu)}\right),$$

with multiplicities

$$s_2 = \frac{1}{2}\left(n - 1 + \frac{(n-1)(\mu - \eta) - 2d}{\sqrt{(\mu - \eta)^2 + 4(d - \mu)}}\right)$$

and

$$s_3 = \frac{1}{2}\left(n - 1 - \frac{(n-1)(\mu - \eta) - 2d}{\sqrt{(\mu - \eta)^2 + 4(d - \mu)}}\right),$$

respectively.

Proof. Let A be the adjacency matrix of A. By the definition of A and the fact that A is symmetric with zeroes on the main diagonal, the (i, j)-entry of the square A^2 counts the number of common neighbors of v_i and v_j in G if $i \neq j$, and is equal to the degree $d(v_i)$ in case $i = j$. The statement that G is $srg(n, d, \eta, \mu)$ is equivalent then to:

(10) $AJ = dJ,$ $A^2 = (d - \mu)I + \mu J + (\eta - \mu)A,$

where J is the n-by-n all-one matrix and I is the n-by-n identity matrix.

Since G is d-regular and connected, we obtain from the Perron–Frobenius Theorem that $\lambda_1 = d$ is an eigenvalue of G with multiplicity 1 and with $e = (1, \ldots, 1)^t$ as the corresponding eigenvector. Let $\lambda \neq d$ be another eigenvalue of G, and let $x \in R^n$ be a corresponding eigenvector. Then x is orthogonal to e, and therefore $Jx = 0$. Applying both sides of the second identity in (10) to x we get the equation: $\lambda^2 x = (d - \mu)x + (\eta - \mu)\lambda x$, which results in the following quadratic equation for λ:

$$\lambda^2 + (\mu - \eta)\lambda + (\mu - d) = 0.$$

This equation has two solutions λ_2 and λ_3 as defined in the proposition formulation. If we denote by s_2 and s_3 the respective multiplicities of λ_2 and λ_3 as eigenvalues of A, we get:

$$1 + s_2 + s_3 = n, \qquad Tr(A) = d + s_2\lambda_2 + s_3\lambda_3 = 0.$$

Solving the above system of linear equations for s_2 and s_3 we obtain the assertion of the proposition. ∎

Using the bound (4) we can derive from the above proposition that if the parameters of a strongly regular graph G satisfy $\eta \approx \mu$ then G has a large eigenvalue gap and is therefore a good pseudo-random graph. We will exhibit several examples of such graphs in the next section.

3. EXAMPLES

Here we present some examples of pseudo-random graphs. Many of them are well known and already appeared, e.g., in [79] and [80], but there also some which have been discovered only recently. Since in the rest of the paper we will mostly discuss properties of (n, d, λ)-graphs, in our examples we emphasize the spectral properties of the constructed graphs. We will also use most of these constructions later to illustrate particular points and to test the strength of the theorems.

Random graphs.

1. Let $G = G(n, p)$ be a random graph with edge probability p. If p satisfies $pn/\log n \to \infty$ and $(1-p)n\log n \to \infty$, then almost surely all the degrees of G are equal to $(1 + o(1))np$. Moreover it was proved

by Füredi and Komlós [44] that the largest eigenvalue of G is a.s. $(1+o(1)) np$ and that $\lambda(G) \leq (2+o(1)) \sqrt{p(1-p)n}$. They stated this result only for constant p but their proof shows that $\lambda(G) \leq O(\sqrt{np})$ also when $p \geq poly \log n/n$.

2. For a positive integer-valued function $d = d(n)$ we define the model $G_{n,d}$ of random regular graphs consisting of all regular graphs on n vertices of degree d with the uniform probability distribution. This definition of a random regular graph is conceptually simple, but it is not easy to use. Fortunately, for small d there is an efficient way to generate $G_{n,d}$ which is useful for theoretical studies. This is the so called *configuration model*. For more details about this model, and random regular graphs in general we refer the interested reader to two excellent monographs [20] and [49], or to a survey [83]. As it turns out, sparse random regular graphs have quite different properties from those of the binomial random graph $G(n,p), p = d/n$. For example, they are almost surely connected. The spectrum of $G_{n,d}$ for a fixed d was studied in [38] by Friedman, Kahn and Szemerédi. Friedman [39] proved that for constant d the second largest eigenvalue of a random d-regular graph is $\lambda = (1+o(1)) 2\sqrt{d-1}$. The approach of Kahn and Szemerédi gives only $O(\sqrt{d})$ bound on λ but continues to work also when d is small power of n. The case $d \gg n^{1/2}$ was recently studied by Krivelevich, Sudakov, Vu and Wormald [61]. They proved that in this case for any two vertices $u, v \in G_{n,d}$ almost surely

$$\left| codeg(u,v) - d^2/n \right| < Cd^3/n^2 + 6d\sqrt{\log n}/\sqrt{n},$$

where C is some constant and $codeg(u,v)$ is the number of common neighbors of u, v. Moreover if $d \geq n/\log n$, then C can be defined to be zero. Using this it is easy to show that for $d \gg n^{1/2}$, the second largest eigenvalue of a random d-regular graph is $o(d)$. The true bound for the second largest eigenvalue of $G_{n,d}$ should be probably $(1+o(1)) 2\sqrt{d-1}$ for all values of d, but we are still far from proving it.

Strongly regular graphs.

3. Let $q = p^\alpha$ be a prime power which is congruent to 1 modulo 4 so that -1 is a square in the finite field $GF(q)$. Let P_q be the graph whose vertices are all elements of $GF(q)$ and two vertices are adjacent if and only if their difference is a quadratic residue in $GF(q)$. This

graph is usually called the *Paley graph*. It is easy to see that P_q is $(q-1)/2$-regular. In addition one can easily compute the number of common neighbors of two vertices in P_q. Let χ be the *quadratic residue character* on $GF(q)$, i.e., $\chi(0) = 0$, $\chi(x) = 1$ if $x \neq 0$ and is a square in $GF(q)$ and $\chi(x) = -1$ otherwise. By definition, $\sum_x \chi(x) = 0$ and the number of common neighbors of two vertices a and b equals

$$\sum_{x \neq a,b} \left(\frac{1 + \chi(a-x)}{2} \right) \left(\frac{1 + \chi(b-x)}{2} \right)$$

$$= \frac{q-2}{4} - \frac{\chi(a-b)}{2} + \frac{1}{4} \sum_{x \neq a,b} \chi(a-x)\chi(b-x).$$

Using that for $x \neq b$, $\chi(b-x) = \chi\big((b-x)^{-1}\big)$, the last term can be rewritten as

$$\sum_{x \neq a,b} \chi(a-x)\chi\big((b-x)^{-1}\big) = \sum_{x \neq a,b} \chi\left(\frac{a-x}{b-x} \right) = \sum_{x \neq a,b} \chi\left(1 + \frac{a-b}{b-x} \right)$$

$$= \sum_{x \neq 0,1} \chi(x) = -1.$$

Thus the number of common neighbors of a and b is $(q-3)/4 - \chi(a-b)/2$. This equals $(q-5)/4$ if a and b are adjacent and $(q-1)/4$ otherwise. This implies that the Paley graph is a strongly regular graph with parameters $\big(q, (q-1)/2, (q-5)/4, (q-1)/4\big)$ and therefore its second largest eigenvalue equals $\big(\sqrt{q}+1\big)/2$.

4. For any odd integer k let H_k denote the graph whose $n_k = 2^{k-1} - 1$ vertices are all binary vectors of length k with an odd number of ones except the all one vector, in which two distinct vertices are adjacent iff the inner product of the corresponding vectors is 1 modulo 2. Using elementary linear algebra it is easy to check that this graph is $(2^{k-2} - 2)$-regular. Also every two nonadjacent vertices vertices in it have $2^{k-3} - 1$ common neighbors and every two adjacent vertices vertices have $2^{k-3} - 3$ common neighbors. Thus H_k is a strongly regular graph with parameters $\big(2^{k-1} - 1, 2^{k-2} - 2, 2^{k-3} - 3, 2^{k-3} - 1\big)$ and with the second largest eigenvalue $\lambda(H_k) = 1 + 2^{\frac{k-3}{2}}$.

5. Let q be a prime power an let $V(G)$ be the elements of the two dimensional vector space over $GF(q)$, so G has q^2 vertices. Partition

the $q+1$ lines through the origin of the space into two sets P and N, where $|P| = k$. Two vertices x and y of the graph G are adjacent if $x - y$ is parallel to a line in P. This example is due to Delsarte and Goethals and to Turyn (see [72]). It is easy to check that G is strongly regular with parameters $\left(k(q-1), (k-1)(k-2) + q - 2, k(k-1)\right)$. Therefore its eigenvalues, besides the trivial one are $-k$ and $q - k$. Thus if k is sufficiently large we obtain that G is $d = k(q-1)$-regular graph whose second largest eigenvalue is much smaller than d.

Graphs arising from finite geometries.

6. For any integer $t \geq 2$ and for any power $q = p^\alpha$ of prime p let $PG(q,t)$ denote the projective geometry of dimension t over the finite field $GF(q)$. The interesting case for our purposes here is that of large q and fixed t. The vertices of $PG(q,t)$ correspond to the equivalence classes of the set of all non-zero vectors $\mathbf{x} = (x_0, \ldots, x_t)$ of length $t+1$ over $GF(q)$, where two vectors are equivalent if one is a multiple of the other by an element of the field. Let G denote the graph whose vertices are the points of $PG(q,t)$ and two (not necessarily distinct) vertices $\mathbf{x}$ and $\mathbf{y}$ are adjacent if and only if $x_0 y_0 + \ldots + x_t y_t = 0$. This construction is well known. In particular, in case $t = 2$ this graph is often called the Erdős–Rényi graph and it contains no cycles of length 4. It is easy to see that the number of vertices of G is $n_{q,t} = \left(q^{t+1} - 1\right)/(q-1) = \left(1 + o(1)\right) q^t$ and that it is $d_{q,t}$-regular for $d_{q,t} = (q^t - 1)/(q-1) = \left(1 + o(1)\right) q^{t-1}$, where $o(1)$ tends to zero as q tends to infinity. It is easy to see that the number of vertices of G with loops is bounded by $2(q^t - 1)/(q-1) = \left(2 + o(1)\right) q^{t-1}$, since for every possible value of $x_0, \ldots, x_{t-1}$ we have at most two possible choices of x_t. Actually using more complicated computation, which we omit, one can determine the exact number of vertices with loops. The eigenvalues of G are easy to compute (see [11]). Indeed, let A be the adjacency matrix of G. Then, by the properties of $PG(q,t)$, $A^2 = AA^T = \mu J + (d_{q,t} - \mu)I$, where $\mu = \left(q^{t-1} - 1\right)/(q-1)$, J is the all one matrix and I is the identity matrix, both of size $n_{q,t} \times n_{q,t}$. Therefore the largest eigenvalue of A is $d_{q,t}$ and the absolute value of all other eigenvalues is $\sqrt{d_{q,t} - \mu} = q^{(t-1)/2}$.

7. The generalized polygons are incidence structures consisting of points $\mathcal{P}$ and lines $\mathcal{L}$. For our purposes we restrict our attention to those in which every point is incident to $q + 1$ lines and every line is incident

to $q+1$ points. A generalized m-gon defines a bipartite graph G with bipartition $(\mathcal{P}, \mathcal{L})$ that satisfies the following conditions. The diameter of G is m and for every vertex $v \in G$ there is a vertex $u \in G$ such that the shortest path from u to v has length m. Also for every $r < m$ and for every two vertices u, v at distance r there exists a unique path of length r connecting them. This immediately implies that every cycle in G has length at least $2m$. For $q \geq 2$, it was proved by Feit and Higman [36] that $(q+1)$-regular generalized m-gons exist only for $m = 3, 4, 6$. A *polarity* of G is a bijection $\pi : \mathcal{P} \cup \mathcal{L} \to \mathcal{P} \cup \mathcal{L}$ such that $\pi(\mathcal{P}) = \mathcal{L}$, $\pi(\mathcal{L}) = \mathcal{P}$ and π^2 is the identity map. Also for every $p \in \mathcal{P}, l \in \mathcal{L}$, $\pi(p)$ is adjacent to $\pi(l)$ if and only if p and l are adjacent. Given π we define a polarity graph G^π to be the graph whose vertices are point in $\mathcal{P}$ and two (not necessarily distinct) points p_1, p_2 are adjacent iff p_1 was adjacent to $\pi(p_2)$ in G. Some properties of G^π can be easily deduced from the corresponding properties of G. In particular, G^π is $(q+1)$-regular and also contains no even cycles of length less than $2m$.

For every q which is an odd power of 2, the incidence graph of the generalized 4-gon has a polarity. The corresponding polarity graph is a $(q+1)$-regular graph with $q^3 + q^2 + q + 1$ vertices. See [23], [62] for more details. This graph contains no cycle of length 6 and it is not difficult to compute its eigenvalues (they can be derived, for example, from the eigenvalues of the corresponding bipartite incidence graph, given in [78]). Indeed, all the eigenvalues, besides the trivial one (which is $q+1$) are either 0 or $\sqrt{2q}$ or $-\sqrt{2q}$. Similarly, for every q which is an odd power of 3, the incidence graph of the generalized 6-gon has a polarity. The corresponding polarity graph is a $(q+1)$-regular graph with $q^5 + q^4 + \cdots + q + 1$ vertices (see again [23], [62]). This graph contains no cycle of length 10 and its eigenvalues can be derived using the same technique as in case of the 4-gon. All these eigenvalues, besides the trivial one are either $\sqrt{3q}$ or $-\sqrt{3q}$ or $\sqrt{q}$ or $-\sqrt{q}$.

Cayley graphs.

8. Let G be a finite group and let S be a set of non-identity elements of G such that $S = S^{-1}$, i.e., for every $s \in S$, s^{-1} also belongs to S. The *Cayley graph* $\Gamma(G, S)$ of this group with respect to the generating set S is the graph whose set of vertices is G and where two vertices g and g' are adjacent if and only if $g'g^{-1} \in S$. Clearly, $\Gamma(G, S)$ is $|S|$-regular

and it is connected iff S is a set of generators of the group. If G is abelian then the eigenvalues of the Cayley graph can be computed in terms of the characters of G. Indeed, let $\chi : G \to C$ be a character of G and let A be the adjacency matrix of $\Gamma(G, S)$ whose rows and columns are indexed by the elements of G. Consider the vector $\mathbf{v}$ defined by $\mathbf{v}(g) = \chi(g)$. Then it is easy to check that $A\mathbf{v} = \alpha\mathbf{v}$ with $\alpha = \sum_{s\in S} \chi(s)$. In addition all eigenvalues can be obtained in this way, since every abelian group has exactly $|G|$ different characters which are orthogonal to each other. Using this fact, one can often give estimates on the eigenvalues of $\Gamma(G, S)$ for abelian groups.

One example of a Cayley graph that has already been described earlier is P_q. In that case the group is the additive group of the finite field $GF(q)$ and S is the set of all quadratic residues modulo q. Next we present a slightly more general construction. Let $q = 2kr + 1$ be a prime power and let Γ be a Cayley graph whose group is the additive group of $GF(q)$ and whose generating set is $S = \{x = y^k \mid$ for some $y \in GF(q)\}$. By definition, Γ is $(q - 1)/k$-regular. On the other hand, this graph is not strongly regular unless $k = 2$, when it is the Paley graph. Let χ be a nontrivial additive character of $GF(q)$ and consider the Gauss sum $\sum_{y\in GF(q)} \chi(y^k)$. Using the classical bound $\left| \sum_{y\in GF(q)} \chi(y^k) \right| \leq (k-1)q^{1/2}$ (see e.g. [63]) and the above connection between characters and eigenvalues we can conclude that the second largest eigenvalue of our graph Γ is bounded by $O(q^{1/2})$.

9. Next we present a surprising construction obtained by Alon [3] of a very dense pseudo-random graph that on the other hand is triangle-free. For a positive integer k, consider the finite field $GF(2^k)$, whose elements are represented by binary vectors of length k. If a, b, c are three such vectors, denote by (a, b, c) the binary vector of length $3k$ whose coordinates are those of a, followed by coordinates of b and then c. Suppose that k is not divisible by 3. Let W_0 be the set of all nonzero elements $\alpha \in GF(2^k)$ so that the leftmost bit in the binary representation of α^7 is 0, and let W_1 be the set of all nonzero elements $\alpha \in GF(2^k)$ for which the leftmost bit of α^7 is 1. Since 3 does not divide k, 7 does not divide $2^k - 1$ and hence $|W_0| = 2^{k-1} - 1$ and $|W_1| = 2^{k-1}$, as when α ranges over all nonzero elements of the field so does α^7. Let G_n be the graph whose vertices are all $n = 2^{3k}$ binary vectors of length $3k$, where two vectors $\mathbf{v}$ and $\mathbf{v}'$ are adjacent if and only if there exist $w_0 \in W_0$ and $w_1 \in W_1$ so that

$\mathbf{v} - \mathbf{v}' = (w_0, w_0^3, w_0^5) + (w_1, w_1^3, w_1^5)$, where here powers are computed in the field $GF(2^k)$ and the addition is addition modulo 2. Note that G_n is the Cayley graph of the additive group $\mathbf{Z}_2^{3k}$ with respect to the generating set $S = U_0 + U_1$, where $U_0 = \{ (w_0, w_0^3, w_0^5) \mid w_0 \in W_0 \}$ and U_1 is defined similarly. A well known fact from Coding Theory (see e.g., [66]), which can be proved using the Vandermonde determinant, is that every set of six distinct vectors in $U_0 \cup U_1$ is linearly independent over $GF(2)$. In particular all the vectors in $U_0 + U_1$ are distinct, $S = |U_0| |U_1|$ and hence G_n is $|S| = 2^{k-1}(2^{k-1} - 1)$-regular. The statement that G_n is triangle free is clearly equivalent to the fact that the sum modulo 2 of any set of 3 nonzero elements of S is not a zero-vector. Let $u_0 + u_1$, $u_0' + u_1'$ and $u_0'' + u_1''$ be three distinct element of S, where $u_0, u_0', u_0'' \in U_0$ and $u_1, u_1', u_1'' \in U_1$. By the above discussion, if the sum of these six vectors is zero, then every vector must appear an even number of times in the sequence $(u_0, u_0', u_0'', u_1, u_1', u_1'')$. However, since U_0 and U_1 are disjoint, this is clearly impossible. Finally, as we already mentioned, the eigenvalues of G_n can be computed in terms of characters of $\mathbf{Z}_2^{3k}$. Using this fact together with the Carlitz-Uchiyama bound on the characters of $\mathbf{Z}_2^{3k}$ it was proved in [3] that the second eigenvalue of G_n is bounded by $\lambda \leq 9 \cdot 2^k + 3 \cdot 2^{k/2} + 1/4$.

10. The construction above can be extended in the obvious way as mentioned in [10]. Let $h \geq 1$ and suppose that k is an integer such that $2^k - 1$ is not divisible by $4h + 3$. Let W_0 be the set of all nonzero elements $\alpha \in GF(2^k)$ so that the leftmost bit in the binary representation of α^{4h+3} is 0, and let W_1 be the set of all nonzero elements $\alpha \in GF(2^k)$ for which the leftmost bit of α^{4h+3} is 1. Since $4h + 3$ does not divide $2^k - 1$ we have that $|W_0| = 2^{k-1} - 1$ and $|W_1| = 2^{k-1}$, as when α ranges over all nonzero elements of the field so does α^{4h+3}. Define G to be the Cayley graph of the additive group $\mathbf{Z}_2^{(2h+1)k}$ with respect to the generating set $S = U_0 + U_1$, where $U_0 = \{ (w_0, w_0^3, \ldots, w_0^{4h+1}) \mid w_0 \in W_0 \}$ and U_1 is defined similarly. Clearly, G is a $2^{k-1}(2^{k-1} - 1)$-regular graph on $2^{(2h+1)k}$ vertices. Using methods from [3], one can show that G contains no odd cycle of length $\leq 2h + 1$ and that the second eigenvalue of G is bounded by $O(2^k)$.

11. Now we describe the celebrated expander graphs constructed by Lubotzky, Phillips and Sarnak [65] and independently by Margulis [68]. Let p and q be unequal primes, both congruent to 1 modulo 4 and such that p is a quadratic residue modulo q. As usual de-

note by $PSL(2,q)$ the factor group of the group of two by two matrices over $GF(q)$ with determinant 1 modulo its normal subgroup consisting of the two scalar matrices $\begin{pmatrix} 1 & 0 \\ 0 & 1 \end{pmatrix}$ and $\begin{pmatrix} -1 & 0 \\ 0 & -1 \end{pmatrix}$. The graphs we describe are Cayley graphs of $PSL(2,q)$. A well known theorem of Jacobi asserts that the number of ways to represent a positive integer n as a sum of 4 squares is $8 \sum_{4 \nmid d,\ d|n} d$. This easily implies that there are precisely $p+1$ vectors $\mathbf{a} = (a_0, a_1, a_2, a_3)$, where a_0 is an odd positive integer, a_1, a_2, a_3 are even integers and $a_0^2 + a_1^2 + a_2^2 + a_3^2 = p$. From each such vector construct the matrix M_a in $PSL(2,q)$ where $M_a = \frac{1}{\sqrt{p}} \begin{pmatrix} a_0 + ia_1 & a_2 + ia_3 \\ -a_2 + ia_3 & a_0 - ia_1 \end{pmatrix}$ and i is an integer satisfying $i^2 = -1 \pmod{q}$. Note that, indeed, the determinant of M_a is 1 and that the square root of p modulo q does exist. Let $G^{p,q}$ denote the Cayley graph of $PSL(2,q)$ with respect to these $p+1$ matrices. In [65] it was proved that if $q > 2\sqrt{p}$ then $G^{p,q}$ is a connected $(p+1)$-regular graph on $n = q(q^2-1)/2$ vertices. Its girth is at least $2\log_p q$ and all the eigenvalues of its adjacency matrix, besides the trivial one $\lambda_1 = p+1$, are at most $2\sqrt{p}$ in absolute value. The bound on the eigenvalues was obtained by applying deep results of Eichler and Igusa concerning the Ramanujan conjecture. The graphs $G^{p,q}$ have very good expansion properties and have numerous applications in Combinatorics and Theoretical Computer Science.

12. The *projective norm graphs* $NG_{p,t}$ have been constructed in [17], modifying an earlier construction given in [52]. These graphs **are not** Cayley graphs, but as one will immediately see, their construction has a similar flavor. The construction is the following. Let $t > 2$ be an integer, let p be a prime, let $GF(p)^*$ be the multiplicative group of the field with p elements and let $GF(p^{t-1})$ be the field with p^{t-1} elements. The set of vertices of the graph $NG_{p,t}$ is the set $V = GF(p^{t-1}) \times GF(p)^*$. Two distinct vertices (X, a) and $(Y, b) \in V$ are adjacent if and only if $N(X+Y) = ab$, where the norm N is understood over $GF(p)$, that is, $N(X) = X^{1+p+\cdots+p^{t-2}}$. Note that $|V| = p^t - p^{t-1}$. If (X, a) and (Y, b) are adjacent, then (X, a) and $Y \neq -X$ determine b. Thus $NG_{p,t}$ is a regular graph of degree $p^{t-1} - 1$. In addition, it was proved in [17], that $NG_{p,t}$ contains no complete bipartite graphs $K_{t,(t-1)!+1}$. These graphs can be also defined in the same manner starting with a prime power instead of

the prime p. It is also not difficult to compute the eigenvalues of this graph. Indeed, put $q = p^{t-1}$ and let A be the adjacency matrix of $NG_{p,t}$. The rows and columns of this matrix are indexed by the ordered pairs of the set $GF(q) \times GF(p)^*$. Let ψ be a character of the additive group of $GF(q)$, and let χ be a character of the multiplicative group of $GF(p)$. Consider the vector $\mathbf{v} : GF(q) \times GF(p)^* \mapsto C$ defined by $\mathbf{v}(X, a) = \psi(X)\chi(a)$. Now one can check (see [14], [76] for more details) that the vector $\mathbf{v}$ is an eigenvector of A^2 with eigenvalue $\left| \sum_{Z \in GF(q), Z \neq 0} \psi(Z)\chi(N(Z)) \right|^2$ and that all eigenvalues of A^2 have this form. Set $\chi'(Z) = \chi(N(Z))$ for all nonzero Z in $GF(q)$. Note that as the norm is multiplicative, χ' is a multiplicative character of the large field. Hence the above expression is a square of the absolute value of the Gauss sum and it is well known (see e.g. [31], [20]) that the value of each such square, besides the trivial one (that is, when either ψ or χ' are trivial), is q. This implies that the second largest eigenvalue of $NG_{p,t}$ is $\sqrt{q} = p^{(t-1)/2}$.

4. PROPERTIES OF PSEUDO-RANDOM GRAPHS

We now examine closely properties of pseudo-random graphs, with a special emphasis on (n, d, λ)-graphs. The majority of them are obtained using the estimate (4) of Theorem 2.11, showing again the extreme importance and applicability of the latter result. It is instructive to compare the properties of pseudo-random graphs, considered below, with the analogous properties of random graphs, usually shown to hold by completely different methods. The set of properties we chose to treat here is not meant to be comprehensive or systematic, but quite a few rather diverse graph parameters will be covered.

4.1. Connectivity and perfect matchings

The *vertex-connectivity* of a graph G is the minimum number of vertices that we need to delete to make G disconnected. We denote this parameter by $\kappa(G)$. For random graphs it is well known (see, e.g., [20]) that the vertex-connectivity is almost surely the same as the minimum degree. Recently it was also proved (see [61] and [30]) that random d-regular graphs are d-vertex-connected. For (n, d, λ)-graphs it is easy to show the following.

Theorem 4.1. *Let G be an (n, d, λ)-graph with $d \leq n/2$. Then the vertex-connectivity of G satisfies:*

$$\kappa(G) \geq d - 36\lambda^2/d.$$

Proof. We can assume that $\lambda \leq d/6$, since otherwise there is nothing to prove. Suppose that there is a subset $S \subset V$ of size less than $d - 36\lambda^2/d$ such that the induced graph $G[V - S]$ is disconnected. Denote by U the set of vertices of the smallest connected component of $G[V - S]$ and set $W = V - (S \cup U)$. Then $|W| \geq (n-d)/2 \geq n/4$ and there is no edge between U and W. Also $|U| + |S| > d$, since all the neighbors of a vertex from U are contained in $S \cup U$. Therefore $|U| \geq 36\lambda^2/d$. Since there are no edges between U and W, by Theorem 2.11, we have that $d|U||W|/n < \lambda\sqrt{|U||W|}$. This implies that

$$|U| < \frac{\lambda^2 n^2}{d^2|W|} = \frac{\lambda}{d}\frac{n}{|W|}\frac{\lambda n}{d} \leq \frac{1}{6} \cdot 4 \cdot \frac{\lambda n}{d} < \frac{\lambda n}{d}.$$

Next note that, by Theorem 2.11, the number of edges spanned by U is at most

$$e(U) \leq \frac{d|U|^2}{2n} + \frac{\lambda|U|}{2} < \frac{\lambda n}{d}\frac{d|U|}{2n} + \frac{\lambda|U|}{2} = \frac{\lambda|U|}{2} + \frac{\lambda|U|}{2} = \lambda|U|.$$

As the degree of every vertex in U is d, it follows that

$$e(U, S) \geq d|U| - 2e(U) > (d - 2\lambda)|U| \geq 2d|U|/3.$$

On the other hand using again Theorem 2.11 together with the facts that $|U| \geq 36\lambda^2/d$, $|S| < d$ and $d \leq n/2$ we conclude that

$$e(U, S) \leq \frac{d|U||S|}{n} + \lambda\sqrt{|U||S|} < \frac{d}{n}d|U| + \lambda\sqrt{d|U|} \leq \frac{d|U|}{2} + \frac{\lambda\sqrt{d|U|}}{\sqrt{|U|}}$$

$$\leq \frac{d|U|}{2} + \frac{\lambda\sqrt{d|U|}}{6\lambda/\sqrt{d}} = \frac{d|U|}{2} + \frac{d|U|}{6} = \frac{2d|U|}{3}.$$

This contradiction completes the proof. ∎

The constants in this theorem can be easily improved and we make no attempt to optimize them. Note that, in particular, for an (n, d, λ)-graph G with $\lambda = O(\sqrt{d})$ we have that $\kappa(G) = d - \Theta(1)$.

Next we present an example which shows that the assertion of Theorem 4.1 is tight up to a constant factor. Let G be any (n, d, λ)-graph with $\lambda = \Theta(\sqrt{d})$. We already constructed several such graphs in the previous section. For an integer k, consider a new graph G_k, which is obtained by replacing each vertex of G by the complete graph of order k and by connecting two vertices of G_k by an edge if and only if the corresponding vertices of G are connected by an edge. Then it follows immediately from the definition that G_k has $n' = nk$ vertices and is d'-regular graph with $d' = dk + k - 1$. Let λ' be the second eigenvalue of G_k. To estimate λ' note that the adjacency matrix of G_k equals to $A_G \otimes J_k + I_n \otimes A_{K_k}$. Here A_G is the adjacency matrix of G, J_k is the all one matrix of size $k \times k$, I_n is the identity matrix of size $n \times n$ and A_{K_k} is the adjacency matrix of the complete graph of order k. Also the tensor product of the $m \times n$ dimensional matrix $A = (a_{ij})$ and the $s \times t$-dimensional matrix $B = (b_{kl})$ is the $ms \times nt$-dimensional matrix $A \otimes B$, whose entry labelled $\big((i, k)(j, l)\big)$ is $a_{ij}b_{kl}$. In case A and B are symmetric matrices with spectrums $\{\lambda_1, \ldots, \lambda_n\}$, $\{\mu_1, \ldots, \mu_t\}$ respectively, it is a simple consequence of the definition that the spectrum of $A \otimes B$ is $\{\lambda_i \mu_k : i = 1, \ldots, n, k = 1, \ldots, t\}$ (see, e.g. [64]). Therefore the second eigenvalue of $A_G \otimes J_k$ is $k\lambda$. On the other hand $I_n \otimes A_{K_k}$ is the adjacency matrix of the disjoint union of k-cliques and therefore the absolute value of all its eigenvalues is at most $k - 1$. Using these two facts we conclude that $\lambda' \leq \lambda k + k - 1$ and that G_k is $(n' = nk, d' = dk + k - 1, \lambda' = \lambda k + k - 1)$-graph. Also it is easy to see that the set of vertices of G_k that corresponds to a vertex in G has exactly dk neighbors outside this set. By deleting these neighbors we can disconnect the graph G_k and thus

$$\kappa(G_k) \leq dk = d' - (k - 1) = d' - \Omega\big((\lambda')^2/d'\big).$$

Sometimes we can improve the result of Theorem 4.1 using the information about co-degrees of vertices in our graph. Such result was used in [61] to determine the vertex-connectivity of dense random d-regular graphs.

Proposition 4.2 [61]. *Let $G = (V, E)$ be a d-regular graph on n vertices such that $\sqrt{n} \log n < d \leq 3n/4$ and the number of common neighbors for every two distinct vertices in G is $(1 + o(1)) d^2/n$. Then the graph G is d-vertex-connected.*

Similarly to vertex-connectivity, define the *edge-connectivity* of a graph G to be the minimum number of edges that we need to delete to make G disconnected. We denote this parameter by $\kappa'(G)$. Clearly the edge-connectivity is always at most the minimum degree of a graph. We also say that G has a *perfect matching* if there is a set of disjoint edges that covers all the vertices of G. Next we show that (n, d, λ)-graphs even with a very weak spectral gap are d-edge-connected and have a perfect matching (if the number of vertices is even).

Theorem 4.3. *Let G be an (n, d, λ)-graph with $d - \lambda \geq 2$. Then G is d-edge-connected. When n is even, it has a perfect matching.*

Proof. Let U be a subset of vertices of G of size at most $n/2$. To prove that G is d-edge-connected we need to show that there are always at least d edges between U and $V(G) - U$. If $1 \leq |U| \leq d$, then every vertex in U has at least $d - (|U| - 1)$ neighbors outside U and therefore $e(U, V(G) - U) \geq |U|(d - |U| + 1) \geq d$. On the other hand if $d \leq |U| \leq n/2$, then using that $d - \lambda \geq 2$ together with Theorem 2.11 we obtain that

$$e\big(U, V(G) - U\big)$$

$$\geq \frac{d|U|(n - |U|)}{n} - \lambda \sqrt{|U|(n - |U|) \left(1 - \frac{|U|}{n}\right) \left(1 - \frac{n - |U|}{n}\right)}$$

$$= (d - \lambda) \frac{(n - |U|)}{n} |U| \geq 2 \cdot \frac{1}{2} \cdot |U| = |U| \geq d,$$

and therefore $\kappa'(G) = d$.

To show that G contains a perfect matching we apply the celebrated Tutte's condition. Since n is even, we need to prove that for every nonempty set of vertices S, the induced graph $G[V - S]$ has at most $|S|$ connected components of odd size. Since G is d-edge-connected we have that there are at least d edges from every connected component of $G[V - S]$ to S. On the other hand there are at most $d|S|$ edges incident with vertices in S. Therefore $G[V - S]$ has at most $|S|$ connected components and hence G contains a perfect matching. ∎

4.2. Maximum cut

Let $G = (V, E)$ be a graph and let S be a nonempty proper subset of V. Denote by $(S, V - S)$ the cut of G consisting of all edges with one end in S and another one in $V - S$. The *size* of the cut is the number of edges in it. The MAX CUT problem is the problem of finding a cut of maximum size in

G. Let $f(G)$ be the size of the maximum cut in G. MAX CUT is one of the most natural combinatorial optimization problems. It is well known that this problem is NP-hard [45]. Therefore it is useful to have bounds on $f(G)$ based on other parameters of the graph, that can be computed efficiently.

Here we describe two such folklore results. First, consider a random partition $V = V_1 \cup V_2$, obtained by assigning each vertex $v \in V$ to V_1 or V_2 with probability $1/2$ independently. It is easy to see that each edge of G has probability $1/2$ to cross between V_1 and V_2. Therefore the expected number of edges in the cut (V_1, V_2) is $m/2$, where m is the number of edges in G. This implies that for every graph $f(G) \geq m/2$. The example of a complete graph shows that this lower bound is asymptotically optimal. The second result provides an upper bound for $f(G)$, for a regular graph G, in terms of the smallest eigenvalue of its adjacency matrix.

Proposition 4.4. *Let G be a d-regular graph (which may have loops) of order n with $m = dn/2$ edges and let $\lambda_1 \geq \lambda_2 \geq \ldots \geq \lambda_n$ be the eigenvalues of the adjacency matrix of G. Then*

$$f(G) \leq \frac{m}{2} - \frac{\lambda_n n}{4}.$$

In particular if G is an (n, d, λ)-graph then $f(G) \leq (d + \lambda)n/4$.

Proof. Let $A = (a_{ij})$ be the adjacency matrix of $G = (V, E)$ and let $V = \{1, \ldots, n\}$. Let $\mathbf{x} = (x_1, \ldots, x_n)$ be any vector with coordinates ± 1. Since the graph G is d-regular we have

$$\sum_{(i,j) \in E} (x_i - x_j)^2 = d \sum_{i=1}^{n} x_i^2 - \sum_{i,j} a_{ij} x_i x_j = dn - \mathbf{x}^t A \mathbf{x}.$$

By the variational definition of the eigenvalues of A, for any vector $z \in R^n$, $z^t A z \geq \lambda_n \|z\|^2$. Therefore

$$(11) \qquad \sum_{(i,j) \in E} (x_i - x_j)^2 = dn - \mathbf{x}^t A \mathbf{x} \leq dn - \lambda_n \|\mathbf{x}\|^2 = dn - \lambda_n n.$$

Let $V = V_1 \cup V_2$ be an arbitrary partition of V into two disjoint subsets and let $e(V_1, V_2)$ be the number of edges in the bipartite subgraph of G with bipartition (V_1, V_2). For every vertex $v \in V(G)$ define $x_v = 1$ if $v \in V_1$ and $x_v = -1$ if $v \in V_2$. Note that for every edge (i, j) of G, $(x_i - x_j)^2 = 4$ if this edge has its ends in the distinct parts of the above partition and is zero otherwise. Now using (11), we conclude that

$$e(V_1, V_2) = \frac{1}{4} \sum_{(i,j) \in E} (x_i - x_j)^2 \leq \frac{1}{4}(dn - \lambda_n n) = \frac{m}{2} - \frac{\lambda_n n}{4}. \qquad \blacksquare$$

This upper bound is often used to show that some particular results about maximum cuts are tight. For example this approach was used in [5] and [8]. In these papers the authors proved that for every graph G with m edges and girth at least $r \geq 4$, $f(G) \geq m/2 + \Omega\left(m^{\frac{r}{r+1}}\right)$. They also show, using Proposition 4.4 and Examples 9, 6 from Section 3, that this bound is tight for $r = 4, 5$.

4.3. Independent sets and the chromatic number

The *independence number* $\alpha(G)$ of a graph G is the maximum cardinality of a set of vertices of G no two of which are adjacent. Using Theorem 2.11 we can immediately establish an upper bound on the size of a maximum independent set of pseudo-random graphs.

Proposition 4.5. *Let G be an (n, d, λ)-graph, then*

$$\alpha(G) \leq \frac{\lambda n}{d + \lambda}.$$

Proof. Let U be an independent set in G, then $e(U) = 0$ and by Theorem 2.11 we have that $d|U|^2/n \leq \lambda|U|(1 - |U|/n)$. This implies that $|U| \leq \lambda n/(d + \lambda)$. $\blacksquare$

Note that even when $\lambda = O\left(\sqrt{d}\right)$ this bound only has order of magnitude $O(n/\sqrt{d})$. This contrasts sharply with the behavior of random graphs where it is known (see [20] and [49]) that the independence number of random graph $G(n, p)$ is only $\Theta\left(\frac{n}{d} \log d\right)$ where $d = (1 + o(1)) np$. More strikingly there are graphs for which the bound in Proposition 4.5 cannot be improved. One such graph is the Paley graph P_q with $q = p^2$ (Example 3 in the previous section). Indeed it is easy to see that in this case all elements of the subfield $GF(p) \subset GF(p^2)$ are quadratic residues in $GF(p^2)$. This implies that for every quadratic non-residue $\beta \in GF(p^2)$ all elements of any multiplicative coset $\beta GF(p)$ form an independent set of size p. As we already mentioned, P_q is an (n, d, λ)-graph with $n = p^2, d = (p^2 - 1)/2$ and $\lambda = (p + 1)/2$. Hence for this graph we get $\alpha(P_q) = \lambda n/(d + \lambda)$.

Next we obtain a lower bound on the independence number of pseudo-random graphs. We present a slightly more general result by Alon et al. [12] which we will need later.

Proposition 4.6 [12]. *Let G be an (n, d, λ)-graph such that $\lambda < d \leq 0.9n$. Then the induced subgraph $G[U]$ of G on any subset U, $|U| = m$, contains an independent set of size at least*

$$\alpha(G[U]) \geq \frac{n}{2(d - \lambda)} \ln \left(\frac{m(d - \lambda)}{n(\lambda + 1)} + 1 \right).$$

In particular,

$$\alpha(G) \geq \frac{n}{2(d - \lambda)} \ln \left(\frac{(d - \lambda)}{(\lambda + 1)} + 1 \right).$$

Sketch of proof. First using Theorem 2.11 it is easy to show that if U is a set of bn vertices of G, then the minimum degree in the induced subgraph $G[U]$ is at most $db + \lambda(1 - b) = (d - \lambda)b + \lambda$. Construct an independent set I in the induced subgraph $G[U]$ of G by the following greedy procedure. Repeatedly choose a vertex of minimum degree in $G[U]$, add it to the independent set I and delete it and its neighbors from U, stopping when the remaining set of vertices is empty. Let a_i, $i \geq 0$ be the sequence of numbers defined by the following recurrence formula:

$$a_0 = m,$$

$$a_{i+1} = a_i - \left(d\frac{a_i}{n} + \lambda(1 - \frac{a_i}{n}) + 1 \right) = \left(1 - \frac{d - \lambda}{n} \right) a_i - (\lambda + 1), \ \forall i \geq 0.$$

By the above discussion, it is easy to see that the size of the remaining set of vertices after i iterations is at least a_i. Therefore the size of the resulting independent set I is at least the smallest index i such that $a_i \leq 0$. By solving the recurrence equation we obtain that this index satisfies:

$$i \geq \frac{n}{2(d - \lambda)} \ln \left(\frac{m(d - \lambda)}{n(\lambda + 1)} + 1 \right). \qquad \blacksquare$$

For an (n, d, λ)-graph G with $\lambda \leq d^{1-\delta}$, $\delta > 0$, this proposition implies that $\alpha(G) \geq \Omega\left(\frac{n}{d} \log d\right)$. This shows that the independence number of a pseudo-random graph with a sufficiently small second eigenvalue is up to a constant factor at least as large as $\alpha\big(G(n,p)\big)$ with $p = d/n$. On the other hand the graph H_k (Example 4, Section 3) shows that even when $\lambda \leq O\big(\sqrt{d}\,\big)$ the independence number of (n, d, λ)-graph can be smaller than $\alpha\big(G(n,p)\big)$ with $p = d/n$. This graph has $n = 2^{k-1} - 1$ vertices, degree $d = \big(1 + o(1)\big) n/2$ and $\lambda = \Theta\big(\sqrt{d}\,\big)$. Also it is easy to see that every independent set in H_k corresponds to a family of orthogonal vectors in $\mathbf{Z}_2^k$ and thus has size at most $k = \big(1 + o(1)\big) \log_2 n$. This is only half of the size of a maximum independent set in the corresponding random graph $G(n, 1/2)$.

A *vertex-coloring* of a graph G is an assignment of a color to each of its vertices. The coloring is *proper* if no two adjacent vertices get the same color. The *chromatic number* $\chi(G)$ of G is the minimum number of colors used in a proper coloring of it. Since every color class in the proper coloring of G forms an independent set we can immediately obtain that $\chi(G) \geq |V(G)|/\alpha(G)$. This together with Proposition 4.5 implies the following result of Hoffman [48].

Corollary 4.7. Let G be an (n, d, λ)-graph. Then the chromatic number of G is at least $1 + d/\lambda$.

On the other hand, using Proposition 4.6, one can obtain the following upper bound on the chromatic number of pseudo-random graphs.

Theorem 4.8 [12]. Let G be an (n, d, λ)-graph such that $\lambda < d \leq 0.9n$. Then the chromatic number of G satisfies

$$\chi(G) \leq \frac{6(d - \lambda)}{\ln\left(\frac{d-\lambda}{\lambda+1} + 1\right)}.$$

Sketch of proof. Color the graph G as follows. As long as the remaining set of vertices U contains at least $n/\ln\left(\frac{d-\lambda}{\lambda+1} + 1\right)$ vertices, by Proposition 4.6 we can find an independent set of vertices in the induced subgraph $G[U]$ of size at least

$$\frac{n}{2(d - \lambda)} \ln\left(\frac{|U|(d - \lambda)}{n(\lambda + 1)} + 1\right) \geq \frac{n}{4(d - \lambda)} \ln\left(\frac{d - \lambda}{\lambda + 1} + 1\right).$$

Color all the members of such a set by a new color, delete them from the graph and continue. When this process terminates, the remaining set of vertices U is of size at most $n/\ln\left(\frac{d-\lambda}{\lambda+1} + 1\right)$ and we used at most $4(d - \lambda)/\ln\left(\frac{d-\lambda}{\lambda+1} + 1\right)$ colors so far. As we already mentioned above, for every subset $U' \subset U$ the induced subgraph $G[U']$ contains a vertex of degree at most

$$(d-\lambda)\frac{|U'|}{n} + \lambda \le (d-\lambda)\frac{|U|}{n} + \lambda \le \frac{d-\lambda}{\ln\left(\frac{d-\lambda}{\lambda+1}+1\right)} + \lambda \le \frac{2(d-\lambda)}{\ln\left(\frac{d-\lambda}{\lambda+1}+1\right)} - 1.$$

Thus we can complete the coloring of G by coloring $G[U]$ using at most $2(d-\lambda)/\ln\left(\frac{d-\lambda}{\lambda+1}+1\right)$ additional colors. The total number of colors used is at most $6(d-\lambda)/\ln\left(\frac{d-\lambda}{\lambda+1}+1\right)$. ∎

For an (n, d, λ)-graph G with $\lambda \le d^{1-\delta}, \delta > 0$ this proposition implies that $\chi(G) \le O\left(\frac{d}{\log d}\right)$. This shows that the chromatic number of a pseudo-random graph with a sufficiently small second eigenvalue is up to a constant factor at least as small as $\chi\big(G(n,p)\big)$ with $p = d/n$. On the other hand, the Paley graph $P_q, q = p^2$, shows that sometimes the chromatic number of a pseudo-random graph can be much smaller than the above bound, even the in case $\lambda = \Theta(\sqrt{d})$. Indeed, as we already mentioned above, all elements of the subfield $GF(p) \subset GF(p^2)$ are quadratic residues in $GF(p^2)$. This implies that for every quadratic non-residue $\beta \in GF(p^2)$ all elements of a multiplicative coset $\beta GF(p)$ form an independent set of size p. Also all additive cosets of $\beta GF(p)$ are independent sets in P_q. This implies that $\chi(P_q) \le \sqrt{q} = p$. In fact P_q contains a clique of size p (all elements of a subfield $GF(p)$), showing that $\chi(P_q) = \sqrt{q} \ll q/\log q$. Therefore the bound in Corollary 4.7 is best possible.

A more complicated quantity related to the chromatic number is the *list-chromatic number* $\chi_l(G)$ of G, introduced in [34] and [82]. This is the minimum integer k such that for every assignment of a set $S(v)$ of k colors to every vertex v of G, there is a proper coloring of G that assigns to each vertex v a color from $S(v)$. The study of this parameter received a considerable amount of attention in recent years, see, e.g., [2], [57] for two surveys. Note that from the definition it follows immediately that $\chi_l(G) \ge \chi(G)$ and it is known that the gap between these two parameters can be arbitrarily large. The list-chromatic number of pseudo-random graphs was studied by Alon, Krivelevich and Sudakov [12] and independently by Vu [84]. In [12] and [84] the authors mainly considered graphs with all degrees $(1+o(1))np$ and all co-degrees $(1+o(1))np^2$. Here we use ideas from these two papers to obtain an upper bound on the list-chromatic number of an (n, d, λ)-graphs. This bound has the same order of magnitude as the list chromatic number of the truly random graph $G(n, p)$ with $p = d/n$ (for more details see [12], [84]).

Theorem 4.9. *Suppose that $0 < \delta < 1$ and let G be an (n, d, λ)-graph satisfying $\lambda \leq d^{1-\delta}$, $d \leq 0.9n$. Then the list-chromatic number of G is bounded by*

$$\chi_l(G) \leq O\left(\frac{d}{\delta \log d}\right).$$

Proof. Suppose that d is sufficiently large and consider first the case when $d \leq n^{1-\delta/4}$. Then by Theorem 2.11 the neighbors of every vertex in G span at most $d^3/n + \lambda d \leq O(d^{2-\delta/4})$ edges. Now we can apply the result of Vu [84] which says that if the neighbors of every vertex in a graph G with maximum degree d span at most $O(d^{2-\delta/4})$ edges then $\chi_l(G) \leq O(d/(\delta \log d))$.

Now consider the case when $d \geq n^{1-\delta/4}$. For every vertex $v \in V$, let $S(v)$ be a list of at least $\frac{7d}{\delta \log n}$ colors. Our objective is to prove that there is a proper coloring of G assigning to each vertex a color from its list. As long as there is a set C of at least $n^{1-\delta/2}$ vertices containing the same color c in their lists we can, by Proposition 4.6, find an independent set of at least $\frac{\delta n}{6d} \log n$ vertices in C, color them all by c, omit them from the graph and omit the color c from all lists. The total number of colors that can be deleted in this process cannot exceed $\frac{6d}{\delta \log n}$ (since in each such deletion at least $\frac{\delta n}{6d} \log n$ vertices are deleted from the graph). When this process terminates, no color appears in more than $n^{1-\delta/2}$ lists, and each list still contains at least $\frac{d}{\delta \log n} > n^{1-\delta/2}$ colors. Therefore, by Hall's theorem, we can assign to each of the remaining vertices a color from its list so that no color is being assigned to more than one vertex, thus completing the coloring and the proof. ∎

4.4. Small subgraphs

We now examine small subgraphs of pseudo-random graphs. Let H be a fixed graph of order s with r edges and with automorphism group $Aut(H)$. Using the second moment method it is not difficult to show that for every constant p the random graph $G(n, p)$ contains

$$(1 + o(1)) \, p^r (1-p)^{\binom{s}{2}-r} \frac{n^s}{|Aut(H)|}$$

induced copies of H. Thomason extended this result to jumbled graphs. He showed in [79] that if a graph G is (p, α)-jumbled and $p^s n \gg 42\alpha s^2$

then the number of induced subgraphs of G which are isomorphic to H is
$(1 + o(1)) p^s (1 - p)^{\binom{s}{2} - r} n^s / |Aut(H)|$.

Here we present a result of Noga Alon [6] that proves that every large subset of the set of vertices of (n, d, λ)-graph contains the "correct" number of copies of any fixed sparse graph. An additional advantage of this result is that its assertion depends not on the number of vertices s in H but only on its maximum degree Δ which can be smaller than s. Special cases of this result have appeared in various papers including [11], [13] and probably other papers as well. The approach here is similar to the one in [13].

Theorem 4.10. [6] *Let H be a fixed graph with r edges, s vertices and maximum degree Δ, and let $G = (V, E)$ be an (n, d, λ)-graph, where, say, $d \leq 0.9n$. Let $m < n$ satisfy $m \gg \lambda \left(\frac{n}{d} \right)^\Delta$. Then, for every subset $V' \subset V$ of cardinality m, the number of (not necessarily induced) copies of H in V' is*

$$\left(1 + o(1) \right) \frac{m^s}{|Aut(H)|} \left(\frac{d}{n} \right)^r.$$

Note that this implies that a similar result holds for the number of induced copies of H. Indeed, if $n \gg d$ and $m \gg \lambda \left(\frac{n}{d} \right)^{\Delta + 1}$ then the number of copies of each graph obtained from H by adding to it at least one edge is, by the above Theorem, negligible compared to the number of copies of H, and hence almost all copies of H in V' are induced. If $d = \Theta(n)$ then, by inclusion-exclusion, the number of induced copies of H in V' as above is also roughly the "correct" number. A special case of the above theorem implies that if $\lambda = O(\sqrt{d})$ and $d \gg n^{2/3}$, then any (n, d, λ)-graph contains many triangles. As shown in Example 9, Section 3, this is not true when $d = \left(\frac{1}{4} + o(1) \right) n^{2/3}$, showing that the assertion of the theorem is not far from being best possible.

Proof of Theorem 4.10. To prove the theorem, consider a random one-to-one mapping of the set of vertices of H into the set of vertices V'. Denote by $A(H)$ the event that every edge of H is mapped on an edge of G. In such a case we say that the mapping is an embedding of H. Note that it suffices to prove that

(12) $$Pr\left(A(H) \right) = \left(1 + o(1) \right) \left(\frac{d}{n} \right)^r.$$

We prove (12) by induction on the number of edges r. The base case ($r = 0$) is trivial. Suppose that (12) holds for all graphs with less than r

edges, and let uv be an edge of H. Let H_{uv} be the graph obtained from H by removing the edge uv (and keeping all vertices). Let H_u and H_v be the induced subgraphs of H on the sets of vertices $V(H) \setminus \{v\}$ and $V(H) \setminus \{u\}$, respectively, and let H' be the induced subgraph of H on the set of vertices $V(H) \setminus \{u, v\}$. Let r' be the number of edges of H' and note that $r - r' \leq 2(\Delta - 1) + 1 = 2\Delta - 1$. Clearly $Pr\big(A(H_{uv})\big) = Pr\big(A(H_{uv}) \mid A(H')\big) \cdot Pr\big(A(H')\big)$. Thus, by the induction hypothesis applied to H_{uv} and to H':

$$Pr\big(A(H_{uv}) \mid A(H')\big) = \big(1 + o(1)\big) \left(\frac{d}{n}\right)^{r-1-r'}.$$

For an embedding f' of H', let $\nu(u, f')$ be the number of extensions of f' to an embedding of H_u in V'; $\nu(v, f')$ denotes the same for v. Clearly, the number of extensions of f' to an embedding of H_{uv} in V' is at least $\nu(u, f')\nu(v, f') - \min\big(\nu(u, f'), \nu(v, f')\big)$ and at most $\nu(u, f')\nu(v, f')$. Thus we have

$$\frac{\nu(u, f')\nu(v, f') - \min\big(\nu(u, f'), \nu(v, f')\big)}{(m - s + 2)(m - s + 1)}$$

$$\leq Pr\big(A(H_{uv}) \mid f'\big) \leq \frac{\nu(u, f')\nu(v, f')}{(m - s + 2)(m - s + 1)}.$$

Taking expectation over all embeddings f' the middle term becomes $Pr\big(A(H_{uv}) \mid A(H')\big)$, which is $\big(1 + o(1)\big)\big(\frac{d}{n}\big)^{r-1-r'}$. Note that by our choice of the parameters and the well known fact that $\lambda = \Omega\big(\sqrt{d}\big)$, the expectation of the term $\min\big(\nu(u, f'), \nu(v, f')\big)$ $(\leq m)$ is negligible and we get

$$E_{f'}\big(\nu(u, f')\nu(v, f') \mid A(H')\big) = \big(1 + o(1)\big)m^2 \left(\frac{d}{n}\right)^{r-1-r'}.$$

Now let f be a random one-to-one mapping of $V(H)$ into V'. Let f' be a fixed embedding of H'. Then

$$Pr_f\big(A(H) \mid f|_{V(H)\setminus\{u,v\}} = f'\big) = \left(\frac{d}{n}\right)\frac{\nu(u, f')\nu(v, f')}{(m - s + 2)(m - s + 1)} + \delta,$$

where $|\delta| \leq \lambda\frac{\sqrt{\nu(u,f')\nu(v,f')}}{(m-s+2)(m-s+1)}$. This follows from Theorem 2.11, where we take the possible images of u as the set U and the possible images of v as the set W. Averaging over embeddings f' we get $Pr\big(A(H) \mid A(H')\big)$ on the left hand side. On the right hand side we get $\left(1 + o(1)\right)\left(\frac{d}{n}\right)^{r-r'}$ from the first term plus the expectation of the error term δ. By Jensen's inequality, the absolute value of this expectation is bounded by

$$\lambda \frac{\sqrt{E\left(\nu(u, f')\nu(v, f')\right)}}{(m - s + 2)(m - s + 1)} = \left(1 + o(1)\right)\frac{\lambda}{m}\left(\frac{d}{n}\right)^{(r-r'-1)/2}.$$

Our assumptions on the parameters imply that this is negligible with respect to the main term. Therefore $Pr\left(A(H)\right) = Pr\left(A(H) \mid A(H')\right) \cdot Pr\left(A(H')\right) = \left(1 + o(1)\right)\left(\frac{d}{n}\right)^r$, completing the proof of Theorem 4.10. ∎

If we are only interested in the existence of one copy of H then one can sometimes improve the conditions on d and λ in Theorem 4.10. For example if H is a complete graph of order r then the following result was proved in [11].

Proposition 4.11 [11]. *Let G be an (n, d, λ)-graph. Then for every integer $r \geq 2$ every set of vertices of G of size more than*

$$\frac{(\lambda + 1)n}{d}\left(1 + \frac{n}{d} + \ldots + \left(\frac{n}{d}\right)^{r-2}\right)$$

contains a copy of a complete graph K_r.

In particular, when $d \geq \Omega(n^{2/3})$ and $\lambda \leq O\left(\sqrt{d}\right)$ then any (n, d, λ)-graph contains a triangle and as shows Example 9 in Section 3 this is tight. Unfortunately we do not know if this bound is also tight for $r \geq 4$. It would be interesting to construct examples of (n, d, λ)-graphs with $d = \Theta\left(n^{1-1/(2r-3)}\right)$ and $\lambda \leq O\left(\sqrt{d}\right)$ which contain no copy of K_r.

Finally we present one additional result about the existence of odd cycles in pseudo-random graphs.

Proposition 4.12. *Let $k \geq 1$ be an integer and let G be an (n, d, λ)-graph such that $d^{2k}/n \gg \lambda^{2k-1}$. Then G contains a cycle of length $2k + 1$.*

Proof. Suppose that G contains no cycle of length $2k + 1$. For every two vertices u, v of G denote by $d(u, v)$ the length of a shortest path from u to v. For every $i \geq 1$ let $N_i(v) = \{u \mid d(u, v) = i\}$ be the set of all vertices in G which are at distance exactly i from v. In [32] Erdős et al. proved that if G contains no cycle of length $2k + 1$ then for any $1 \leq i \leq k$ the induced graph $G\left[N_i(v)\right]$ contains an independent set of size $\left|N_i(v)\right|/(2k - 1)$. This result together with Proposition 4.5 implies that for every vertex v and for every $1 \leq i \leq k$, $\left|N_i(v)\right| \leq (2k-1)\lambda n/d$. Since $d^{2k}/n \gg \lambda^{2k-1}$ we have that $\lambda = o(d)$. Therefore by Theorem 2.11

$$e\big(N_i(v)\big) \leq \frac{d}{2n}\left|N_i(v)\right|^2 + \lambda\left|N_i(v)\right| \leq \frac{d}{n}\frac{(2k-1)\lambda n}{2d}\left|N_i(v)\right| + \lambda\left|N_i(v)\right|$$

$$< 2k\lambda\left|N_i(v)\right| = o\big(d\left|N_i(v)\right|\big).$$

Next we prove by induction that for every $1 \leq i \leq k$, $\frac{\left|N_{i+1}(v)\right|}{\left|N_i(v)\right|} \geq \left(1 - o(1)\right)d^2/\lambda^2$. By the above discussion the number of edges spanned by $N_1(v)$ is $o(d^2)$ and therefore $e\big(N_1(v), N_2(v)\big) = d^2 - o(d^2) = \left(1 - o(1)\right)d^2$. On the other hand, by Theorem 2.11

$$e\big(N_1(v), N_2(v)\big) \leq \frac{d}{n}\left|N_1(v)\right|\left|N_2(v)\right| + \lambda\sqrt{\left|N_1(v)\right|\left|N_2(v)\right|}$$

$$\leq \frac{d}{n}d\frac{(2k-1)\lambda n}{d} + \lambda\sqrt{d\left|N_2(v)\right|}$$

$$= \lambda d\sqrt{\frac{\left|N_2(v)\right|}{d}} + O(\lambda d) = \lambda d\sqrt{\frac{\left|N_2(v)\right|}{\left|N_1(v)\right|}} + o(d^2).$$

Therefore $\frac{\left|N_2(v)\right|}{\left|N_1(v)\right|} \geq \left(1 - o(1)\right)d^2/\lambda^2$. Now assume that $\frac{\left|N_i(v)\right|}{\left|N_{i-1}(v)\right|} \geq \left(1 - o(1)\right)d^2/\lambda^2$. Since the number of edges spanned by $N_i(v)$ is $o\big(d\left|N_i(v)\right|\big)$ we obtain

$$e\big(N_i(v), N_{i+1}(v)\big) = d\left|N_i(v)\right| - 2e\big(N_i(v)\big) - e\big(N_{i-1}(v), N_i(v)\big)$$

$$\geq d\left|N_i(v)\right| - o\big(d\left|N_i(v)\right|\big) - d\left|N_{i-1}(v)\right|$$

$$\geq \left(1 - o(1)\right)d\left|N_i(v)\right| - \left(1 + o(1)\right)d(\lambda^2/d^2)\left|N_i(v)\right|$$

$$= \left(1 - o(1)\right)d\left|N_i(v)\right| - o\big(d\left|N_i(v)\right|\big)$$

$$= \left(1 - o(1)\right)d\left|N_i(v)\right|.$$

On the other hand, by Theorem 2.11

$$e\big(N_i(v), N_{i+1}(v)\big) \leq \frac{d}{n}\big|N_i(v)\big|\,\big|N_{i+1}(v)\big| + \lambda\sqrt{\big|N_i(v)\big|\,\big|N_{i+1}(v)\big|}$$

$$\leq \frac{d}{n}\frac{(2k-1)\lambda n}{d}\big|N_i(v)\big| + \lambda\sqrt{\big|N_i(v)\big|\,\big|N_{i+1}(v)\big|}$$

$$= O\big(\lambda\big|N_i(v)\big|\big) + \lambda\big|N_i(v)\big|\sqrt{\frac{\big|N_{i+1}(v)\big|}{\big|N_i(v)\big|}}$$

$$= \lambda\big|N_i(v)\big|\sqrt{\frac{\big|N_{i+1}(v)\big|}{\big|N_i(v)\big|}} + o\big(d\big|N_i(v)\big|\big).$$

Therefore $\frac{|N_{i+1}(v)|}{|N_i(v)|} \geq \big(1 - o(1)\big)d^2/\lambda^2$ and we proved the induction step.

Finally note that

$$\big|N_k(v)\big| = d\prod_{i=1}^{k-1}\frac{\big|N_{i+1}(v)\big|}{\big|N_i(v)\big|} \geq \big(1 + o(1)\big)d\left(\frac{d^2}{\lambda^2}\right)^{k-1}$$

$$= \big(1 + o(1)\big)\frac{d^{2k-1}}{\lambda^{2k-2}} \gg (2k-1)\frac{\lambda n}{d}.$$

This contradiction completes the proof. ∎

This result implies that when $d \gg n^{\frac{2}{2k+1}}$ and $\lambda \leq O\big(\sqrt{d}\big)$ then any (n, d, λ)-graph contains a cycle of length $2k + 1$. As shown by Example 10 of the previous section this result is tight. It is worth mentioning here that it follows from the result of Bondy and Simonovits [22] that any d-regular graph with $d \gg n^{1/k}$ contains a cycle of length $2k$. Here we do not need to make any assumption about the second eigenvalue λ. This bound is known to be tight for $k = 2, 3, 5$ (see Examples 6,7, Section 3).

4.5. Extremal properties

Turán's theorem [81] is one of the fundamental results in Extremal Graph Theory. It states that among n-vertex graphs not containing a clique of size t the complete $(t - 1)$-partite graph with (almost) equal parts has the maximum number of edges. For two graphs G and H we define the Turán number $ex(G, H)$ of H in G, as the largest integer e, such that there is an H-free subgraph of G with e edges. Obviously $ex(G, H) \leq |E(G)|$, where $E(G)$ denotes the edge set of G. Turán's theorem, in an asymptotic form, can be restated as

$$ex(K_n, K_t) = \left(\frac{t-2}{t-1} + o(1) \right) \binom{n}{2},$$

that is the largest K_t-free subgraph of K_n contains approximately $\frac{t-2}{t-1}$-fraction of its edges. Here we would like to describe an extension of this result to (n, d, λ)-graphs.

For an arbitrary graph G on n vertices it is easy to give a lower bound on $ex(G, K_t)$ following Turán's construction. One can partition the vertex set of G into $t-1$ parts such that the degree of each vertex within its own part is at most $\frac{1}{t-1}$-times its degree in G. Thus the subgraph consisting of the edges of G connecting two different parts has at least a $\frac{t-2}{t-1}$-fraction of the edges of G and is clearly K_t-free. We say that a graph (or rather a family of graphs) is t-*Turán* if this trivial lower bound is essentially an upper bound as well. More precisely, G is t-Turán if $ex(G, K_t) = \left(\frac{t-2}{t-1} + o(1) \right) |E(G)|$.

It has been shown that for any fixed t, there is a number $m(t, n)$ such that almost all graphs on n vertices with $m \geq m(t, n)$ edges are t-Turán (see [77], [51] for the most recent estimate for $m(t, n)$). However, these results are about random graphs and do not provide a deterministic sufficient condition for a graph to be t-Turán. It appears that such a condition can be obtained by a simple assumption about the spectrum of the graph. This was proved by Sudakov, Szabó and Vu in [75]. They obtained the following result.

Theorem 4.13 [75]. *Let $t \geq 3$ be an integer and let $G = (V, E)$ be an (n, d, λ)-graph. If $\lambda = o(d^{t-1}/n^{t-2})$ then*

$$ex(G, K_t) = \left(\frac{t-2}{t-1} + o(1) \right) |E(G)|.$$

Note that this theorem generalizes Turán's theorem, as the second eigenvalue of the complete graph K_n is 1.

Let us briefly discuss the sharpness of Theorem 4.13. For $t = 3$, one can show that its condition involving n, d and λ is asymptotically tight. Indeed, in this case the above theorem states that if $d^2/n \gg \lambda$, then one needs to delete about half of the edges of G to destroy all the triangles. On the other hand, by taking the example of Alon (Section 3, Example 9) whose parameters are: $d = \Theta(n^{2/3})$, $\lambda = \Theta(n^{1/3})$, and blowing it up (which means replacing each vertex by an independent set of size k and connecting two vertices in the new graph if and only if the corresponding vertices of G are connected by an edge) we get a graph $G(k)$ with the following properties:

$$\left| V\big(G(k)\big) \right| = n_k = nk; \quad G(k) \text{ is } d_k = dk\text{-regular}; \quad G(k) \text{ is triangle-free};$$

$$\lambda\big(G(k)\big) = k\lambda \quad \text{and} \quad \lambda\big(G(k)\big) = \Omega\big(d_k^2/n_k\big).$$

The above bound for the second eigenvalue of $G(k)$ can be obtained by using well known results on the eigenvalues of the tensor product of two matrices, see [59] for more details. This construction implies that for $t = 3$ and any sensible degree d the condition in Theorem 4.13 is not far from being best possible.

4.6. Factors and fractional factors

Let H be a fixed graph on n vertices. We say that a graph G on n vertices has an *H-factor* if G contains n/h vertex disjoint copies of H. Of course, a trivial necessary condition for the existence of an H-factor in G is that h divides n. For example, if H is just an edge $H = K_2$, then an H-factor is a perfect matching in G.

One of the most important classes of graph embedding problems is to find sufficient conditions for the existence of an H-factor in a graph G, usually assuming that H is fixed while the order n of G grows. In many cases such conditions are formulated in terms of the minimum degree of G. For example, the classical result of Hajnal and Szemerédi [47] asserts that if the minimum degree $\delta(G)$ satisfies $\delta(G) \geq \left(1 - \frac{1}{r}\right)n$, then G contains $\lfloor n/r \rfloor$ vertex disjoint copies of K_r. The statement of this theorem is easily seen to be tight.

It turns our that pseudo-randomness allows in many cases to significantly weaken sufficient conditions for H-factors and to obtain results which fail to hold for general graphs of the same edge density.

Consider first the case of a constant edge density p. In this case the celebrated Blow-up Lemma of Komlós, Sárközy and Szemerédi [54] can be used to show the existence of H-factors. In order to formulate the Blow-up Lemma we need to introduce the notion of a super-regular pair. Given $\varepsilon > 0$ and $0 < p < 1$, a bipartite graph G with bipartition (V_1, V_2), $|V_1| = |V_2| = n$, is called *super (p, ε)-regular* if

1. For all vertices $v \in V(G)$,

$$(p - \varepsilon)n \le d(v) \le (p + \varepsilon)n \; ;$$

2. For every pair of sets (U, W), $U \subset V_1$, $W \subset V_2$, $|U|, |W| \ge \varepsilon n$,

$$\left| \frac{e(U, W)}{|U||W|} - \frac{|E(G)|}{n^2} \right| \le \varepsilon.$$

Theorem 4.14 [54]. *For every choice of integers r and Δ and a real $0 < p < 1$ there exist an $\varepsilon > 0$ and an integer $n_0(\varepsilon)$ such that the following is true. Consider an r-partite graph G with all partition sets $V_1, \ldots, V_r$ of order $n > n_0$ and all $\binom{r}{2}$ bipartite subgraphs $G[V_i, V_j]$ super (p, ε)-regular. Then for every r-partite graph H with maximum degree $\Delta(H) \le \Delta$ and all partition sets $X_1, \ldots, X_r$ of order n, there exists an embedding f of H into G with each set X_i mapped onto V_i, $i = 1, \ldots, r$.*

(The above version of the Blow-up Lemma, due to Rödl and Ruciński [71], is somewhat different from and yet equivalent to the original formulation of Komlós et al. We use it here as it is somewhat closer in spirit to the notion of pseudo-randomness).

The Blow-up Lemma is a very powerful embedding tool. Combined with another "big cannon", the Szemerédi Regularity Lemma, it can be used to obtain approximate versions of many of the most famous embedding conjectures. We suggest the reader to consult a survey of Komlós [53] for more details and discussions.

It is easy to show that if G is an (n, d, λ)-graph with $d = \Theta(n)$ and $\lambda = o(n)$, and h divides n, then a random partition of $V(G)$ into h equal parts $V_1, \ldots, V_h$ produces almost surely $\binom{h}{2}$ super $(d/n, \varepsilon)$-regular pairs. Thus the Blow-up Lemma can be applied to the obtained h-partite subgraph of G and we get:

Corollary 4.15. *Let G be an (n, d, λ)-graph with $d = \Theta(n)$, $\lambda = o(n)$. If h divides n, then G contains an H-factor, for every fixed graph H on h vertices.*

The case of a vanishing edge density $p = o(1)$ is as usual significantly more complicated. Here a sufficient condition for the existence of an H-factor should depend heavily on the graph H, as there may exist quite dense pseudo-random graphs without a single copy of H, see, for example, the Alon graph (Example 9 of Section 3). When $H = K_2$, already a very weak pseudo-randomness condition suffices to guarantee an H-factor, or a perfect matching, as provided by Theorem 4.3. We thus consider the case $H = K_3$, the task here is to guarantee a *triangle factor*, i.e. a collection of $n/3$ vertex disjoint triangles. This problem has been treated by Krivelevich, Sudakov and Szabó [59] who obtained the following result:

Theorem 4.16 [59]. *Let G be an (n, d, λ)-graph. If n is divisible by 3 and*

$$\lambda = o\left(\frac{d^3}{n^2 \log n}\right),$$

then G has a triangle factor.

For best pseudo-random graphs with $\lambda = \Theta(\sqrt{d})$ the condition of the above theorem is fulfilled when $d \gg n^{4/5} \log^{2/5} n$.

To prove Theorem 4.16 Krivelevich et al. first partition the vertex set $V(G)$ into three parts V_1, V_2, V_3 of equal cardinality at random. Then they choose a perfect matching M between V_1 an V_2 at random and form an auxiliary bipartite graph Γ whose parts are M and V_3, and whose edges are formed by connecting $e \in M$ and $v \in V_3$ if both endpoints of e are connected by edges to v in G. The existence of a perfect matching in Γ is equivalent to the existence of a triangle factor in G. The authors of [59] then proceed to show that if M is chosen at random then the Hall condition is satisfied for Γ with positive probability.

The result of Theorem 4.16 is probably not tight. In fact, the following conjecture is stated in [59]:

Conjecture 4.17 [59]. *There exists an absolute constant $c > 0$ so that every d-regular graph G on $3n$ vertices, satisfying $\lambda(G) \leq cd^2/n$, has a triangle factor.*

If true the above conjecture would be best possible, up to a constant multiplicative factor. This is shown by taking the example of Alon (Section 3, Example 9) and blowing each of its vertices by an independent set of size k. As we already discussed in the previous section (see also [59]), this gives a triangle-free d_k-regular graph $G(k)$ on n_k vertices which satisfies $\lambda(G(k)) = \Omega(d_k^2/n_k)$.

Krivelevich, Sudakov and Szabó considered in [59] also the fractional version of the triangle factor problem. Given a graph $G = (V, E)$, denote by $T = T(G)$ the set of all triangles of G. A function $f : T \to \mathbb{R}_+$ is called a *fractional triangle factor* if for every $v \in V(G)$ one has $\sum_{v \in t} f(t) = 1$. If G contains a triangle factor T_0, then assigning values $f(t) = 1$ for all $t \in T_0$, and $f(t) = 0$ for all other $t \in T$ produces a fractional triangle factor. This simple argument shows that the existence of a triangle factor in G implies the existence of a fractional triangle factor. The converse statement is easily seen to be invalid in general.

The fact that a fractional triangle factor f can take non-integer values, as opposed to the characteristic vector of a "usual" (i.e. integer) triangle factor, enables to invoke the powerful machinery of Linear Programming to prove a much better result than Theorem 4.16.

Theorem 4.18 [59]. *Let $G = (V, E)$ be a (n, d, λ)-graph. If $\lambda \leq 0.1 d^2/n$ then G has a fractional triangle factor.*

This statement is optimal up to a constant factor – see the discussion following Conjecture 4.17.

Already for the next case $H = K_4$ analogs of Theorem 4.16 and 4.18 are not known. In fact, even an analog of Conjecture 4.17 is not available either, mainly due to the fact that we do not know the weakest possible spectral condition guaranteeing a single copy of K_4, or K_r in general, for $r \geq 4$.

Finally it would be interesting to show that for every integer Δ there exist a real M and an integer n_0 so that the following is true. If $n \geq n_0$ and G is an (n, d, λ)-graph for which $\lambda \leq d(d/n)^M$, then G contains a copy of any graph H on at most n vertices with maximum degree $\Delta(H) \leq \Delta$. This can be considered as a sparse analog of the Blow-up Lemma.

4.7. Hamiltonicity

A *Hamilton cycle* in a graph is a cycle passing through all the vertices of this graph. A graph is called *Hamiltonian* if it has at least one Hamilton cycle. For background information on Hamiltonian cycles the reader can consult a survey of Chvátal [28].

The notion of Hamilton cycles is one of the most central in modern Graph Theory, and many efforts have been devoted to obtain sufficient conditions for Hamiltonicity. The absolute majority of such known conditions (for example, the famous theorem of Dirac asserting that a graph on n vertices with minimal degree at least $n/2$ is Hamiltonian) deal with graphs which are fairly dense. Apparently there are very few sufficient conditions for the existence of a Hamilton cycle in sparse graphs.

As it turns out spectral properties of graphs can supply rather powerful sufficient conditions for Hamiltonicity. Here is one such result, quite general and yet very simple to prove, given our knowledge of properties of pseudo-random graphs.

Proposition 4.19. *Let G be an (n, d, λ)-graph. If*

$$d - 36\frac{\lambda^2}{d} \geq \frac{\lambda n}{d + \lambda},$$

then G is Hamiltonian.

Proof. According to Theorem 4.1 G is $(d - 36\lambda^2/d)$-vertex-connected. Also, $\alpha(G) \leq \lambda n/(d + \lambda)$, as stated in Proposition 4.5. Finally, a theorem of Chvátal and Erdős [29] asserts that if the vertex-connectivity of a graph G is at least as large as its independence number, then G is Hamiltonian. ∎

The Chvátal–Erdős Theorem has also been used by Thomason in [79], who proved that a (p, α)-jumbled graph G with minimal degree $\delta(G) = \Omega(\alpha/p)$ is Hamiltonian. His proof is quite similar in spirit to that of the above proposition.

Assuming that $\lambda = o(d)$ and $d \to \infty$, the condition of Proposition 4.19 reads then as: $\lambda \leq (1 - o(1)) d^2/n$. For best possible pseudo-random graphs, where $\lambda = \Theta(\sqrt{d})$, this condition starts working when $d = \Omega(n^{2/3})$.

One can however prove a much stronger asymptotical result, using more sophisticated tools for assuring Hamiltonicity. The authors prove such a result in [58]:

Theorem 4.20 [58]. *Let G be an (n, d, λ)-graph. If n is large enough and*

$$\lambda \leq \frac{(\log \log n)^2}{1000 \log n (\log \log \log n)} d,$$

then G is Hamiltonian.

The proof of Theorem 4.20 is quite involved technically. Its main instrument is the famous rotation-extension technique of Posa [70], or rather a version of it developed by Komlós and Szemerédi in [56] to obtain the exact threshold for the appearance of a Hamilton cycle in the random graph $G(n, p)$. We omit the proof details here, referring the reader to [58].

For reasonably good pseudo-random graphs, in which $\lambda \leq d^{1-\varepsilon}$ for some $\varepsilon > 0$, Theorem 4.20 starts working already when the degree d is only poly-logarithmic in n – quite a progress compared to the easy Proposition 4.19! It is possible though that an even stronger result is true as given by the following conjecture:

Conjecture 4.21 [58]. There exists a positive constant C such that for large enough n, any (n, d, λ)-graph that satisfies $d/\lambda > C$ contains a Hamilton cycle.

This conjecture is closely related to another well known problem on Hamiltonicity. The *toughness* $t(G)$ of a graph G is the largest real t so that for every positive integer $x \geq 2$ one should delete at least tx vertices from G in order to get an induced subgraph of it with at least x connected components. G is t-tough if $t(G) \geq t$. This parameter was introduced by Chvátal in [27], where he observed that Hamiltonian graphs are 1-tough and conjectured that t-tough graphs are Hamiltonian for large enough t. Alon showed in [4] that if G is an (n, d, λ)-graph, then the toughness of G satisfies $t(G) > \Omega(d/\lambda)$. Therefore the conjecture of Chvátal implies the above conjecture.

Krivelevich and Sudakov used Theorem 4.20 in [58] to derive Hamiltonicity of sparse random Cayley graphs. Given a group G of order n, choose a set S of s non-identity elements uniformly at random and form a Cayley graph $\Gamma(G, S \cup S^{-1})$ (see Example 8 in Section 3 for the definition of a Cayley graph). The question is how large should be the value of $t = t(n)$ so as to guarantee the almost sure Hamiltonicity of the random Cayley graph no matter which group G we started with.

Theorem 4.22 [58]. *Let G be a group of order n. Then for every $c > 0$ and large enough n a Cayley graph $X(G, S \cup S^{-1})$, formed by choosing a set S of $c \log^5 n$ random generators in G, is almost surely Hamiltonian.*

Sketch of proof. Let λ be the second largest by absolute value eigenvalue of $X(G, S)$. Note that the Cayley graph $X(G, S)$ is d-regular for $d \geq c \log^5 n$. Therefore to prove Hamiltonicity of $X(G, S)$, by Theorem 4.20 it is enough to show that almost surely $\lambda/d \leq O(\log n)$. This can be done by applying an approach of Alon and Roichman [16] for bounding the second eigenvalue of a random Cayley graph. ∎

We note that a well known conjecture claims that every connected Cayley graph is Hamiltonian. If true the conjecture would easily imply that as few as $O(\log n)$ random generators are enough to give almost sure connectivity and thus Hamiltonicity.

4.8. Random subgraphs of pseudo-random graphs

There is a clear tendency in recent years to study random graphs different from the classical by now model $G(n, p)$ of binomial random graphs. One of the most natural models for random graphs, directly generalizing $G(n, p)$, is defined as follows. Let $G = (V, E)$ be a graph and let $0 < p < 1$. The *random subgraph* G_p if formed by choosing every edge of G independently and with probability p. Thus, when G is the complete graph K_n we get back the probability space $G(n, p)$. In many cases the obtained random graph G_p has many interesting and peculiar features, sometimes reminiscent of those of $G(n, p)$, and sometimes inherited from those of the host graph G.

In this subsection we report on various results obtained on random subgraphs of pseudo-random graphs. While studying this subject, we study in fact not a single probability space, but rather a family of probability spaces, having many common features, guaranteed by those of pseudo-random graphs. Although several results have already been achieved in this direction, overall it is much less developed than the study of binomial random graphs $G(n, p)$, and one can certainly expect many new results on this topic to appear in the future.

We start with Hamiltonicity of random subgraphs of pseudo-random graphs. As we learned in the previous section spectral condition are in many cases sufficient to guarantee Hamiltonicity. Suppose then that a host graph G is a Hamiltonian (n, d, λ)-graph. How small can the edge probability $p = p(n)$ be chosen so as to guarantee almost sure Hamiltonicity of the random subgraph G_p? This question has been studied by Frieze and the first author in [42]. They obtained the following result.

Theorem 4.23 [42]. *Let G be an (n, d, λ)-graph. Assume that $\lambda = o\left(\frac{d^{5/2}}{n^{3/2}(\log n)^{3/2}}\right)$. Form a random subgraph G_p of G by choosing each edge of G independently with probability p. Then for any function $\omega(n)$ tending to infinity arbitrarily slowly:*

1. *if $p(n) = \frac{1}{d}\left(\log n + \log\log n - \omega(n)\right)$, then G_p is almost surely not Hamiltonian;*

2. if $p(n) = \frac{1}{d}\left(\log n + \log \log n + \omega(n)\right)$, then G_p is almost surely Hamiltonian.

Just as in the case of $G(n,p)$ (see, e.g. [20]) it is quite easy to predict the critical probability for the appearance of a Hamilton cycle in G_p. An obvious obstacle for its existence is a vertex of degree at most one. If such a vertex almost surely exists in G_p, then G_p is almost surely non-Hamiltonian. It is a straightforward exercise to show that the smaller probability in the statement of Theorem 4.23 gives the almost sure existence of such a vertex. The larger probability can be shown to be sufficient to eliminate almost surely all vertices of degree at most one in G_p. Proving that this is sufficient for almost sure Hamiltonicity is much harder. Again as in the case of $G(n,p)$ the rotation-extension technique of Posa [70] comes to our rescue. We omit technical details of the proof of Theorem 4.23, referring the reader to [42].

One of the most important events in the study of random graphs was the discovery of the sudden appearance of the giant component by Erdős and Rényi [33]. They proved that all connected components of $G(n,c/n)$ with $0 < c < 1$ are almost surely trees or unicyclic and have size $O(\log n)$. On the other hand, if $c > 1$, then $G(n,c/n)$ contains almost surely a unique component of size linear in n (the so called *giant component*), while all other components are at most logarithmic in size. Thus, the random graph $G(n,p)$ experiences the so called *phase transition* at $p = 1/n$.

Very recently Frieze, Krivelevich and Martin showed [43] that a very similar behavior holds for random subgraphs of many pseudo-random graphs. To formulate their result, for $\alpha > 1$ we define $\bar{\alpha} < 1$ to be the unique solution (other than α) of the equation $xe^{-x} = \alpha e^{-\alpha}$.

Theorem 4.24 [43]. *Let G be an (n,d,λ)-graph. Assume that $\lambda = o(d)$. Consider the random subgraph $G_{\alpha/d}$, formed by choosing each edge of G independently and with probability $p = \alpha/d$. Then:*

(a) *If $\alpha < 1$ then almost surely the maximum component size is $O(\log n)$.*

(b) *If $\alpha > 1$ then almost surely there is a unique giant component of asymptotic size $\left(1 - \frac{\bar{\alpha}}{\alpha}\right)n$ and the remaining components are of size $O(\log n)$.*

Let us outline briefly the proof of Theorem 4.24. First, bound (4) and known estimates on the number of k-vertex trees in d-regular graphs are used to get estimates on the expectation of the number of connected components of size k in G_p, for various values of k. Using these estimates it is proved then that almost surely G_p has no connected components of size between $(1/\alpha\gamma)\log n$ and γn for a properly chosen $\gamma = \gamma(\alpha)$. Define $f(\alpha)$ to be 1 for all $\alpha \leq 1$, and to be $\bar{\alpha}/\alpha$ for $\alpha > 1$. One can show then that almost surely in $G_{\alpha/d}$ the number of vertices in components of size between 1 and $d^{1/3}$ is equal to $nf(\alpha)$ up to the error term which is $O(n^{5/6}\log n)$. This is done by first calculating the expectation of the last quantity, which is asymptotically equal to $nf(\alpha)$, and then by applying the Azuma–Hoeffding martingale inequality.

Given the above, the proof of Theorem 4.24 is straightforward. For the case $\alpha < 1$ we have $nf(\alpha) = n$ and therefore all but at most $n^{5/6}\log n$ vertices lie in components of size at most $(1/\alpha\gamma)\log n$. The remaining vertices should be in components of size at least γn, but there is no room for such components. If $\alpha > 1$, then $(\bar{\alpha}/\alpha)n + O(n^{5/6}\log n)$ vertices belong to components of size at most $(1/\alpha\gamma)\log n$, and all remaining vertices are in components of size at least γn. These components are easily shown to merge quickly into one giant component of a linear size. The detail can be found in [43] (see also [7] for some related results).

One of the recent most popular subjects in the study of random graphs is proving sharpness of thresholds for various combinatorial properties. This direction of research was spurred by a powerful theorem of Friedgut–Bourgain [37], providing a sufficient condition for the sharpness of a threshold. The authors together with Vu apply this theorem in [60] to show sharpness of graph connectivity, sometimes also called *network reliability*, in random subgraphs of a wide class of graphs. Here are the relevant definitions. For a connected graph G and edge probability p denote by $f(p) = f(G, p)$ the probability that a random subgraph G_p is connected. The function $f(p)$ can be easily shown to be strictly monotone. For a fixed positive constant $x \leq 1$ and a graph G, let p_x denote the (unique) value of p where $f(G, p_x) = x$. We say that a family $(G_i)_{i=1}^{\infty}$ of graphs satisfies the *sharp threshold* property if for any fixed positive $\varepsilon \leq 1/2$

$$\lim_{i \to \infty} \frac{p_\varepsilon(G_i)}{p_{1-\varepsilon}(G_i)} \to 1.$$

Thus the threshold for connectivity is sharp if the width of the transition interval is negligible compared to the critical probability. Krivelevich, Sudakov and Vu proved in [60] the following theorem.

Theorem 4.25 [60]. *Let $(G_i)_{i=1}^{\infty}$ be a family of distinct graphs, where G_i has n_i vertices, maximum degree d_i and it is k_i-edge-connected. If*

$$\lim_{i \to \infty} \frac{k_i \ln n_i}{d_i} = \infty,$$

then the family $(G_i)_{i=1}^{\infty}$ has a sharp connectivity threshold.

The above theorem extends a celebrated result of Margulis [67] on network reliability (Margulis' result applies to the case where the critical probability is a constant).

Since (n, d, λ) graphs are $d(1 - o(1))$-connected as long as $\lambda = o(d)$ by Theorem 4.1, we immediately get the following result on the sharpness of the connectivity threshold for pseudo-random graphs.

Corollary 4.26. *Let G be an (n, d, λ)-graph. If $\lambda = o(d)$, then the threshold for connectivity in the random subgraph G_p is sharp.*

Thus already weak connectivity is sufficient to guarantee sharpness of the threshold. This result has potential practical applications as discussed in [60].

Finally we consider a different probability space created from a graph $G = (V, E)$. This space is obtained by putting random weights on the edges of G independently. One can then ask about the behavior of optimal solutions for various combinatorial optimization problems.

Beveridge, Frieze and McDiarmid treated in [19] the problem of estimating the weight of a random minimum length spanning tree in regular graphs. For each edge e of a connected graph $G = (V, E)$ define the length X_e of e to be a random variable uniformly distributed in the interval $(0, 1)$, where all X_e are independent. Let $mst(G, \mathbf{X})$ denote the minimum length of a spanning tree in such a graph, and let $mst(G)$ be the expected value of $mst(G, \mathbf{X})$. Of course, the value of $mst(G)$ depends on the connectivity structure of the graph G. Beveridge et al. were able to prove however that if the graph G is assumed to be almost regular and has a modest edge expansion, then $mst(G)$ can be calculated asymptotically:

Theorem 4.27 [19]. *Let $\alpha = \alpha(d) = O(d^{-1/3})$ and let $\rho(d)$ and $\omega(d)$ tend to infinity with d. Suppose that the graph $G = (V, E)$ satisfies*

$$d \leq d(v) \leq (1 + \alpha)d \quad \text{for all } v \in V(G),$$

and

$$\frac{e(S, V \setminus S)}{|S|} \geq \omega d^{2/3} \log d \quad \text{for all } S \subset V \text{ with } d/2 < |S| \leq \min\left\{\rho d, |V|/2\right\}.$$

Then

$$mst(G) = \left(1 + o(1)\right) \frac{|V|}{d} \zeta(3),$$

where the $o(1)$ term tends to 0 as $d \to \infty$, and $\zeta(3) = \sum_{i=1}^{\infty} i^{-3} = 1.202\ldots$.

The above theorem extends a celebrated result of Frieze [40], who proved it in the case of the complete graph $G = K_n$.

Pseudo-random graphs supply easily the degree of edge expansion required by Theorem 4.27. We thus get:

Corollary 4.28. *Let G be an (n, d, λ)-graph. If $\lambda = o(d)$ then*

$$mst(G) = \left(1 + o(1)\right) \frac{n}{d} \zeta(3).$$

Beveridge, Frieze and McDiarmid also proved that the random variable $mst(G, \mathbf{X})$ is sharply concentrated around its mean given by Theorem 4.27.

Comparing between the very well developed research of binomial random graphs $G(n, p)$ and few currently available results on random subgraphs of pseudo-random graphs, we can say that many interesting problems in the latter subject are yet to be addressed, such as the asymptotic behavior of the independence number and the chromatic number, connectivity, existence of matchings and factors, spectral properties, to mention just a few.

4.9. Enumerative aspects

Pseudo-random graphs on n vertices with edge density p are quite similar in many aspects to the random graph $G(n, p)$. One can thus expect that counting statistics in pseudo-random graphs will be close to those in truly random graphs of the same density. As the random graph $G(n, p)$ is a product probability space in which each edge behaves independently, computing the expected number of most subgraphs in $G(n, p)$ is straightforward. Here are just a few examples:

- The expected number of perfect matchings in $G(n, p)$ is $\frac{n!}{(n/2)! 2^{n/2}} p^{n/2}$ (assuming of course that n is even);

- The expected number of spanning trees in $G(n, p)$ is $n^{n-2}p^{n-1}$;

- The expected number of Hamilton cycles in $G(n, p)$ is $\frac{(n-1)!}{2}p^n$.

In certain cases it is possible to prove that the actual number of subgraphs in a pseudo-random graph on n vertices with edge density $p = p(n)$ is close to the corresponding expected value in the binomial random graph $G(n, p)$.

Frieze in [41] gave estimates on the number of perfect matchings and Hamilton cycles in what he calls super ε-regular graphs. Let $G = (V, E)$ be a graph on n vertices with $\binom{n}{2}p$ edges, where $0 < p < 1$ is a constant. Then G is called *super (p, ε)-regular*, for a constant $\varepsilon > 0$, if

1. For all vertices $v \in V(G)$,

$$(p - \varepsilon)n \le d(v) \le (p + \varepsilon)n \, ;$$

2. For all $U, W \subset V$, $U \cap W = \emptyset$, $|U|, |W| \ge \varepsilon n$,

$$\left| \frac{e(U, W)}{|U| \, |W|} - p \right| \le \varepsilon.$$

Thus, a super (p, ε)-regular graph G can be considered a non-bipartite analog of the notion of a super-regular pair defined above. In our terminology, G is a weakly pseudo-random graph of constant density p, in which *all* degrees are asymptotically equal to pn. Assume that $n = 2\nu$ is even. Let $m(G)$ denote the number of perfect matchings in G and let $h(G)$ denote the number of Hamilton cycles in G, and let $t(G)$ denote the number of spanning trees in G.

Theorem 4.29 [41]. *If ε is sufficiently small and n is sufficiently large then*

(a)

$$(p - 2\varepsilon)^\nu \frac{n!}{\nu! 2^\nu} \le m(G) \le (p + 2\varepsilon)^\nu \frac{n!}{\nu! 2^\nu} \, ;$$

(b)

$$(p - 2\varepsilon)^n n! \le h(G) \le (p + 2\varepsilon)^n n! \, ;$$

Theorem 4.29 thus implies that the numbers of perfect matchings and of Hamilton cycles in super ε-regular graphs are quite close asymptotically to the expected values of the corresponding quantities in the random graph $G(n, p)$. Part (b) of Theorem 4.29 improves significantly Corollary 2.9 of Thomason [79] which estimates from below the number of Hamilton cycles in jumbled graphs.

Here is a very brief sketch of the proof of Theorem 4.29. To estimate the number of perfect matchings in G, Frieze takes a random partition of the vertices of G into two equal parts A and B and estimates the number of perfect matchings in the bipartite subgraph of G between A and B. This bipartite graph is almost surely super 2ε-regular, which allows to apply bounds previously obtained by Alon, Rödl and Ruciński [15] for such graphs.

Since each Hamilton cycle is a union of two perfect matchings, it follows immediately that $h(G) \leq m^2(G)/2$, establishing the desired upper bound on $h(G)$. In order to prove a lower bound, let f_k be the number of 2-factors in G containing exactly k cycles, so that $f_1 = h(G)$. Let also A be the number of ordered pairs of edge disjoint perfect matchings in G. Then

$$
(13) \qquad A = \sum_{i=1}^{\lfloor n/3 \rfloor} 2^k f_k.
$$

For a perfect matching M in G let a_M be the number of perfect matchings of G disjoint from M. Since deleting M disturbs ε-regularity of G only marginally, one can use part (a) of the theorem to get $a_M \geq (p - 2\varepsilon)^\nu \frac{n!}{\nu! 2^\nu}$. Thus

$$
(14) \qquad A = \sum_{M \in G} a_M \geq \left((p - 2\varepsilon)^\nu \frac{n!}{\nu! 2^\nu} \right)^2 \geq (p - 2\varepsilon)^n n! \cdot \frac{1}{3 n^{1/2}}.
$$

Next Frieze shows that the ratio f_{k+1}/f_k can be bounded by a polynomial in n for all $1 \leq k \leq k_1 = O(p^{-2})$, $f_k \leq 5^{-(k-k_0)/2} \max\{f_{k_0+1}, f_{k_0}\}$ for all $k \geq k_0 + 2$, $k_0 = \Theta(p^{-3} \log n)$ and that the ratio $\left(f_{k_1+1} + \ldots + f_{\lfloor n/3 \rfloor} \right)/f_{k_1}$ is also bounded by a polynomial in n. Then from (13), $A \leq O_p(1) \sum_{k=1}^{k_0+1} f_k$ and thus $A \leq n^{O(1)} f_1$. Plugging (14) we get the desired lower bound.

One can also show (see [1]) that the number of spanning trees $t(G)$ in super (p, ε)-regular graphs satisfies:

$$
(p - 2\varepsilon)^{n-1} n^{n-2} \leq t(G) \leq (p + 2\varepsilon)^{n-1} n^{n-2},
$$

for small enough $\varepsilon > 0$ and large enough n. In order to estimate from below the number of spanning trees in G, consider a random mapping $f :$ $V(G) \rightarrow V(G)$, defined by choosing for each $v \in V$ its neighbor $f(v)$ at random. Each such f defines a digraph $D_f = (V, A_f)$, $A_f = \{(v, f(v)) :$ $v \in V\}$. Each component of D_f consists of cycle C with a rooted forest whose roots are all in C. Suppose that D_f has k_f components. Then a spanning tree of G can be obtained by deleting the lexicographically first edge of each cycle in D_f, and then extending the k_f components to a spanning tree. Showing that D_f has typically $O(\sqrt{n})$ components implies that most of the mappings f create a digraph close to a spanning tree of G, and therefore:

$$t(G) \geq n^{-O(\sqrt{n})}|f : V \rightarrow V| \geq n^{-O(\sqrt{n})}(p - \varepsilon)n^n.$$

For the upper bound on $t(G)$ let $\Omega^* = \{f : V \rightarrow V : (v, f(v)) \in E(G)$ for $v \neq 1$ and $f(1) = 1\}$. Then

$$t(G) \leq |\Omega^*| \leq ((p + \varepsilon)n)^{n-1} \leq (p + 2\varepsilon)^{n-1}n^{n-2}.$$

To see this consider the following injection from the spanning trees of G into Ω^*: orient each edge of a tree T towards vertex 1 and set $f(1) = 1$. Note that this proof does not use the fact that the graph is pseudo-random. Surprisingly it shows that all nearly regular connected graphs with the same density have approximately the same number of spanning trees.

For sparse pseudo-random graphs one can use Theorem 4.23 to estimate the number of Hamilton cycles. Let G be an (n, d, λ)-graph satisfying the conditions of Theorem 4.23. Consider the random subgraph G_p of G, where $p = (\log n + 2 \log \log n)/d$. Let X be the random variable counting the number of Hamilton cycles in G_p. According to Theorem 4.23, G_p has almost surely a Hamilton cycle, and therefore $E[X] \geq 1 - o(1)$. On the other hand, the probability that a given Hamilton cycle of G appears in G_p is exactly p^n. Therefore the linearity of expectation implies $E[X] = h(G)p^n$. Combining the above two estimates we derive:

$$h(G) \geq \frac{1 - o(1)}{p^n} = \left(\frac{d}{(1 + o(1)) \log n}\right)^n.$$

We thus get the following corollary:

Corollary 4.30 [42]. *Let G be an (n, d, λ)-graph with*

$$\lambda = o\big(d^{5/2}/\big(n^{3/2}(\log n)^{3/2}\big)\big).$$

Then G contains at least $\left(\frac{d}{(1+o(1))\log n}\right)^n$ Hamilton cycles.

Note that the number of Hamilton cycles in any d-regular graph on n vertices obviously does not exceed d^n. Thus for graphs satisfying the conditions of Theorem 4.23 the above corollary provides an asymptotically tight estimate on the exponent of the number of Hamilton cycles.

5. CONCLUSION

Although we have made an effort to provide a systematic coverage of the current research in pseudo-random graphs, there are certainly quite a few subjects that were left outside this survey, due to the limitations of space and time (and of the authors' energy). Probably the most notable omission is a discussion of diverse applications of pseudo-random graphs to questions from other fields, mostly Extremal Graph Theory, where pseudo-random graphs provide the best known bounds for an amazing array of problems. We hope to cover this direction in one of our future papers. Still, we would like to believe that this survey can be helpful in mastering various results and techniques pertaining to this field. Undoubtedly many more of them are bound to appear in the future and will make this fascinating subject even more deep, diverse and appealing.

Acknowledgment. The authors would like to thank Noga Alon for many illuminating discussions and for kindly granting us his permission to present his Theorem 4.10 here. The proofs of Theorems 4.1, 4.3 were obtained in discussions with him.

REFERENCES

[1] N. Alon, *The number of spanning trees in regular graphs*, Random Structures and Algorithms **1** (1990), 175–181.

[2]   N. Alon, *Restricted colorings of graphs*, in: "Surveys in Combinatorics", Proc. 14th British Combinatorial Conference, London Math. Society Lecture Notes Series 187, ed. by K. Walker, Cambridge University Press, 1993, 1–33.

[3]   N. Alon, *Explicit Ramsey graphs and orthonormal labelings*, The Electronic J. Combinatorics **1** (1994), R12, 8pp.

[4]   N. Alon, *Tough Ramsey graphs without short cycles*, J. Algebraic Combinatorics **4** (1995), 189–195.

[5]   N. Alon, *Bipartite subgraphs*, Combinatorica **16** (1996), 301–311.

[6]   N. Alon, private communication.

[7]   N. Alon, I. Benjamini and A. Stacey, *Percolation on finite graphs and isoperimetric inequalities*, Ann. Probab. **32** (2004), 1727–1745.

[8]   N. Alon, B. Bollobás, M. Krivelevich and B. Sudakov, *Maximum cuts and judicious partitions in graphs without short cycles*, J. Combinatorial Theory Ser. B **88** (2003), 329–346.

[9]   N. Alon, R. Duke, H. Lefmann, V. Rödl and R. Yuster, *The algorithmic aspects of the regularity lemma*, J. Algorithms **16** (1994), 80–109.

[10] N. Alon and N. Kahale, *Approximating the independence number via the θ-function*, Math. Programming **80** (1998), 253–264.

[11] N. Alon and M. Krivelevich, *Constructive bounds for a Ramsey-type problem*, Graphs and Combinatorics **13** (1997), 217–225.

[12] N. Alon, M. Krivelevich and B. Sudakov, *List coloring of random and pseudorandom graphs*, Combinatorica **19** (1999), 453–472.

[13] N. Alon and P. Pudlak, *Constructive lower bounds for off-diagonal Ramsey numbers*, Israel J. Math. **122** (2001), 243–251.

[14] N. Alon and V. Rödl, *Asymptotically tight bounds for some multicolored Ramsey numbers*, Combinatorica **25** (2005), 125–141.

[15] N. Alon, V. Rödl and A. Ruciński, *Perfect matchings in ε-regular graphs*, Electronic J. Combinatorics, Vol. 5 (1998), publ. R13.

[16] N. Alon and Y. Roichman, *Random Cayley graphs and expanders*, Random Structures and Algorithms **5** (1994), 271–284.

[17] N. Alon, L. Rónyai and T. Szabó, *Norm-graphs: variations and applications*, J. Combinatorial Theory Ser. B **76** (1999), 280–290.

[18] N. Alon and J. Spencer, *The probabilistic method*, 2nd Ed., Wiley, New York 2000.

[19] A. Beveridge, A. Frieze and C. McDiarmid, *Random minimum length spanning trees in regular graphs*, Combinatorica **18** (1998), 311–333.

[20] B. Bollobás, *Random graphs*, 2nd Ed., Cambridge University Press, 2001.

[21] R. C. Bose, *Strongly regular graphs, partial geometries, and partially balanced designs*, Pacific J. Math. **13** (1963), 389–419.

[22] A. Bondy and M. Simonovits, *Cycles of even length in graphs*, J. Combin. Theory Ser. B **16** (1974), 97–105.

[23] A. E. Brouwer, A. M. Cohen and A. Neumaier, *Distance-Regular Graphs*, Springer-Verlag, Berlin, 1989.

[24] A. E. Brouwer and J. H. van Lint, *Strongly regular graphs and partial geometries*, in: Enumeration and design, D. M. Jackson and S. A. Vanstone, Eds., Academic Press, 1984, 85–122.

[25] F. R. K. Chung and R. Graham, *Sparse quasi-random graphs*, Combinatorica **22** (2002), 217–244.

[26] F. R. K. Chung, R. L. Graham and R. M. Wilson, *Quasi-random graphs*, Combinatorica **9** (1989), 345–362.

[27] V. Chvátal, *Tough graphs and hamiltonian circuits*, Discrete Mathematics **5** (1973), 215–218.

[28] V. Chvátal, *Hamiltonian cycles*, in: The traveling salesman problem: a guided tour of combinatorial optimization, E. L. Lawler, J. K. Lenstra, A. H. G. Rinnooy Kan and D. B. Shmoys, Eds., Wiley 1985, 403–429.

[29] V. Chvátal and P. Erdős, *A note on Hamiltonian circuits*, Discrete Math. **2** (1972), 111–113.

[30] C. Cooper, A. Frieze and B. Reed, *Random regular graphs of non-constant degree: connectivity and Hamilton cycles*, Combinatorics, Probability and Computing **11** (2002), 249–262.

[31] H. Davenport, *Multiplicative Number Theory*, 2nd edition, Springer Verlag, New York, 1980.

[32] P. Erdős, R. Faudree, C. Rousseau and R. Schelp, *On cycle-complete graph Ramsey numbers*, J. Graph Theory **2** (1978), 53–64.

[33] P. Erdős and A. Rényi, *On the evolution of random graphs*, Publ. Math. Inst. Hungar. Acad. Sci. **5** (1960), 17–61.

[34] P. Erdős, A. L. Rubin and H. Taylor, *Choosability in graphs*, Proc. West Coast Conf. on Combinatorics, Graph Theory and Computing, Congressus Numerantium XXVI, 1979, 125–157.

[35] P. Erdős and J. Spencer, *Imbalances in k-colorations*, Networks **1** (1972), 379–385.

[36] W. Feit and G. Higman, *The nonexistence of certain generalized polygons*, J. Algebra **1** (1964), 114–131.

[37] E. Friedgut, *Sharp thresholds of graph properties, and the k-sat problem*. With an appendix by Jean Bourgain, Journal Amer. Math. Soc. **12** (1999), 1017–1054.

[38] J. Friedman, J. Kahn and E. Szemerédi, *On the second eigenvalue in random regular graphs*, Proc. of 21th ACM STOC (1989), 587–598.

[39] J. Friedman, *A proof of Alon's second eigenvalue conjecture*, preprint.

[40] A. Frieze, *On the value of a random minimum spanning tree problem*, Discrete Applied Math. **10** (1985), 47–56.

[41] A. Frieze, *On the number of perfect matchings and Hamilton cycles in ε-regular non-bipartite graphs*, Electronic J. Combinatorics Vol. 7 (2000), publ. R57.

[42] A. Frieze and M. Krivelevich, *Hamilton cycles in random subgraphs of pseudo-random graphs,* Discrete Math. **256** (2002), 137–150.

[43] A. Frieze, M. Krivelevich and R. Martin, *The emergence of a giant component in random subgraphs of pseudo-random graphs,* Random Structures and Algorithms **24** (2004), 42–50.

[44] Z. Füredi and J. Komlós, *The eigenvalues of random symmetric matrices,* Combinatorica **1** (1981), 233–241.

[45] M. R. Garey and D. S. Johnson, *Computers and Intractability: A Guide to the Theory of NP-Completeness,* W. H. Freeman, 1979.

[46] C. Godsil and G. Royle, *Algebraic graph theory,* Springer Verlag, New York, 2001.

[47] A. Hajnal and E. Szemerédi, Proof of a conjecture of Erdős, in: *Combinatorial Theory and its applications,* Vol. II, P. Erdős, A. Rényi and V. T. Sós, Eds., Colloq. Math. Soc. J. Bolyai 4, North Holland, Amsterdam, 1970, pp. 601–623.

[48] A. Hoffman, *On eigenvalues and colorings of graphs,* in: Graph Theory and its Applications, Academic Press, New York, 1970, 79–91.

[49] S. Janson, T. Luczak and A. Ruciński, *Random graphs,* Wiley, New York, 2000.

[50] Y. Kohayakawa, V. Rödl and P. Sissokho, *Embedding graphs with bounded degree in sparse pseudo-random graphs,* Israel Journal of Mathematics **139** (2004), 93–137.

[51] Y. Kohayakawa, V. Rödl and M. Schacht, *The Turán theorem for random graphs,* Combinatorics, Probability, and Computing **13** (2004), 61–91.

[52] J. Kollár, L. Rónyai and T. Szabó, *Norm-graphs and bipartite Turán numbers,* Combinatorica **16** (1996), 399–406.

[53] J. Komlós, *The blow-up lemma,* Combinatorics, Probability and Computing **8** (1999), 161–176.

[54] J. Komlós, G. N. Sárközy and E. Szemerédi, *The blow-up lemma,* Combinatorica **17** (1997), 109–123.

[55] J. Komlós and M. Simonovits, *Szemerédi Regularity Lemma and its applications in Extremal Graph Theory,* in: Paul Erdős is 80 II, Bolyai Soc. Math. Stud. 2, Budapest 1996, 295–352.

[56] J. Komlós and E. Szemerédi, *Limit distributions for the existence of Hamilton circuits in a random graph,* Discrete Mathematics **43** (1983), 55–63.

[57] J. Kratochvil, Zs. Tuza and M. Voigt, *New trends in the theory of graph colorings: choosability and list coloring,* Contemporary Trends in Discrete Mathematics (R. L. Graham et al., eds.), DIMACS Series in Discrete Math. and Theor. Computer Science 49, Amer. Math. Soc., Providence, RI, 1999, 183–197.

[58] M. Krivelevich and B. Sudakov, *Sparse pseudo-random graphs are Hamiltonian,* J. Graph Theory **42** (2003), 17–33.

[59] M. Krivelevich, B, Sudakov and T. Szabó, *Triangle factors in pseudo-random graphs,* Combinatorica **24** (2004), 403–426.

[60] M. Krivelevich, B. Sudakov and V. Vu, *A sharp threshold for network reliability,* Combinatorics, Probability and Computing **11** (2002), 465–474.

[61] M. Krivelevich, B. Sudakov, V. Vu and N. Wormald, *Random regular graphs of high degree,* Random Structures and Algorithms **18** (2001), 346–363.

[62] F. Lazebnik, V. A. Ustimenko and A. J. Woldar, *Polarities and 2k-cycle-free graphs,* Discrete Math. **197/198** (1999), 503–513.

[63] R. Lidl and H. Niederreiter, *Finite fields,* Cambridge University Press, Cambridge, 1997.

[64] L. Lovász, *Combinatorial problems and exercises,* 2nd Ed., North Holland, Amsterdam, 1993.

[65] A. Lubotzky, R. Phillips and P. Sarnak, *Ramanujan graphs,* Combinatorica **8** (1988), 261–277.

[66] F. J. MacWilliams and N. J. A. Sloane, *The Theory of Error-Correcting Codes,* North Holand, Amsterdam, 1977.

[67] G. Margulis, *Probabilistic characteristics of graphs with large connectivity,* Problems Info. Transmission **10** (1974), 101–108.

[68] G. Margulis, *Explicit group-theoretic constructions of combinatorial schemes and their applications in the construction of expanders and concentrators* (in Russian) Problemy Peredachi Informatsii **24** (1988), 51–60; translation in: Problems Inform. Transmission 24 (1988), 39–46.

[69] A. Nilli, *On the second eigenvalue of a graph,* Discrete Math. **91** (1991), 207–210.

[70] L. Posa, *Hamiltonian circuits in random graphs,* Discrete Math. **14** (1976), 359–364.

[71] V. Rödl and A. Ruciński, *Perfect matchings in ϵ-regular graphs and the blow-up lemma,* Combinatorica **19** (1999), 437–452.

[72] J. Seidel, *A survey of two-graphs,* in: Colloquio Internazionale sulle Teorie Combinatorie (Rome, 1973), vol I, Atti dei Convegni Lincei, No. 17, Accad. Naz. Lincei, Rome, 1976, 481–511.

[73] M. Simonovits and V. T. Sós, *Szemerédi's partition and quasirandomness,* Random Structures and Algorithms **2** (1991), 1–10.

[74] M. Simonovits and V. T. Sós, *Hereditary extended properties, quasi-random graphs and not necessarily induced subgraphs,* Combinatorica **17** (1997), 577–596.

[75] B. Sudakov, T. Szabó and V. H. Vu, *A generalization of Turán's theorem,* Journal of Graph Theory **49** (2005), 187–195.

[76] T. Szabó, *On the spectrum of projective norm-graphs,* Information Processing Letters **86** (2003) 71–74.

[77] T. Szabó and V. H. Vu, *Turán's theorem in sparse random graphs,* Random Structures and Algorithms **23** (2003) 225–234.

[78] R. M. Tanner, *Explicit concentrators from generalized N-gons,* SIAM J. Algebraic Discrete Methods **5** (1984), 287–293.

[79] A. Thomason, *Pseudo-random graphs,* in: Proceedings of Random Graphs, Poznań 1985, M. Karoński, ed., Annals of Discrete Math. **33** (North Holland 1987), 307–331.

[80] A. Thomason, *Random graphs, strongly regular graphs and pseudo-random graphs,* Surveys in Combinatorics, 1987, C. Whitehead, ed., LMS Lecture Note Series **123** (1987), 173–195.

[81] P. Turán, *Egy gráfelméleti szélsőértékfeladatról* (in Hungarian), Mat. Fiz. Lapok **48** (1941), 436–452.

[82] V. G. Vizing, *Coloring the vertices of a graph in prescribed colors* (in Russian), Diskret. Analiz. No. 29, Metody Diskret. Anal. v. Teorii Kodov i Shem **101** (1976), 3–10.

[83] N. C. Wormald, *Models of random regular graphs,* in: Surveys in Combinatorics, 1999, J. D. Lamb and D. A. Preece, eds, pp. 239–298.

[84] V. Vu, *On some degree conditions which guarantee the upper bound on chromatic (choice) number of random graphs,* J. Graph Theory **31** (1999), 201–226.

Michael Krivelevich

Department of Mathematics
Raymond and Beverly Sackler
Faculty of Exact Sciences
Tel Aviv University
Tel Aviv 69978
Israel

krivelev@post.tau.ac.il

Benny Sudakov

Department of Mathematics
Princeton University
Princeton, NJ 08544
U.S.A.

bsudakov@math.princeton.edu

BOLYAI SOCIETY
MATHEMATICAL STUDIES, 15

Conference on Finite
and Infinite Sets
Budapest, pp. 263–283.

BOUNDS AND EXTREMA FOR CLASSES OF GRAPHS AND FINITE STRUCTURES

J. NEŠETŘIL[*]

We consider the homomorphism (or colouring) order $\mathcal{C}$ induced by all finite structures (of a given type; for example graphs) and the existence of a homomorphism between them. This ordering may be seen as a lattice which is however far from being complete. In this paper we study (upper) bounds, suprema and maximal elements in $\mathcal{C}$ of some frequently studied classes of structures (such as classes of structures with bounded degree of its vertices, degenerated and classes determined by a finite set of forbidden substructures). We relate these extrema to cuts and duality theorems for $\mathcal{C}$. Some of these results hold for general finite relational structures. In view of combinatorial problems related to coloring problems this should be regarded as a surprise. We support this view also by showing both analogies and striking differences between undirected and oriented graphs (i.e. for the easiest types) This is based on our recent work with C. Tardif.

We also consider minor closed classes of graphs and we survey recent results obtained by P. Ossona de Mendez and author. We show how the order setting captures Hadwiger conjecture and suggests some new problems.

1. INTRODUCTION

Graph theory receives its mathematical motivation mostly from two areas of mathematics: algebra and geometry (topology) and it is fair to say that graphical notions stood at the birth of algebraic topology (in the beginning called combinatorial topology). Consequently, various operations and relations for graphs stress either its algebraic aspects (e.g. colourings and

---
[*]Supported by Grants LN00A56 and 1M0021620808 of the Czech Ministary of Education.

homomorphisms, various products and spaces associated with graphs) or its geometrical aspects (e.g. drawings, contractions, embeddings). It is only natural that the key place in modern graph theory is played by (fortunate) mixtures of both approaches as exhibited best by various modifications of the notion of graph minor. From the algebraic point of view perhaps the most natural notion which captures comparision of two graphs is that of a homomorphism.

A *homomorphism* $G \rightarrow H$ is a mapping $f : V(G) \rightarrow V(H)$ which satisfies $f(u)f(v) \in E(H)$ for any edge $uv \in E(G)$. (We shall consider both directed and undirected graphs. This will be always clearly specified. Some of the results hold for general finite structures. Section 5 is devoted entirely to them.)

The central notion of this paper is the quasiorder (and partial order) induced by the existence of a homomorphism. This notion and its context is illustrated on the example of graphs.

Given graphs G, H we denote by $G \leq H$ the existence of a homomorphism $G \rightarrow H$. Clearly $\leq$ is a quasiorder. If we consider isomorphism types of minimal retracts (or *cores*, see [16]) then we obtain a partial order. This quasiorder (and partial order) is called *colouring order* (or *homomorphism order*, [16]) and it is denoted by $\mathcal{C}$. We denote by $G \sim H$ the equivalence given by $G \leq H \leq G$. We also denote by $<$ the strict version of $\leq$ (thus $G < H$ iff $G \leq H$ and $G \not\sim H$). For a graph H we denote by $\mathcal{C}_H$ the principal ideal determined by H: $\mathcal{C}_H = \{G;\ G \leq H\}$. $\mathcal{C}_H$ is also called a *colour class*. This name is justified by interpreting homomorphisms as generalized colourings: Indeed, for undirected graphs a homomorphism $G \rightarrow K_k$ is a just a (proper) k-colouring of graph G and thus a homomorphism $G \rightarrow H$ is also called a *H-colouring*. Consequently, $\mathcal{C}_H$ is the class of all H-colourable graphs; hence the name colour class. It follows that the question whether $G \leq H$ is difficult to decide (and it is NP-complete in a very strong sense). We refer to [14, 16, 6, 9] as a background information. Our graph–theory terminology is standard.

It is perhaps surprising how many fine combinatorial questions are captured by *order-theoretic* properties of the colouring order $\mathcal{C}$. In this paper we concentrate on extremal elements of this order: greatest and maximal elements, suprema and (upper) bounds in general. It appears that these extremal graphs capture various problems which are as remote as *duality theorems* ([23]) and celebrated Hadwiger conjecture. These interpretations lead also to some, hopefully interesting, problems.

Given a class $\mathcal{K}$ of graphs it is usually a difficult question to find a graph H which is maximal (or greatest, or even supremum) of $\mathcal{K}$ in $\mathcal{C}$. Among other things such result yields maximal chromatic number of a graph in $\mathcal{K}$. As these concepts are the subject of this paper we recall the corresponding definitions in the setting of colouring order $\mathcal{C}$:

A graph H is said to be *maximal* of $\mathcal{K}$ if $H \in \mathcal{K}$ and no graph $G \in \mathcal{K}$ satisfies $H < G$.

A graph H is said to be an *(upper) bound* of $\mathcal{K}$ if every graph $G \in \mathcal{K}$ satisfies $H \leq G$. If in addition $H \in \mathcal{K}$ then H is said to be *greatest* graph in $\mathcal{K}$ (or *maximum* of $\mathcal{K}$).

A graph H is said to be *supremum* of $\mathcal{K}$ if $G \leq H$ for every $G \in \mathcal{K}$ and if for every graph $H' < H$ there exists a graph $G \in \mathcal{K}$ such that $G \not\leq H'$.

For example, in this setting, the 4-colour theorem says that K_4 is the greatest graph in the class of all planar graphs. This obviously cannot be improved. On the other hand, Grötzsch's theorem says that K_3 is an upper bound of the class of all planar K_3-free graphs. However, as we will see, this may be improved as K_3 fails to be a supremum of this class.

This is our motivating example. By proving that k is the maximal chromatic number of a class $\mathcal{K}$ we claim that the graph K_k is an upper bound for $\mathcal{K}$ (in the coloring order $\mathcal{C}$) while the graph K_{k-1} fails to be such upper bound. But the homomorphism order is dense and thus it is natural to ask if there exists a smaller upper bound.

In this paper we determine suprema and greatest elements of some of the frequently studied classes of graphs (compare [6]). These include classes Forb $(\mathcal{F})$ where $\mathcal{F}$ is a finite set of connected graphs: We denote by Forb $(\mathcal{F})$ the class of all finite graphs G which satisfy $F \not\leq G$ for every $F \in \mathcal{F}$. Alternatively, Forb $(\mathcal{F})$ is the class of all graphs which do not contain a homomorphic image of a graph from $\mathcal{F}$. Or we could say that Forb $(\mathcal{F})$ is the class of all $\mathcal{F}$-free graphs. In yet another way we can say that Forb $(\mathcal{F})$ is the class of all graphs defined by forbidden homomorphisms from a finite set of graphs $\mathcal{F}$. In our context these are natural classes of graphs. A bit surprisingly all related questions can be solved for classes Forb $(\mathcal{F})$. For undirected graphs this is much easier than for directed graphs where we use strong results obtained jointly with C. Tardif [23, 24].

As an approximation to the minor closed classes we also consider extrema relativized by classes of bounded degree graphs and classes of d-degenerated graphs. While for degenerated graphs we have a full discussion of extremal properties for bounded degrees this seems to be a very difficult problem.

In a way this line of research presents a development of *inverse program* for graph colouring problems: while in the usual setting one investigates chromatic number and similar characteristics of a given class of graphs here we are interested in the structural properties of the bounds themselves. Random graphs (and high-chromatic sparse) graphs provide us with rigidity properties.

In this paper we also generalize some of the algebraic constructions to finite relational models in the full generality. Such a generalizations is not for its own sake. Homomorphisms of relational structures and the corresponding H-coloring problem is equivalent to CSP problems and this general approach led recently to a new approach to classical problems such as *dichotomy conjecture*. For more on this see [9, 1, 13]. In Section 2 we briefly introduce this general framework.

The paper is organized as follows: In Section 2 we start with general systems of type Δ and in Section 3 we prove non-existence of proper suprema for classes of bounded degree Δ-systems in the full generality. Sections 4, and 5 deal with graphs. In Section 4 we consider d-degenerated graphs and we display the striking difference between classes of d-degenerated graphs and classes of bounded degree graphs. We determine the suprema of degenerated classes in every color class. In Section 5 we consider minor closed classes and show the relevance of a recent result [20] to Hadwiger conjecture via our order-theoretic setting. It is also here where we introduce the cuts and their characterization problem. In Section 6 we return to Δ-systems and completely characterize extrema for classes of type $\mathrm{Forb}\,(\mathcal{F})$ for a finite set $\mathcal{F}$ of connected graphs. This is related to our joint work with C. Tardif ([23, 24]). In Section 5 we consider oriented graphs and we prove the main result on classes $\mathrm{Forb}\,(\mathcal{F})$ and in Section 6 we conclude with some remarks and problems.

2. RELATIONAL STRUCTURES

A relational structure of a given type generalizes the notion of a relation and of a graph to more relations and to higher (non-binary) arities.

A *type* Δ is a sequence $(\delta_i;\ i \in I)$ of positive integers. A *relational system* A of type Δ is a pair $\big(X, (R_i;\ i \in I)\big)$ where X is a set and $R_i \in X^{\delta_i}$; that is R_i is a δ_i-nary relation on X. In this paper we shall always assume that X is a finite set (thus we consider finite relational systems only).

The type $\Delta = (\delta_i;\ i \in I)$ will be fixed throughout this paper. Note that for the type $\Delta = (2)$ relational systems of type Δ correspond to oriented graphs, the case $\Delta = (2, 2)$ corresponds to oriented graphs with blue-green colored edges. Relational systems (of type Δ) will be denoted by capital letters $A, B, C, \ldots$. A relational system of type Δ is also called a Δ-*system*. If $A = \big(X, (R_i;\ i \in I)\big)$ we also denote the base set X as $\underline{A}$ and the relation R_i by $R_i(A)$. Let $A = \big(X, (R_i;\ i \in I)\big)$ and $B = \big(Y, (S_i;\ i \in I)\big)$ be Δ-systems. A mapping $f : X \to Y$ is called a *homomorphism* if for each $i \in I$ holds: $(x_1, \ldots, x_{\delta_i}) \in R_i$ implies $\big(f(x_1), \ldots, f(x_{\delta_i})\big) \in S_i$.

In other words a homomorphism f is any mapping $F : \underline{A} \to \underline{B}$ which satisfies $f\big(R_i(A)\big) \subset R_i(B)$ for each $i \in I$. (Here we extended the definition of f by putting $f(x_1, \ldots, x_t) = \big(f(x_1), \ldots, f(x_t)\big)$.)

For Δ-systems A and B we write $A \to B$ if there exists a homomorphism from A to B. Hence the symbol $\to$ denotes a relation that is defined on the class of all Δ-systems. This relation is clearly reflexive and transitive, thus induces a quasi-ordering of all Δ-systems. As is usual with quasi-orderings, it is convenient to reduce it to a partial order on classes of equivalent objects: Two Δ-systems A and B are called *homomorphically equivalent* if we have both $A \to B$ and $B \to A$; we then write $A \sim B$.

The relation $\to$ induces an order on the classes of homomorphically equivalent Δ-system, which we call the *homomorphism order* and we denote it again by $\mathcal{C}$ (suppressing type Δ which will be clear from the context). Other categorical notions which were introduced in Section 1 are defined analogously as in the case of graphs. Particularly the core of a Δ-system is defined analogously as for graphs. The operations of sum, product and exponentiation reveal the rich categorical structure of the homomorphism order:

- The *sum* $A + B$ of A and B has the property that for any Δ-system C, we have $A + B \to C$ if and only if $A \to C$ and $B \to C$.

- The *product* $A \times B$ of A and B has the property that for any Δ-system C, we have $C \to A \times B$ if and only if $C \to A$ and $C \to B$.

It follows that the homomorphism order is a distributive lattice. This categorical description will be more relevant to us that the actual (i.e. inner) description of sums and products, which is bit technical though standard.

Some further notions for graphs and their classes may be translated to relational systems. So we shall speak about cores, classes Forb $(\mathcal{F})$, bounds, suprema etc. for Δ-systems. The type Δ will be always properly understood from the context. We add the following two notions which relate Δ-systems and graphs:

Let $A = \big(X, (R_i; \ i \in I)\big)$ be a Δ-system. The *graph-shadow* of A is the graph (X, E) where $xy \in E$ providing there exists $i \in I$ and $(x_1, \ldots, x_t) \in R_i$ such that $x = x_a$ and $y = x_b$ for *distinct* indices $a \neq b$. The graph shadow of A will be denoted $sh(A)$. Note that $sh(A)$ may have loops. We say that Δ-system A is *connected* if its shadow $sh(A)$ is connected. Alternatively, A is connected if it cannot be written as a sum $B + C$. *Degree* of a vertex $x \in \underline{A}$ is the degree of x in $sh(A)$. (This may sound as slightly unusual definition but we are interested in bounded degrees of Δ- systems for a fixed type Δ and thus our definition suffices.)

Given Δ-system $A = \big(X, (R_i; \ i \in I)\big)$ we also define *incidence graph* $\mathrm{Inc}\,(A)$ as the bipartite graph $\big(X \bigcup \sum_{i \in I} R_i, E\big)$ where $xe \in E$ iff $x \in e \in R_i$ for some $i \in I$. Here we denoted by $\sum_{i \in I} R_i$ the disjoint union of sets $R_i; \ i \in I$.

3. BOUNDED DEGREES

Let type Δ be fixed throughout this section. Denote by Δ_d the class of all Δ-systems A with maximal degree $\leq d$. For a finite set $\mathcal{F} = \{F_1, \ldots, F_t\}$ of connected graphs denote also $\Delta_d(\mathcal{F})$ the class of all Δ-systems $A \in \Delta_d$ with $F_i \nrightarrow G$ for $i = 1, \ldots, t$. Thus $\Delta_d(\mathcal{F}) = \Delta_d \cap \mathrm{Forb}\,(\mathcal{F})$.

Celebrated Brooks theorem states that while K_{d+1} is a bound (and indeed greatest element) of the class $(2)_d$ by forbidding K_{d+1} this may be improved to a better bound K_d. It follows that K_{d+1} fails to be supremum of the class $(2)_d\big(\{K_{d+1}\}\big)$ (which here means just graphs). This is not an accident and a similar statement holds in general thus yielding a whole hierarchy of Brook's type bounds.

We say that a A is a *proper supremum* of a class $\mathcal{K}$ of Δ-systems if A is supremum and $A \notin \mathcal{K}$ (as we are working with equivalence classes the later condition of course means that $A \nsim B$ for every Δ-system $B \in \mathcal{K}$). The following is the main result of this section:

Theorem 3.1. *The class $\Delta_d(\mathcal{F})$ has no proper suprema for any $d \geq 1$ and any finite set $\mathcal{F}$ of connected Δ-systems.*

Motivated by the above interpretation of Brooks theorem Theorem 3.1 means that one cannot hope to prove the "best Brooks-type" bound. We shall later see that this statement is in a sharp contrast with properties of all structures and even of classes of d-degenerated graphs (see Section 3).

Advancing the proof of Theorem 3.1 we state first the following easy:

Lemma 3.2. *Let B, B' be bounds of a class $\mathcal{K}$ of Δ-systems. Then $B \times B'$ is also a bound.*

The key of the proof of Theorem 3.1 is the following construction which extends [3, 8]:

Proposition 3.3. *The class $\Delta_d(\mathcal{F})$ has a bound $B \in \mathrm{Forb}\,(\mathcal{F})$.*

Proof. Put $\mathcal{F} = \{F_1, \ldots, F_t\}$, let D denotes the maximal size of $F_i, i \in I$ and let δ denotes maximal δ_i; $i \in I$ (i.e. the maximal arity of relation in our Δ-systems). Put $a = \delta D$ and $b = d^{2a+1}$. Let $A = (X, (R_i; \ i \in I)) \in \Delta_d(\mathcal{F})$. By our assumption all vertices of X have degree $\leq d$. Consider the (auxiliar) graph defined by $xy \in E$ iff the distance of x and y in $sh(A)$ is at most $2a$. It follows that that the graph (X, E) has all vertices of degree $< b$ and thus it may be properly colored by b colors. For the shadow graph $sh(A)$ this in turn means that there exists a coloring $c : X \rightarrow \{1, \ldots, b\}$ such that any two vertices x, z of $sh(A)$ at distance at most a get different color. Particularly any subgraph of $sh(A)$ induced by all vertices of $sh(A)$ at distance at most a from a fixed vertex is colored by distinct colors only. This property may be used for a construction of a bound $C \in \mathrm{Forb}\,(\mathcal{F})$:

(A $2a-ball$ is a Δ-structure A together with a fixed vertex r such that all other vertices of A have distance $< 2a$ from r in $sh(A)$.)

The vertices of C are all $2a$-balls $(B, r) \in \Delta_d(\mathcal{F})$ where $B \subset \{1, \ldots, b\}$. We put $((B_1, r_1), \ldots, (B_{\delta_i}, r_{\delta_i})) \in R_i(C)$ iff the following holds (for every $i \in I$):

    i. $(r_1, \ldots, r_{\delta_i}) \in R_i(B_j)$ for all $j = 1, \ldots, \delta_i$.

    ii. For any $j, 1 \leq j \leq \delta_i$ all the vertices of B_j of distance $< a$ from r_j belong as well to any other $B_{j'}, 1 \leq j' \leq \delta_i$;

Clearly C is a Δ-system. We already indicated all the essential fact which yield that C is a bound of the class $\Delta_d(\mathcal{F})$: Given a system $A \in \Delta_d(\mathcal{F})$ and a mapping $c : X \rightarrow \{1, \ldots, b\}$ as above we simply define the mapping

$f : X \to \underline{B}$ by putting $f(x) = c(B_{2a}, x)$ where (B_{2a}, x) is the $2a$−ball in A induced by all vertices of distance $< 2a$ from x and by $c(B_{2a}, x)$ we denoted its image by the map c (i.e. isomorphic copy of (B_{2a}, x) induced on the set $\{c(y);\ y \in \underline{B}_{2a}\}$). It follows from conditions i. that f is a homomorphism $A \to C$.

Now we prove that $C \in \mathrm{Forb}\,(\mathcal{F})$. Towards this end let $f : F \to C$ be a homomorphism for an $F \in \mathcal{F}$. Put $\underline{F} = \{u_1, \ldots, u_t\}$ and $f(u_j) = (A_j, r_j)$. According to the definition each of the vertices u_i has distance $< D$ from u_1 and thus each of the vertice $r_1, \ldots, r_t$ is reached from r_1 by a walk of length $< \delta D$. According to the ii. we get that $\{r_1, \ldots, r_t\}$ is a subset of $\underline{A}_1$ and thus (by ii. of the definition of edges of C) we get that f induces a homomorphism $F \to A_1$ which is a contradiction. ∎

The statement of Theorem 3.1 will be proved in the following more technical form:

Theorem 3.4. *Let $\mathcal{F}$ be a finite set of connected Δ-structures. Let C be a bound for the class $\Delta_d(\mathcal{F})$, $C \notin \Delta_d(\mathcal{F})$. Then there exists a bound C' for $\Delta_d(\mathcal{F})$ with $C' < C$.*

Proof. In the situation of Theorem 3.4 let C be a bound. If C is connected than we can consider the system $\mathcal{F} \cup \{C\} = \mathcal{F}'$. It is easy to see that $\mathrm{Forb}\,(\mathcal{F}) = \mathrm{Forb}\,(\mathcal{F}')$ and using Proposition 3.3 there exists a bound C' of $\Delta_d(\mathcal{F}')$ in $\mathrm{Forb}\,(\mathcal{F}')$. But then obviously $C' < C$.

Thus assume that C is a disconnected core. Let K be a component of C such that $K \notin \Delta_d(\mathcal{F})$. As before put $\mathcal{F} \cup \{K\} = \mathcal{F}'$. Applying Proposition 3.3 and Lemma 3.2, there exists a Δ-system C' such that C' is a bound of $\Delta_d(\mathcal{F}')$, $C' \in \mathrm{Forb}\,(\mathcal{F}')$, and $C' \le C$. But the definition of $\mathcal{F}'$ in fact implies $C' < C$. Thus it suffices to prove that C' is also a bound for the class $\Delta_d(\mathcal{F})$. In fact we prove again $\Delta_d(\mathcal{F}) = \Delta_d(\mathcal{F}')$. One direction is clear: $\Delta_d(\mathcal{F}) \supset \Delta_d(\mathcal{F}')$ (as $\mathcal{F} \subset \mathcal{F}'$). In the reverse direction assume contrary: let $A \in \Delta_d(\mathcal{F}) \setminus \Delta_d(\mathcal{F}')$. It is $K \le A \le C$ and thus there exists a component L of A such that $K \le L$. Now $A \le C$ and thus $K \le L \le K'$ for a component K' of C. However we assumed that C is a core and thus $K = K'$ and thus also $K = L$. This is a contradiction as $K \notin \Delta_d(\mathcal{F})$ while A and thus also $L \in \Delta_d(\mathcal{F})$. ∎

It follows that for classes of form $\Delta_d \cap \mathrm{Forb}\,(\mathcal{F})$ a supremum exists only if there exists $H \in \Delta_d \cap \mathrm{Forb}\,(\mathcal{F})$ such that $\Delta_d \cap \mathrm{Forb}\,(\mathcal{F}) \subset \mathcal{C}_H$; this means that H is the greatest element of $\Delta_d(\mathcal{F})$. However the structure of classes $\Delta_d \cap \mathcal{C}_H$ is far from obvious. For example the following two problems have been isolated;

Problem 3.5 (Independence problem). Let $d \geq 3$. Is it true that for every graph $G \in \Delta_d$, $G < K_d$ there exists a graph $G' \in \Delta_d$ such that neither $G \leq G'$ nor $G' \leq G$ (i.e. graphs G and G' are incomparable graphs in Δ_d)?

This problem is related to the complexity of H-colourings of bounded degree graphs which have been studied e.g. in [5, 11].

Problem 3.6 (Pentagon problem). Does there exists an g such that any cubic (i.e. 3-regular) graph G with girth g is homomorphic to C_5 (i.e. is C_5-colourable)?

Partial results related to this problem were obtained in [28, 10, 11, 7]. One should note that for C_{2k+1}, $k > 2$, (instead of C_5) the answer is negative.

4. DEGENERATED CLASSES OF GRAPHS

In this section we consider undirected graphs only. The degenerated graphs are low-density graphs and as such they serve as an approximation for coloring problems of bounded degree- and minor-restricted classes of graphs. We shall see that their homomorphism behaviour differs very much from these classes (which are discussed in Sections 3 and 5). Recall that a graph $G = (V, E)$ is said to be *d-degenerated* if there exists a linear ordering $v_1 < v_2 < \cdots < v_n$ of vertices of G such that for every i holds

$$\left| \{v_j;\ j < i,\ v_j v_i \in E\} \right| \leq d$$

Alternatively, a d-degenerated graph can be defined by the condition

$$\delta(G') \leq d$$

for every subgraph G' of G ($\delta(G)$ denotes the minimal degree of a vertex of G). (Yet another way is to define d-degenerated graphs by the hereditary edge-density.)

The class DEG_1 is just the class of all forests. For $d > 1$ these classes are more interesting. Similarily as in the previous section we denote by $DEG_d(\mathcal{F})$ the class of all d-degenerated graphs which belong to the class Forb $(\mathcal{F})$. While these definitions are formally similar the extremal properties of these classes are strikingly different.

Theorem 4.1. *Let $d \geq 2$. Then the following holds:*

i. K_{d+1} is the greatest graph in DEG_d.

ii. For every finite set $\mathcal{F}$ of non-bipartite connected graphs the class $DEG_d(\mathcal{F})$ has supremum K_{d+1}.

Corollary 4.2. *For any $d \geq 2$ and any proper subclass $DEG_d(\mathcal{F})$ of Δ_d (where $\mathcal{F}$ is a finite set of non-bipartite connected graphs) the class $DEG_d(\mathcal{F})$ fails to be bounded by an $\mathcal{F}$-free graph.*

Note that for sets $\mathcal{F}$ which contain a bipartite graph the situation is much simpler and different - K_1 is a bound.

Proof. Clearly it suffices to prove *ii*. Let $d \geq 2$ and $\mathcal{F}$ be as assumed. Let $\mathcal{F}'$ denotes the set of all non-bipartite blocks of graphs belonging to $\mathcal{F}$. As any graph $F \in \mathcal{F}$ contains a non-bipartite block it follows that the class $DEG_d(\mathcal{F}')$ is a subclass of the class $DEG_d(\mathcal{F})$. Put l the maximal number of vertices of a graph belonging to $\mathcal{F}'$. Now let H be a graph, $H < K_{d+1}$. Put $k = |V(H)|$. We shall construct a graph G with the following properties:

1. G has girth $> l$ (and thus particularly $F \not\leq G$ for any $F \in \mathcal{F}'$ and consequently also $F \not\leq G$ for any $F \in \mathcal{F}$).

2. G is d-degenerated;

3. Any homomorphic image of G with at most k vertices contains K_{d+1}.

It follows from 3. that $G \not\leq H$ and thus H fails to be a bound of $DEG_d(\mathcal{F})$.

The graph G will be constructed by means of Descartes–Tutte-type of construction as follows (compare [10]):

We shall construct graphs $G_1, G_2, \ldots, G_{d+1}$; G_{d+1} will be the desired graph G. Put $G_1 = K_1$ and $G_2 = K_2$. In the induction step assume that G_i is constructed. Put $|V(G_i)| = p_i$ and let $(X_{i+1}, \mathcal{M}_{i+1})$ be p_i-uniform hypergraph without cycles of length $\leq l$ and with chromatic number $> k$ (it exists by [4, 12]). For every $M \in \mathcal{M}_{i+1}$ take an isomorphic copy G_i^M of G_i and assume $V(G_i^M) \cap X_{i+1} = \emptyset$, $V(G_i^M) \cap V(G_i^{M'}) = \emptyset$, for all $M \neq M' \in \mathcal{M}_{i+1}$. Finally for every $M \in \mathcal{M}_{i+1}$ fix a bijection $\pi_{i+1}^M : V(G_i^M) \to M$. Define the graph $G_{i+1} = (V_{i+1}, E_{i+1})$ as follows:

$$V_{i+1} = X_{i+1} \cup \bigcup \left(V(G_i^M); M \in \mathcal{M}_{i+1} \right)$$

$$E_{i+1} = \bigcup \left(E(G_i^M); M \in \mathcal{M}_{i+1} \right) \cup \left\{ v\pi_{i+1}^M(v); \ v \in V(G_i^M), \ M \in \mathcal{M}_{i+1} \right\}.$$

G_{i+1} does not contain cycles of length $\leq l$ (in fact, by our choice of G_1 and G_2 it does not contain cycles of length $\leq 3l$; we do not optimize here). We also prove by induction for $i = 1, 2, \ldots, d+1$ that G_i is an $(i-1)$-degenerated graph. In the induction step assume that G_i has an $(i-1)$-degenerated ordering. For $V(G_{i+1})$ choose such an ordering which satisfies $x < v$ for all $x \in X_{i+1}$ and $v \in V(G_i^M)$ and coincides on any set $V(G_i^M)$ with $(i-1)$-degenerated ordering (of G_i^M). Clearly this is an i-degenerated ordering of G_{i+1}.

Finally, let $f : V(G_{d+1}) \to H$ be a homomorphism, $|V(H)| \leq k$. By the downward induction for $j = d+1, d, \ldots, 1$ we prove that for every j there exists $M_j \in \mathcal{M}_j$ such that f restricted to the set M_j is a constant. However this is nearly obvious as the building blocks of our construction – the hypergraphs $(X_i, \mathcal{M}_j)$ – have all chromatic number $> k$. As every M_j is joined by an edge to all vertices of $V(G_{j-1}^{M_j})$ we get that the homomorphic image of G under f contains K_{d+1}, which is a contradiction. ∎

We use properties of degenerated classes of graphs again in Section 6.

5. Minor closed classes of graphs

A class of graphs $\mathcal{K}$ is said to be *minor closed* if it contains all minors of any of its member. We say that $\mathcal{K}$ is *proper* if it does not contain all graphs. Note that all graphs in a proper minor closed class of graphs are d-degenerated for a $d = d(\mathcal{K})$ (by Mader's Theorem). Consequently any minor closed class of graphs is bounded (in $\mathcal{C}$). However extremal graphs are much more difficult for minor closed classes then for bounded degree and d-degenerated classes. One of the few general results was obtained recently [20] as a culmination of previous efforts [17, 18, 19]. It is the analogy of Theorem 3.1 (for minor closed classes instead of classes of bounded degrees):

Theorem 5.1. *Let $\mathcal{K}$ be any proper minor closed class of graphs. Let $\mathcal{F}$ be a finite set of connected graphs. Then the class* Forb $(\mathcal{F}) \cap \mathcal{K}$ *(of all $\mathcal{F}$-free graphs from $\mathcal{K}$) is bounded by a graph from* Forb $(\mathcal{F})$ *(i.e. by a $\mathcal{F}$-free graph).*

Explicitly, there exists a graph $H = H(\mathcal{F}, \mathcal{K})$ with the following properties:

i. $F \not\leq H$ for every $F \in \mathcal{F}$;

ii. $G \leq H$ for any $G \in \mathcal{K} \cap \mathrm{Forb}\,(\mathcal{F})$.

Additionally we may assume that the chromatic number of H is equal to the maximal chromatic number of a graph in $\mathcal{K}$.

The proof of this statement is presently not easy and in fact the proof does not yield an explicit bound H. Let us just remark that Theorem 5.1 (and its special case proved already in [19]) implies that the Grötzsch's theorem (which asserts that K_3 is a bound for all triangle-free planar graphs) does not yield the best bound: By virtue of Theorem 5.1 there exists a bound H satisfying $H < K_3$. The bounds for minor closed classes are related to the *Hadwiger conjecture* which we state in three ways: *i.* is the usual formulation and *ii.* is a formulation in the spirit of this paper. We also add a localised version *iii.* of *ii.*. (A class of graphs is said to be *principal ideal* in the minor order if the class consists from minors of a graph.)

Conjecture 5.2 (Hadwiger).

i. For every graph G holds $\chi(G) \leq h(G)$ where $h(G)$ is the maximal complete graph which is a minor of G.

ii. Any proper minor closed class $\mathcal{K}$ of graphs has the greatest element which is a complete graph.

iii. Any principal ideal of the minor quasiorder has greatest graph in the homomorphism order and it is a complete graph.

iv. Any principal ideal of the minor quasiorder has greatest graph in the homomorphism order.

It is easy to see that the first three forms of Hadwiger conjecture are indeed equivalent: *ii.* $\Rightarrow$ *i.* holds as for any graph G we can apply *i.* to the corresponding principal ideal. If H is the greatest element of $\mathcal{K}$ then $\chi(G) \leq \chi(H)$ and thus *ii.* implies $\chi(G) \leq h(G)$.

Converesely, assume *i.* and let $\mathcal{K}$ be a proper minor closed class. Let H be a graph in $\mathcal{K}$ with the maximal chromatic number, put $k = \chi(H)$. Then $\mathcal{K}$ is bounded by K_k and by *ii.* applied to the graph H we know that $K_k \in \mathcal{K}$. The equivalence of *i.* and *iii.* follows similarly. The equivalence of *i.* and *iv.* was recently (independently) observed by Ossona de Mendez and author [19] and by Nasseraser and Nigussie [15]; see also [9].

In view of results of the previous two sections perhaps one could consider the following weaker problems:

Does a proper minor closed class of graphs has a greatest element?

Does a proper minor closed class of graphs has a supremum?

At some instances these weaker forms are more accessible and true. However in the full generality there is an evidence that they are as difficult as the Hadwiger conjecture. We formulate this more precisely in the following statement. We say that a proper minor closed class $\mathcal{K}$ is *connected* if it is determined by a set of connected forbidden minors. We then have

Proposition 5.3. *For every connected proper minor closed class $\mathcal{K}$ are the following three statements equivalent:*

i. Hadwiger conjecture holds for $\mathcal{K}$;

ii. $\mathcal{K}$ has a greatest element (in the coloring order $\mathcal{C}$);

iii. $\mathcal{K}$ has a supremum (in the coloring order $\mathcal{C}$).

Proof. Clearly *i.* implies both *ii.* and *iii.*. The equivalence of *i.* and *ii.* is a recent result of Nasseraser and Nigussie [15] and of [19]. We prove that *iii.* implies *i.*. This will follow from the following which is perhaps of an independent interest:

Claim. Every connected proper minor closed class $\mathcal{K}$ does not have a proper supremum (in $\mathcal{C}$).

We show that this is a consequence of Theorem 5.1: Assume contrary, let H be a proper supremum of $\mathcal{K}$, let K be a connected component of H, $K \notin \mathcal{K}$. (K exists as $\mathcal{K}$ is a connected minor closed class). Put $\mathcal{K}' = \mathcal{K} \cap \mathrm{Forb}\left(\{K\}\right)$. According to Theorem 5.1 there exists an K-free bound H' of $\mathcal{K}'$ with $H' \leq H$. But then of course $H' < H$. It remains to check that H' is also a bound for $\mathcal{K}$. However $\mathcal{K} = \mathcal{K}'$. (This is similar to the proof of Theorem 3.4: Assume there exists $G \in \mathcal{K} \setminus \mathcal{K}'$, by our assumptions we may assume that G is connected. However then $K \leq G \leq H$ but the only possibility for the second inequality is $G \leq K$. Thus $K \sim G$ and $K \in \mathcal{K}$ a contradicion.) ∎

Let us also note that by Theorem 5.1 any proper minor closed class of graphs is bounded by a graph H with clique number $\omega(H) = h(\mathcal{K})$ where $h(\mathcal{K})$ is the largest clique contained in $\mathcal{K}$ which may be interpreted as yet another approximation to Hadwiger conjecture.

Another interpretation of restricted extrema for classes of graph is by means of cuts which are defined as follows:

Let $\mathcal{K}$ be a class of graphs. A finite subset C of $\mathcal{K}$ is said to be a *cut* if for every graph $G \in \mathcal{K}$ there exists a graph $H \in C$ such that either $G \leq H$ or $H \leq G$ and C is minimal with this property. In the other words a set

C is a cut if any graph in $\mathcal{K}$ is comparable with at least one element of C. If $|C| = 1$ then C is called 1-*cut*. It is easy to see that $\{K_1\}$ and $\{K_2\}$ are the only finite minimal cuts for the class of all (undirected) graphs. This we state in the following form as

Theorem 5.4. *Let $G_1, G_2, \ldots, G_t$ be a set of non-bipartite graphs. Then there exists a graph G such that G and G_i, $i = 1, 2, \ldots, t$ are incomparable.*

Proof. Let l denotes the maximal number of vertices of graphs G_i, $i = 1, 2, \ldots, t$. It suffices to consider any graph G with $\chi(G) > l$ and with the girth $> l$. ∎

In this context one should also mention the following result for countable graph proved recently in [22]:

Theorem 5.5. *K_1, K_2 and the infinite complete graph K_ω are the only minimal 1-cuts for the class of all countable graphs.*

As opposed to the finite case countable graphs allow finite cuts of any size. And contrary to the 1-cuts, the minimal cuts of size $t > 1$ are abundant:

Theorem 5.6. *For every $t > 1$ there are (for countable graphs) infinitely many minimal cuts each of size t.*

Proof. Let $t > 1$ be fixed. Let $F_1, F_2, \ldots, F_{t-1}$ be finite connected graphs which are pairwise incomparable in $\mathcal{C}$. We can use Theorem 5.4, a random $(t-1)$-tuple of graphs will do as well. Now we can apply a result of [2] which gives the existence of a countable graph H which is universal for the class $\mathrm{Forb}\,(F_1, F_2, \ldots, F_{t-1})$ (when considered as the class of all countable graphs). Explicitly: H is a graph such that $F_i \not\leq H$ for every $i = 1, \ldots, t-1$ and if G is a countable graph satisfying $F_i \not\leq G$ for every $i = 1, \ldots, t-1$ then G is an induced subgraph of H. However then the set $C = \{F_1, F_2, \ldots, F_{t-1}, H\}$ is obviously a cut in the class of all countable graphs. It is also easy to check that C is a minimal cut. ∎

This proof is perhaps more interesting than the statement of Theorem 5.6: presently there are no other known minimal cuts for infinite graphs. This suggest the following problem (which is also supported by some results for oriented graphs (see Section 4):

Problem 5.7. Is it true that any minimal cut of size at least 2 for the class of all countable graphs contains always a finite graph?

6. Bounds, suprema and dualities for finite structures

In the previous two sections we considered undirected graphs only. It is a special feature of this area that there is a big gap between directed and undirected graphs. We briefly review some recent results for directed graph which are relevant to the context of this paper. At the end we return to the Δ-systems and prove some analogous results for this case.

First we consider classes Forb $(\mathcal{F})$ (of all directed graphs G which do not contain any $F \in$ Forb $(\mathcal{F})$ with $F \leq G$). While for the undirected graphs these classes are bounded in trivial instances only for directed graphs we have a much richer an interesting spectrum of results. Recall that an oriented graph G is said to be *balanced* iff every cycle in G has the same number of forwarding and backwarding arcs. In terms of homomorphisms this is the same as to say that there exists a homomorphism $G \to \vec{P_n}$ where $\vec{P_n}$ is the directed path of length n (i.e. with $n+1$ vertices). For a balanced graph G we also put $al(G) = \min \{n; \ G \to \vec{P_n}\}$ (*algebraic length* of G).

We start with the following:

Theorem 6.1. *For a finite set $\mathcal{F}$ of graphs the following statements are equivalent:*

  i. The class Forb $(\mathcal{F})$ *is bounded;*

  ii. At lest one of the graphs $F \in \mathcal{F}$ is balanced.

Proof. This is yet another version of sparse high chromatic graphs. *ii.* implies *i.* as the chromatic number of graphs in Forb $(\mathcal{F})$ is bounded by $1 + al(F)$ for a balanced $F \in \mathcal{F}$. Conversely, suppose that no $F \in \mathcal{F}$ is balanced. Alternatively we know that any homomorphic image of any $F \in \mathcal{F}$ contains a cycle. Consider any orientation $\vec{G}$ of a high chromatic graph G without short cycles. It follows that $\vec{G} \in$ Forb $(\mathcal{F})$ and thus there is no bound for this class. ∎

The characterization of classes of form Forb $(\mathcal{F})$ with a greatest element is a more difficult result:

Theorem 6.2. *For a finite set $\mathcal{F}$ of graphs the following statements are equivalent:*

  i. The class Forb $(\mathcal{F})$ *has greatest element;*

  ii. $F \in \mathcal{F}$ is a set of oriented trees.

Theorem 6.2 is proved in [23] in a different context which we now outline: Let $\mathcal{F} = \{F_1, F_2, \ldots, F_t\}$ be a finite set of graphs and suppose that H is the greatest element of the class Forb $(\mathcal{F})$. These facts may be expressed equivalently by the validity of the following statement:

For every graph G holds

$$F_i \nrightarrow G, \quad i = 1, \ldots, t \quad \Leftrightarrow \quad G \rightarrow H.$$

Such statement is called a *Homomorphism Duality*. H is called the *dual* of the set $\{F_1, \ldots, F_t\}$ (up to the homomorphism equivalence the dual is uniquely determined). The main result of [23] characterizes all finite sets of graphs which have dual graph – these are just sets of finite trees (and sets which are homomorphically equivalent to them).

Let us remark that Theorem 6.2 may be seen as characterization of all Gallai–Roy–Vitaver (and Hasse) – type theorems. (Gallai–Hasse–Roy–Vitaver theorem corresponds to the case $\mathcal{F} = \{\vec{P}_n\}$. In this case the dual graph is the transitive tournament with n vertices.) This point of view is taken in [24].

Let us finally discuss the existence of suprema for the classes Forb $(\mathcal{F})$. Here we have also a full solution which is perhaps surprising (and combines several techniques described above):

Theorem 6.3. *For a finite set $\mathcal{F}$ of connected graphs the following statements are equivalent:*

    i. The class Forb $(\mathcal{F})$ *has supremum;*

    ii. At least one of the graphs $F \in \mathcal{F}$ is balanced.

In the other words every bounded class Forb $(\mathcal{F})$ *of oriented graphs has a supremum.*

Proof. Clearly it suffices to prove *ii.* $\Rightarrow$ *i.* Put $\mathcal{F} = \{F_1, F_2, \ldots, F_t\}$. Denote also by $\mathcal{F}'$ the class of all homomorphic images of graphs F_i which are trees. $\mathcal{F}'$ is a non-empty set. Consider the class Forb $(\mathcal{F}')$ and let $H = H_{\mathcal{F}'}$ be the greatest element of Forb $(\mathcal{F}')$ (i.e. H is the dual of $\mathcal{F}'$). We prove that H is supremum of Forb $(\mathcal{F})$. Clearly Forb $(\mathcal{F})$ is a subclass of Forb $(\mathcal{F}')$ and thus $G \leq H$ for every $G \in$ Forb $(\mathcal{F})$. Now suppose that A is a graph satisfying $A < H$; let k be the number of vertices of A. Let G be the graph with the following properties:

    1. $G \rightarrow H$;

    2. $G \nrightarrow A$;

3. G does not contain cycles of length $\leq k$.

The existence of graph G will not be proved here as it follows from (oriented version of) Sparse Incomparability Lemma which has been isolated in several papers, see e.g. [21, 25, 16]. It suffices to prove that $G \in \text{Forb}(\mathcal{F})$. (This shows that A is not a bound of $\text{Forb}(\mathcal{F})$ and thus H is indeed supremum of the class $\text{Forb}(\mathcal{F})$.) Assume contrary: let $f : F_i \to G$. The homomorphic image $f(F_i)$ has at most k vertices and thus it induces a tree in G. Therefore $f(F_i) \in \mathcal{F}'$. It follows that also $G \notin \text{Forb}(\mathcal{F}')$ and thus (by homomorphism duality) $G \twoheadrightarrow H$. This is a contradiction. ∎

At the end of this paper let us return to the general Δ-systems.

Let $\Delta = (\delta_i; i \in I)$ be a fixed type, assume without loss of generality (of this section) that $\delta_i \geq 2$. A special role is played by the following Δ-system: *all-loop* system is the system L where $\underline{L} = \{1\}$ and $R_i(L) = \{(1, \ldots, 1)\}$ for every $i \in I$ (the all-loop system will be always denoted by L). The all-loop system is the only (up to homomorphism equivalence) absolute retract for Δ-systems:

Proposition 6.4. *$A \sim L$ if and only if A contains L as a subsystem.*

We say that a class $\mathcal{K}$ is *bounded* if there exists a Δ-system $C \approx L$ such that $A \leq C$ for every $A \in \mathcal{K}$. Which classes $\mathcal{K}$ are bounded? This is a non-trivial question even for finite undirected graphs (see e.g. problems stated in [6, 11]). We can completely solve these questions for classes $\text{Forb}(\mathcal{F})$ where $\mathcal{F}$ is a finite set of connected Δ-systems. We shall need a generalization of a balanced graph: We say that a Δ-system $A = (X, (R_i; i \in I))$ is *balanced* if it is homomorphic to a Δ-tree. (Recall, A is said to be a tree if its incidence system $\text{Inc}(A)$ is a tree. (Clearly for oriented graphs both definitions coincide. One can devise also more explicit definition of a balanced Δ-system but this will not be needed.)

Theorem 6.5. *For a finite set $\mathcal{F}$ of Δ-systems the following statements are equivalent:*

  i. The class $\text{Forb}(\mathcal{F})$ is bounded;

  ii. At least one of the systems $F \in \mathcal{F}$ is balanced.

Proof. *i.* implies *ii.* similarly as in the proof of Theorem 6.1: Suppose contrary, let every $F \in \mathcal{F}$ be an unbalanced Δ-system. Put $n = \max\{|F|;\ F \in \mathcal{F}\}$ and let the class $\text{Forb}(\mathcal{F})$ be bounded by a system B with N vertices. Denote by k the sum of all $\delta_i, i \in I$, and consider a k-uniform hypergraph $(X, \mathcal{M})$ with chromatic number $> N$ not containing cycles with at most n vertices. Modify $(X, \mathcal{M})$ to a Δ-system $A = \big(X, (R_i;\ i \in I)\big)$ by inserting for every edge $M \in \mathcal{M}$ a collection of $|I|$ mutually disjoint δ_i-tuples, $i \in I$. Clearly $A \nrightarrow B$ (as any homomorphism $f : A \to C$ to a a system C with at most N points implies that C contains an all-loop). On the other hand any subsystem of A with at most n points is necessarily a system without cycles. Thus $A \in \mathrm{Forb}\,(\mathcal{F})$, a contradiction.

Conversely assume that the class $\mathrm{Forb}\,(\mathcal{F})$ is unbounded. We use the following family $K(\Delta, X, i_0)$ of complete systems as our *scale class*: the vertices of $K = K(\Delta, X, i_0)$ is the set X and $R_i(K) = X^{\delta_i}$ for all $i \neq i_0$ while $R_{i_0}(K) = X^{\delta_{i_0}} \setminus \big\{(x, \ldots, x);\ x \in X\big\}$. (I.e. $R_{i_0}(K)$ is the set of all non-constant δ_{i_0}-tuples). For $X = \{1, 2, \ldots, N\}$ we put briefly $K(N, i)$ for $K(\Delta, X, i)$. None of the systems $K(N, i)$ is a bound of $\mathrm{Forb}\,(\mathcal{F})$. Thus for every N, i there exists a Δ-system $A_{(N,i)} \in \mathrm{Forb}\,(\mathcal{F})$ such that $A_{(N,i)} \nrightarrow K(N, i)$. This implies that the shadow graph $sh\big(A_{(N,i)}\big)$ has chromatic number $> N$ but we shall need more. Let $i \in I$ be fixed. Fix two indices $1 \leq a < b \leq \delta_i$. By $A_{(N,i)}(a, b)$ denote the oriented graph $(X, E_{a,b})$ where $E_{a,b}$ consists from all pairs (x_a, x_b) which appear in a δ_i-tuple $(x_1, \ldots, x_{\delta_i}) \in R_i(A_{(N,i)})$. It follows from $A_{(N,i)} \nrightarrow K(N, i)$ that $\chi\big(A_{(N,i)}(a, b)\big) > N$. From this follows that the graph $A_N(a, b)$ fails to be degenerated and thus it contains a subgraph with all out- and in-degrees $> N/2$. As N was arbitrary we can repeat this argument and find (for every n) a subsystem A' of $A_{(N(n),i)}$ for which all the graphs $A'(a, b)$ have large in- and out-degrees. We then repeat this argument for all $i \in I$. This may be then used to find any Δ-tree with at most n vertices. Particularly, every balanced $F \in \mathcal{F}$ will for some A satisfying $A \nrightarrow K(N, i)$, $i \in I$, satisfy $F \to A$ which is again a contradiction. $\blacksquare$

It is important that the main result of [23] solves the existence of greatest elements in classes of type $\mathrm{Forb}\,(\mathcal{F})$ and this in turn can be used to characterize those classes $\mathrm{Forb}\,(\mathcal{F})$ of Δ-structures which have suprema. We combine these statement to a single statement:

Theorem 6.6.

I. For a finite set $\mathcal{F}$ of graphs the following statements are equivalent:

i. The class $\mathrm{Forb}\,(\mathcal{F})$ has greatest element;

ii. $F \in \mathcal{F}$ is a set of Δ-trees.

II. For a finite set $\mathcal{F}$ of connected graphs the following statements are equivalent:

i. The class Forb ($\mathcal{F}$) *has supremum;*

ii. At least one of the graphs $F \in \mathcal{F}$ *is balanced.*

As our above proofs were categorical we can use general results of [23] and proceed analogously. We omit the details.

7. Summary and Concluding Remarks

The purpose of this paper is to initiate the study of graph bounds in a homomorphism and partial order setting. From this point of view greatest elements and suprema present tight bounds (which cannot be "improved"). We have proved (Theorems 6.6) that classes which are defined by forbidden homomorphisms from a finite set of connected systems have suprema if and only if they are bounded. On the other hand the same classes when relativized by small degrees are bounded but do not have suprema at all (with a few isolated cases; see Theorem 3.1). Similar negative results were obtained for minor closed classes. This is in a sharp contrast with the situation for degenerated graphs where suprema are easy to describe (and they form a chain). This perhaps sheds some light on questions like Hadwiger conjecture which can be expressed in the same vein. Most of the questions, theorems and proofs considered in this paper can be carried over to more general situations. This provides a connection with universal algebra and model theory. We hope to return to this in near future.

What we propose here is a global approach to extremal–theory estimates (such as bounds for chromatic number) by means of coloring (homomorphism) order. We studied some local properties of the coloring order (such as suprema and greatest elements). To present a good bound (i.e. supremum) for a class of graphs is equivalent to finding a smallest finite homomorphism universal graph. Whether this hom-universal graph can have the same local properties as the class itself is one of the central questions of this paper. We gave instances with both positive and negative answer. A satisfactory solution we could provide for classes which are defined by finitely many homomorphism obstructions. We relativized these results by bounded degree-, degeneracy- and minor closed-restrictions. This leads to some seemingly difficult problems but it also shows how these questions are relevant and that global structure of colorings can capture some of the key combinatorial conjectures.

Acknowledgement. I thank to Patrice Ossona de Mendez for several remarks which improved quality of this paper.

REFERENCES

[1] A. A. Bulatov, P. G. Jeavons and A. A. Krokhin, Constraint satisfaction problems and finite algebras, in: *Proceedings of the 27th ICALP'00,* LNCS 1853, Springer Verlag 2000, pp. 414–425.

[2] G. Cherlin, S. Shelah and N. Shi, Universal Graphs with Forbidden Subgraphs and Algebraic Closure, *Advances in Applied Math.,* **22** (1999), 454–491.

[3] P. Dreyer, Ch. Malon and J. Nešetřil, Universal H-colourable graphs without a given configuration, *Discrete Math.,* **250** (2002), 245–252.

[4] P. Erdős and A. Hajnal, On chromatic number of set systems, *Acta Math. Acad. Sci. Hung.,* **17** (1966), 61–99.

[5] A. Galluccio, P. Hell and J. Nešetřil, The complexity of H-colouring of counded degree graphs, *Discrete Math.,* **222** (2000), 101–109.

[6] A. Gyarfás, Problems from the world surrounding perfect graphs, *Zostos. Mat.,* **19** (1987), 413–441.

[7] H. Hatani, Random cubic graphs are not homomorphic to the cycle of length 7 (to appear in *J. Comb. Th. B.*).

[8] R. Häggkvist and P. Hell, Universality of A-mote graphs, *European J. Comb.* (1993), 23–27.

[9] P. Hell and J. Nešetřil, *Graphs and Homomorphisms,* Oxford University Press, 2004.

[10] Y. H. Kim and J. Nešetřil, *On colourings of bounded degree graphs* (in preparation).

[11] A. Kostochka, J. Nešetřil and P. Smolíková, Colouring and homomorphisms of degenerated and bounded degree graphs, *Discrete Math.,* **233, 1–3** (2001), 257–276.

[12] L. Lovász, On the chromatic number of finite set systems, *Acta Math. Acad. Sci. Hung.,* **19** (1968), 59–67.

[13] T. Luczak and J. Nešetřil, A probabilistic approach to the dichotomy problem, *ITI Series,* 2003-152 (submitted).

[14] J. Matoušek and J. Nešetřil, *Invitation to Discrete Mathematics,* Oxford Univ. Press, 1998.

[15] R. Nasserasr and Y. Nigussie, *On the new reformulation of Hadwiger's conjecture* (in preparation).

[16] J. Nešetřil, Aspects of Structural Combinatorics (Graph Homomorphisms and their Use), *Taiwanese J. Math.* **3, 4** (1999), 381–424.

[17] P. Ossona de Mendez and J. Nešetřil, Colorings and Homomorphisms of Minor Closed Classes, in: *J. Goodman and R. Pollack Festschrift* (J. Pach, ed.), Springer 2003, 651–664.

[18] P. Ossona de Mendez and J. Nešetřil, Folding, to appear in *J. Comb. Th. B*.

[19] P. Ossona de Mendez and J. Nešetřil, Cuts and Bounds for Graphs, *KAM–DIMATIA Series* 2002-592 (to appear in Discrete Math. – ACCOTA volume).

[20] P. Ossona de Mendez and J. Nešetřil, Tree depth, orientation and coloring, *ITI Series* 2004-179, Charles University, Prague (to appear in *European J. Comb.*).

[21] J. Nešetřil and V. Rödl, Chromatically optimal rigid graphs, *J. Comb. Th. B*, **46** (1989), 133–141.

[22] J. Nešetřil and S. Shelah, On the Order of Countable Graphs, *European J. Comb.,* **24** (2003), 649–663.

[23] J. Nešetřil and C. Tardif, Duality Theorems for Finite Structures (Characterizing Gaps and Good Characterizations), *J. Comb. Th. B*, **80** (2000), 80–97.

[24] J. Nešetřil and C. Tardif, On maximal finite antichains in the homomorphism order of directed graphs, *Discussiones Math. Graph Theory*, **23, 2** (2003), 325–332.

[25] J. Nešetřil and X. Zhu, On Sparse graphs with Given colourings and Homomorphisms, *J. Comb. Th. B*, **90** (2004), 161–172..

[26] N. Robertson and P. Seymour, Graph Minors, *J. Comb. Th. B* (since 1985).

[27] L. M. Vitaver, Determination of minimal coloring of vertices of a graph by means of Boolean powers of the incidence matrix, *Dokl. Akad. Nauk SSSR*, **147** (1962), 758–759.

[28] I. M. Wanless and N. C. Wormald, Graphs with no homomorphisms onto cycles, *J. Comb. Th. B*, **82** (2001), 155–160.

Jaroslav Nešetřil

Department of Applied Mathematics
and
Institute of Thoretical Computer sciences
(ITI)
Charles University
Malostranské nám. 25
11800 Praha
Czech Republic

nesetril@kam.ms.mff.cuni.cz

BOLYAI SOCIETY
MATHEMATICAL STUDIES, 15

Conference on Finite
and Infinite Sets
Budapest, pp. 285–300.

RELAXING PLANARITY FOR TOPOLOGICAL GRAPHS

J. PACH, R. RADOIČIĆ and G. TÓTH*

According to Euler's formula, every planar graph with n vertices has at most $O(n)$ edges. How much can we relax the condition of planarity without violating the conclusion? After surveying some classical and recent results of this kind, we prove that every graph of n vertices, which can be drawn in the plane without three pairwise crossing edges, has at most $O(n)$ edges. For straight-line drawings, this statement has been established by Agarwal et al., using a more complicated argument, but for the general case previously no bound better than $O(n^{3/2})$ was known.

1. INTRODUCTION

A *geometric graph* is a graph drawn in the plane so that its vertices are represented by points in general position (i.e., no three are collinear) and its edges by straight-line segments connecting the corresponding points. *Topological graphs* are defined similarly, except that now each edge can be represented by any simple (non-selfintersecting) Jordan arc passing through no vertices other than its endpoints. Throughout this paper, we assume that if two edges of a topological graph G share an interior point, then at this point they properly cross. We also assume, for simplicity, that no three edges cross at the same point and that any two edges cross only a finite number of times. If any two edges of G have at most one point in common (including their endpoints), then G is said to be a *simple* topological graph.

---

*János Pach has been supported by NSF Grant CCR-00-98245, by PSC-CUNY Research Award 63352-0036, and by OTKA T-032458. Géza Tóth has been supported by OTKA-T-038397 and by an award from the New York University Research Challenge Fund.

Clearly, every geometric graph is simple. Let $V(G)$ and $E(G)$ denote the vertex set and edge set of G, respectively. We will make no notational distinction between the vertices (edges) of the underlying abstract graph, and the points (arcs) representing them in the plane.

It follows from Euler's Polyhedral Formula that every simple planar graph with n vertices has at most $3n - 6$ edges. Equivalently, every topological graph with n vertices and more than $3n - 6$ edges has a pair of crossing edges. What happens if, instead of a crossing pair of edges, we want to guarantee the existence of some larger configurations involving several crossings? What kind of *unavoidable* substructures must occur in every geometric (or topological) graph G having n vertices and more than Cn edges, for an appropriate large constant $C > 0$?

In the next four sections, we approach this question from four different directions, each leading to different answers. In the last section, we prove that any topological graph with n vertices and no three pairwise crossing edges has at most $O(n)$ edges. For *simple* topological graphs, this result was first established by Agarwal–Aronov–Pach–Pollack–Sharir [1], using a more complicated argument.

2. Ordinary and topological minors

A graph H is said to be a *minor* of another graph G if H can be obtained from a subgraph of G by a series of edge contractions. If a subgraph of G can be obtained from H by replacing its edges with independent paths between their endpoints, then H is called a *topological minor* of G. Clearly, a topological minor of G is also its (ordinary) minor.

If a graph G with n vertices has no minor isomorphic to K_5 or to $K_{3,3}$, then by Kuratowski's theorem it is planar and its number of edges cannot exceed $3n - 6$. It follows from an old result of Wagner that the same conclusion holds under the weaker assumption that G has no K_5 minor. A few years ago Mader [16] proved the following famous conjecture of Dirac:

Theorem 2.1 (Mader). *Every graph of n vertices with no topological K_5 minor has at most $3n - 6$ edges.*

If we only assume that G has no topological K_r minor for some $r > 5$, we can still conclude that G is *sparse*, i.e., its number of edges is at most linear in n.

Theorem 2.2 (Komlós–Szemerédi [10], Bollobás–Thomason [4]). *For any positive integer r, every graph of n vertices with no topological K_r minor has at most cr^2n edges.*

Moreover, Komlós and Szemerédi showed that the above statement is true with any positive constant $c > 1/4$, provided that r is large enough. Apart from the value of the constant, this theorem is sharp, as is shown by the union of pairwise disjoint copies of a complete bipartite graph of size roughly r^2.

We have a better bound on the number of edges, under the stronger assumption that G has no K_r minor.

Theorem 2.3 (Kostochka [11], Thomason [31]). *For any positive integer r, every graph of n vertices with no K_r minor has at most $cr\sqrt{\log r}\,n$ edges.*

The best value of the constant c for which the theorem holds was asymptotically determined in [33]. The theorem is sharp up to the constant. (Warning! The letters c and C used in several statements will denote *unrelated* positive constants.)

Reversing Theorem 2.3, we obtain that every graph with n vertices and more than $cr\sqrt{\log r}\,n$ edges has a K_r minor. This immediately implies that if the chromatic number $\chi(G)$ of G is at least $2cr\sqrt{\log r} + 1$, then G has a K_r minor. According to Hadwiger's notorious conjecture, for the same conclusion it is enough to assume that $\chi(G) \geq r$. This is known to be true for $r \leq 6$ (see [28]).

3. QUASI-PLANAR GRAPHS

A graph is planar if and only if it can be drawn as a topological graph with no crossing edges. What happens if we relax this condition and we allow r crossings per edge, for some fixed $r \geq 0$?

Theorem 3.1 [25]. *Let r be a natural number and let G be a simple topological graph of n vertices, in which every edge crosses at most r others. Then, for any $r \leq 4$, we have $|E(G)| \leq (r + 3)(n - 2)$.*

The case $r = 0$ is Euler's theorem, which is sharp. In the case $r = 1$, studied in [25] and independently by Gärtner, Thiele, and Ziegler (personal communication), the above bound can be attained for all $n \geq 12$. The result is also sharp for $r = 2$, provided that $n \equiv 5 \pmod{15}$ is sufficiently large (see Fig. 1).

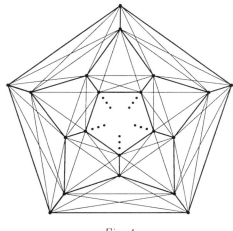

Fig. 1

However, for $r = 3$, we have recently proved that $\left| E(G) \right| \leq 5.5(n - 2)$, and this bound is best possible up to an additive constant [23]. For very large values of r, a much better upper bound can be deduced from the following theorem of Ajtai–Chvátal–Newborn–Szemerédi [2] and Leighton [14]: any topological graph with n vertices and $e > 4n$ edges has at least constant times e^3/n^2 crossings.

Corollary 3.2 [23]. *Any topological graph with n vertices, whose each edge crosses at most r others, has at most $4\sqrt{r}n$ edges.*

One can also obtain a linear upper bound for the number of edges of a topological graph under the weaker assumption that no edge can cross more than r other edges *incident to the same vertex*. This can be further generalized, as follows.

Theorem 3.3 [20]. *Let G be a topological graph with n vertices which contains no $r + s$ edges such that the first r are incident to the same vertex and each of them crosses the other s edges. Then we have $\left| E(G) \right| \leq C_s rn$, where C_s is a constant depending only on s.*

In particular, it follows that if a topological graph contains no large *gridlike* crossing pattern (two large sets of edges such that every element of the first set crosses all elements of the second), its number of edges is at most linear in n. It is a challenging open problem to decide whether the same assertion remains true for all topological graphs containing no large *complete* crossing pattern.

For any positive integer r, we call a topological graph r-*quasi-planar* if it has no r pairwise crossing edges. A topological graph is x-*monotone* if all of its edges are x-monotone curves, i.e., every vertical line crosses them at most once. Clearly, every geometric graph is x-monotone, because its edges are straight-line segments (that are assumed to be non-vertical). If the vertices of a geometric graph are in convex position, then it is said to be a *convex* geometric graph.

Theorem 3.4 [7]. *The maximum number of edges of any r-quasi-planar convex geometric graph with $n \geq 2r$ edges is*

$$2(r-1)n - \binom{2r-1}{2}.$$

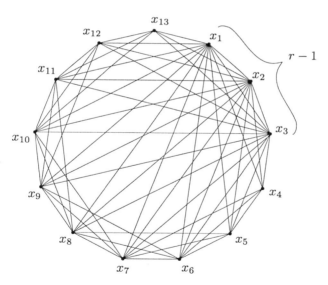

Fig. 2. Construction showing that Theorem 3.4 is sharp ($n = 13$, $r = 4$)

Theorem 3.5 (Valtr [34]). *Every r-quasi-planar x-monotone topological graph with n vertices has at most $C_r n \log n$ edges, for a suitable constant C_r depending on r.*

Theorem 3.6 [24]. *For any $r \geq 4$, every r-quasi-planar simple topological graph G with n vertices has at most $C_r n (\log n)^{2(r-3)}$ edges, for a suitable constant C_r depending only on r.*

In Section 6, we will point out that Theorem 3.6 remains true even if we drop the assumption that G is simple, i.e., two edges may cross more than once.

For 3-quasi-planar topological graphs we have a linear upper bound.

Theorem 3.7 [1]. *Every 3-quasi-planar simple topological graph G with n vertices has at most Cn edges, for a suitable constant C.*

In Section 7, we give a short new proof of the last theorem, showing that here, too, one can drop the assumption that no two edges cross more than once (i.e., that G is simple). In this case, previously no bound better than $O(n^{3/2})$ was known. Theorem 3.7 can also be extended in another direction: it remains true for every topological graph G with no $r + 2$ edges such that each of the first r edges crosses the last two and the last two edges cross each other. Of course, the constant C in the theorem now depends on r [22].

All theorems in this section provide (usually linear) upper bounds on the number of edges of topological graphs satisfying certain conditions. In each case, one may ask whether a stronger statement is true. Is it possible that the graphs in question can be decomposed into a small number planar graphs? For instance, the following stronger form of Theorem 3.7 may hold:

Conjecture 3.8. There is a constant k such that the edges of every 3-quasi-planar topological graph G can be colored by k colors so that no two edges of the same color cross each other.

McGuinness [18] proved that Conjecture 3.8 is true for simple topological graphs, provided that there is a closed Jordan curve crossing every edge of G precisely once. The statement is also true for r-quasi-planar convex geometric graphs, for any fixed r (see [12], [13]).

4. GENERALIZED THRACKLES AND THEIR RELATIVES

Two edges are said to be *adjacent* if they share an endpoint. We say that a graph drawn in the plane is a *generalized thrackle* if any two edges meet an odd number of times, counting their common endpoints, if they have any.

That is, a graph is a generalized thrackle if and only if it has no two adjacent edges that cross an odd number of times and no two non-adjacent edges that cross an even number of times. In particular, a generalized thrackle cannot have two non-adjacent edges that are disjoint. Although at first glance this property may appear to be the exact opposite of planarity, surprisingly, the two notions are not that different. In particular, for bipartite graphs, they are equivalent.

Theorem 4.1 [15]. *A bipartite graph can be drawn in the plane as a generalized thrackle if and only if it is planar.*

Using the fact that every graph G has a bipartite subgraph with at least $|E(G)|/2$ edges, we obtain that if a graph G of n vertices can be drawn as a generalized thrackle, then $|E(G)| = O(n)$.

Theorem 4.2 (Cairns–Nikolayevsky [6]). *Every generalized thrackle with n vertices has at most $2n - 2$ edges. This bound is sharp.*

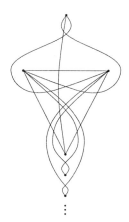

Fig. 3. A generalized thrackle with n vertices and $2n - 2$ edges

A geometric graph G is a generalized thrackle if and only if it has no two disjoint edges. (The edges are supposed to be *closed* sets, so that two disjoint edges are necessarily non-adjacent.) One can relax this condition by assuming that G has no r pairwise disjoint edges, for some fixed $r \geq 2$. For $r = 2$, it was proved by Hopf–Pannwitz [9] that every graph satisfying this property has at most n edges, and that this bound is sharp. For $r = 3$, the first linear bound on the number of edges of such graphs was established by

Alon–Erdős [3], which was later improved to $3n$ by Goddard–Katchalski–Kleitman [8]. For general r, the first linear bound was established in [26]. The best currently known estimate is the following:

Theorem 4.3 (Tóth [32]). *Every geometric graph with n vertices and no r pairwise disjoint edges has at most $2^9(r-1)^2 n$ edges.*

It is likely that the dependence of this bound on r can be further improved to linear. If we want to prove the analogue of Theorem 4.3 for topological graphs, we have to make some additional assumptions on the structure of G, otherwise it is possible that any two edges of G cross each other.

Conjecture 4.4 (Conway's Thrackle Conjecture). Let G be a simple topological graph of n vertices. If G has no two disjoint edges, then $|E(G)| \leq n$.

For many related results, consult [15], [6], [35]. The next interesting open question is to decide whether the maximum number of edges of a simple topological graph with n vertices and no three pairwise disjoint edges is $O(n)$.

5. Locally planar graphs

For any $r \geq 3$, a topological graph G is called *r-locally planar* if G has no selfintersecting path of length at most r. Roughly speaking, this means that the embedding of the graph is planar in a neighborhood of radius $r/2$ around any vertex. In [21], we showed that there exist 3-locally planar geometric graphs with n vertices and with at least constant times $n \log n$ edges. Somewhat surprisingly (to us), Tardos [30] managed to extend this result to any fixed $r \geq 3$. He constructed a sequence of r-locally planar geometric graphs with n vertices and a superlinear number of edges (approximately n times the $\lfloor r/2 \rfloor$ times iterated logarithm of n). Moreover, these graphs are bipartite and all of their edges can be stabbed by the same line.

The following positive result is probably very far from being sharp.

Theorem 5.1 [21]. *The maximum number of edges of a 3-locally planar topological graph with n vertices is $O(n^{3/2})$.*

For geometric graphs, much stronger results are known.

Theorem 5.2 [21]. *The maximum number of edges of a 3-locally planar x-monotone topological graph with n vertices is $O(n \log n)$. This bound is asymptotically sharp.*

For 5-locally planar x-monotone topological graphs, we have a slightly better upper bound on the number of edges: $O(n \log n / \log \log n)$. This bound can be further improved under the additional assumption that all edges of the graph cross the y-axis.

Theorem 5.3 [21]. *Let G be an x-monotone r-locally planar topological graph of n vertices all of whose edges cross the y-axis. Then, we have $|E(G)| \leq crn(\log n)^{1/\lfloor r/2 \rfloor}$ for a suitable constant c.*

6. STRENGTHENING THEOREM 3.6

In this section, we outline the proof of

Theorem 6.1. *Every r-quasi-planar topological graph with n vertices has at most*
$$f_r(n) := C_r n (\log n)^{4(r-3)}$$
edges, where $r \geq 2$ and C_r is a suitable positive constant depending on r.

Let G be a graph with vertex set $V(G)$ and edge set $E(G)$. The *bisection width* $b(G)$ of G is defined as the minimum number of edges, whose removal splits the graph into two roughly equal subgraphs. More precisely, $b(G)$ is the minimum number of edges running between V_1 and V_2, over all partitions of the vertex set of G into two disjoint parts $V_1 \cup V_2$ such that $|V_1|, |V_2| \geq |V(G)|/3$. The *pair-crossing number* PAIR-CR (G) of a graph G is the minimum number of crossing pairs of edges in any drawing of G.

We need a recent result of Kolman and Matoušek [17], whose analogue for ordinary crossing numbers was proved in [24] and [29].

Lemma 6.2 (Matoušek). *Let G be a graph of n vertices with degrees $d_1, d_2, \ldots, d_n$. Then we have*
$$b^2(G) \leq c(\log n)^2 \left(\text{PAIR-CR}\,(G) + \sum_{i=1}^{n} d_i^2 \right),$$
where c is a suitable constant.

We follow the idea of the original proof of Theorem 3.6. We establish Theorem 6.1 by double induction on r and n. By Theorem 7.1 (in the next section), the statement is true for $r = 3$ and for all n. It is also true for any $r > 2$ and $n \leq n_r$, provided that C_r is sufficiently large in terms of n_r, because then the stated bound exceeds $\binom{n}{2}$. (The integers n_r can be specified later so as to satisfy certain simple technical conditions.)

Assume that we have already proved Theorem 6.1 for some $r \geq 3$ and all n. Let $n \geq n_{r+1}$, and suppose that the theorem holds for $r + 1$ and for all topological graphs having fewer than n vertices.

Let G be an $(r + 1)$-quasi-planar topological graph of n vertices. For simplicity, we use the same letter G to denote the underlying abstract graph. For any edge $e \in E(G)$, let $G_e \subset G$ denote the topological graph consisting of all edges of G that cross e. Clearly, G_e is r-quasi-planar. Thus, by the induction hypothesis, we have

$$\text{PAIR-CR}\,(G) \leq \frac{1}{2} \sum_{e \in E(G)} |E(G_e)| \leq \frac{1}{2} |E(G)| f_r(n).$$

Using the fact that $\sum_{i=1}^{n} d_i^2 \leq 2|E(G)|n$ holds for every graph G with degrees $d_1, d_2, \ldots, d_n$, Lemma 6.2 implies that

$$b(G) \leq \left(c(\log n)^2 |E(G)| f_r(n) \right)^{1/2}.$$

Consider a partition of $V(G)$ into two parts of sizes $n_1, n_2 \leq 2n/3$ such that the number of edges running between them is $b(G)$. Obviously, both subgraphs induced by these parts are $(r + 1)$-quasi-planar. Thus, we can apply the induction hypothesis to obtain

$$|E(G)| \leq f_{r+1}(n_1) + f_{r+1}(n_2) + b(G).$$

Comparing the last two inequalities, the result follows by some routine calculation.

7. STRENGTHENING THEOREM 3.7

The aim of this section is to prove the following stronger version of Theorem 3.7.

Theorem 7.1. *Every 3-quasi-planar topological graph with n vertices has at most Cn edges, for a suitable constant C.*

Let G be a 3-quasi-planar topological graph with n vertices. Redraw G, if necessary, without creating 3 pairwise crossing edges so that the number of crossings in the resulting topological graph $\tilde{G}$ is as small as possible. Obviously, no edge of $\tilde{G}$ crosses itself, otherwise we could reduce the number of crossings by removing the loop. Suppose that $\tilde{G}$ has two distinct edges that cross at least twice. A region enclosed by two pieces of the participating edges is called a *lens*. Suppose there is a lens ℓ that contains no vertex of $\tilde{G}$. Consider a *minimal* lens $\ell' \subseteq \ell$, by containment. Notice that by swapping the two sides of ℓ', we could reduce the number of crossings without creating any new pair of crossing edges. In particular, $\tilde{G}$ remains 3-quasi-planar. Therefore, we can conclude that

Claim 1. Each lens of $\tilde{G}$ contains a vertex.

We may assume without loss of generality that the underlying abstract graph of G is connected, because otherwise we can prove Theorem 7.1 by induction on the number of vertices. Let $e_1, e_2, \ldots, e_{n-1} \in E(G)$ be a sequence of edges such that $e_1, e_2, \ldots, e_i$ form a tree $T_i \subseteq G$ for every $1 \leq i \leq n-1$. In particular, $e_1, e_2, \ldots, e_{n-1}$ form a spanning tree of G.

First, we construct a sequence of crossing-free topological graphs (trees), $\tilde{T}_1, \tilde{T}_2, \ldots, \tilde{T}_{n-1}$. Let $\tilde{T}_1$ be defined as a topological graph of two vertices, consisting of the single edge e_1 (as was drawn in $\tilde{G}$). Suppose that $\tilde{T}_i$ has already been defined for some $i \geq 1$, and let v denote the endpoint of e_{i+1} that does not belong to T_i. Now add to $\tilde{T}_i$ the piece of e_{i+1} between v and its first crossing with $\tilde{T}_i$. More precisely, follow the edge e_{i+1} from v up to the point v' where it hits $\tilde{T}_i$ for the first time, and denote this piece of e_{i+1} by $\tilde{e}_{i+1}$. If v' is a vertex of $\tilde{T}_i$, then add v and $\tilde{e}_{i+1}$ to $\tilde{T}_i$ and let $\tilde{T}_{i+1}$ be the resulting topological graph. If v' is in the interior of an edge e of $\tilde{T}_i$, then introduce a new vertex at v'. It divides e into two edges, e' and e''. Add both of them to $\tilde{T}_i$, and delete e. Also add v and $\tilde{e}_{i+1}$, and let $\tilde{T}_{i+1}$ be the resulting topological graph.

After $n-2$ steps, we obtain a topological tree $\tilde{T} := \tilde{T}_{n-1}$, which (1) is crossing-free, (2) has fewer than $2n$ vertices, (3) contains each vertex of $\tilde{G}$, and (4) has the property that each of its edges is either a *full edge*, or a *piece of an edge* of $\tilde{G}$.

Let D denote the open region obtained by removing from the plane every point belonging to $\tilde{T}$. Define a *convex* geometric graph H, as follows.

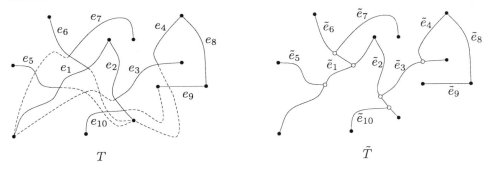

Fig. 4. Constructing $\tilde{T}$ from T

Travelling around the boundary of D in clockwise direction, we encounter two kinds of different "features": vertices and edges of $\tilde{T}$. Represent each such feature by a different vertex x_i of H, in clockwise order in convex position. Note that the same feature will be represented by several x_i's: every edge will be represented twice, because we visit both of its sides, and every vertex will be represented as many times as its degree in $\tilde{T}$. It is not hard to see that the number of vertices $x_i \in V(H)$ does not exceed $8n$.

Next, we define the edges of H. Let E be the set of edges of $\tilde{G}\setminus T$. Every edge $e \in E$ may cross $\tilde{T}$ at several points. These crossing points divide e into several pieces, called *segments*. Let S denote the set of all segments of all edges $e \in E$. With the exception of its endpoints, every segment $s \in S$ runs in the region D. The endpoints of s belong to two features along the boundary of D, represented by two vertices x_i and x_j of H. Connect x_i and x_j by a straight-line edge of H. Notice that H has no loops, because if $x_i = x_j$, then, using the fact that $\tilde{T}$ is connected, one can easily conclude that the lens enclosed by s and by the edge of $\tilde{T}$ corresponding to x_i has no vertex of G in its interior. This contradicts Claim 1.

Of course, several different segments may give rise to the same edge $x_i x_j \in E(H)$. Two such segments are said to be of the *same type*. Observe that two segments of the same type cannot cross. Indeed, as no edge intersects itself, the two crossing segments would belong to distinct edges $e_1, e_2 \in E$. Since any two vertices of G are connected by at most one edge, at least one of x_i and x_j corresponds to an edge (and not to a vertex) of $\tilde{T}$, which together with e_1 and e_2 would form a pairwise intersecting triple of edges, contradicting our assumption that G is 3-quasi-planar.

Claim 2. *H is a 3-quasi-planar convex geometric graph.*

To establish this claim, it is sufficient to observe that if two edges of H cross each other, then the "features" of $\tilde{T}$ corresponding to their endpoints alternate in the clockwise order around the boundary of D. Therefore, any three pairwise crossing edges of H would correspond to three pairwise crossing segments, which is a contradiction.

A *segment* s is said to be *shielded* if there are two other segments, s_1 and s_2, of the same type, one on each side of s. Otherwise, s is called *exposed*. An *edge* $e \in E$ is said to be *exposed* if at least one of its segments is exposed. Otherwise, e is called a *shielded edge*.

In view of Claim 2, we can apply Theorem 3.4 [7] to H. We obtain that $\left|E(H)\right| \leq 4\left|V(H)\right| - 10 < 32n$, that is, there are fewer than $32n$ different *types* of segments. There are at most two exposed segments of the same type, so the total number of exposed segments is smaller than $64n$, and this is also an upper bound on the number of exposed edges in E.

It remains to bound the number of shielded edges in E.

Claim 3. There are no shielded edges.

Suppose, in order to obtain a contradiction, that there is a shielded edge $e \in E$. Orient e arbitrarily, and denote its segments by $s_1, s_2, \ldots, s_m \in S$, listed according to this orientation. For any $1 \leq i \leq m$, let $t_i \in S$ be the (unique) segment of the same type as s_i, running closest to s_i on its left side.

Since there is no self-intersecting edge and empty lens in $\tilde{G}$, the segments t_i and t_{i+1} belong to the same edge $f \in E$, for every $i < m$ (see Fig. 5). However, this means that both endpoints of e and f coincide, which is impossible.

We can conclude that E has fewer than $64n$ elements, all of which are exposed. Thus, taking into account the $n - 1$ edges of the spanning tree T, the total number of edges of $\tilde{G}$ is smaller than $65n$.

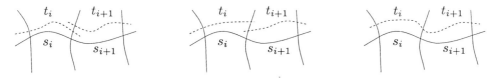

Fig. 5. t_i and t_{i+1} belong to the same edge

REFERENCES

[1] P. K. Agarwal, B. Aronov, J. Pach, R. Pollack and M. Sharir, Quasi-planar graphs have a linear number of edges, *Combinatorica,* **17** (1997), 1–9.

[2] M. Ajtai, V. Chvátal, M. Newborn and E. Szemerédi, Crossing-free subgraphs, in: *Theory and Practice of Combinatorics, North-Holland Math. Stud.,* **60,** North-Holland, Amsterdam–New York, 1982, 9–12.

[3] N. Alon and P. Erdős, Disjoint edges in geometric graphs, *Discrete Comput. Geom.,* **4** (1989), 287–290.

[4] Bollobás and A. Thomason, Proof of a conjecture of Mader, Erdős and Hajnal on topological complete subgraphs, *European J. Combin.,* **19** (1998), 883–887.

[5] P. Braß, G. Károlyi and P. Valtr, A Turán-type extremal theory for convex geometric graphs, in: *Discrete and Computational Geometry – The Goodman-Pollack Festschrift* (B. Aronov et al., eds.), Springer Verlag, Berlin, 2003, to appear.

[6] G. Cairns and Y. Nikolayevsky, Bounds for generalized thrackles, *Discrete Comput. Geom.,* **23** (2000), 191–206.

[7] V. Capoyleas and J. Pach, A Turán-type theorem on chords of a convex polygon, *Journal of Combinatorial Theory, Series B,* **56** (1992), 9–15.

[8] W. Goddard, M. Katchalski and D. J. Kleitman, Forcing disjoint segments in the plane, *European J. Combin.,* **17** (1996), 391–395.

[9] H. Hopf and E. Pannwitz, Aufgabe Nr. 167, *Jahresbericht der deutschen Mathematiker-Vereinigung,* **43** (1934), 114.

[10] J. Komlós and E. Szemerédi, Topological cliques in graphs II, *Combin. Probab. Comput.,* **5** (1996), 79–90.

[11] A. V. Kostochka, Lower bound of the Hadwiger number of graphs by their average degree, *Combinatorica,* **4** (1984), 307–316.

[12] A. V. Kostochka, Upper bounds on the chromatic number of graphs (in Russian), *Trudy Inst. Mat. (Novosibirsk), Modeli i Metody Optim.,* **10** (1988), 204–226.

[13] A. V. Kostochka and J. Kratochvíl, Covering and coloring polygon-circle graphs, *Discrete Math.,* **163** (1997), 299–305.

[14] F. T. Leighton, New lower bound techniques for VLSI, *Math. Systems Theory,* **17** (1984), 47–70.

[15] L. Lovász, J. Pach and M. Szegedy, On Conway's thrackle conjecture, *Discrete and Computational Geometry,* **18** (1997), 369–376.

[16] W. Mader, $3n - 5$ edges do force a subdivision of K_5, *Combinatorica,* **18** (1998), 569–595.

[17] P. Kolman and J. Matoušek, Crossing number, pair-crossing number, and expansion, *J. Comb. Theory,* Ser. B, **92** (2004), 99–113.

[18] S. McGuinness, Colouring arcwise connected sets in the plane I, *Graphs & Combin.,* **16** (2000), 429–439.

[19]   J. Pach, Geometric graph theory, in: *Surveys in Combinatorics, 1999 (J. D. Lamb and D. A. Preece, eds.), London Mathematical Society Lecture Notes,* **267**, Cambridge University Press, Cambridge, 1999, 167–200.

[20]   J. Pach, R. Pinchasi, M. Sharir and G. Tóth, Topological graphs with no large grids, Special Issue dedicated to Victor Neumann-Lara, *Graphs and Combinatorics* (accepted).

[21]   J. Pach, R. Pinchasi, G. Tardos and G. Tóth, Geometric graphs with no self-intersecting path of length three, in: *Graph Drawing* (M. T. Goodrich, S. G. Kobourov, eds.), Lecture Notes in Computer Science **2528**, Springer-Verlag (Berlin, 2002), 295–311. Also in: *European Journal of Combinatorics,* **25** (2004), 793–811.

[22]   J. Pach, R. Radoičić and G. Tóth, A generalization of quasi-planarity, in: *Towards a Theory of Geometric Graphs* (J. Pach, ed.), Contemporary Mathematics **342**, AMS (2004), 177–183.

[23]   J. Pach, R. Radoičić, G. Tardos and G. Tóth, Improving the Crossing Lemma by finding more crossings in sparse graphs, *Proceedings of the 20th Annual Symposium on Computational Geometry* (2004), 68–75. Also in: *Discrete and Computational Geometry,* accepted.

[24]   J. Pach, F. Shahrokhi and M. Szegedy, Applications of the crossing number, *Algorithmica,* **16** (1996), 111–117.

[25]   J. Pach and G. Tóth, Graphs drawn with few crossings per edge, *Combinatorica,* **17** (1997), 427–439.

[26]   J. Pach and J. Törőcsik, Some geometric applications of Dilworth's theorem, *Discrete Comput. Geom.,* **12** (1994), 1–7.

[27]   R. Pinchasi and R. Radoičić, Topological graphs with no self-intersecting cycle of length 4, *Proceedings of the 19th Annual Symposium on Computational Geometry* (2003), 98–103. Also in: *Towards a Theory of Geometric Graphs* (J. Pach, ed.), Contemporary Mathematics **342**, AMS (2004), 233–243.

[28]   N. Robertson, P. Seymour and R. Thomas, Hadwiger's conjecture for K_6-free graphs, *Combinatorica,* **13** (1993), 279–361.

[29]   O. Sýkora and I. Vrt'o, On VLSI layouts of the star graph and related networks, *Integration, The VLSI Journal,* **17** (1994), 83–93.

[30]   G. Tardos, *Construction of locally plane graphs,* manuscript.

[31]   A. Thomason, An extremal function for contractions of graphs, *Math. Proc. Cambridge Philos. Soc.,* **95** (1984), 261–265.

[32]   G. Tóth, Note on geometric graphs, *J. Combin. Theory, Ser. A,* **89** (2000), 126–132.

[33]   A. Thomason, The extremal function for complete minors, *J. Combin. Theory Ser. B,* **81** (2001), 318–338.

[34]   P. Valtr, On geometric graphs with no k pairwise parallel edges, *Discrete and Computational Geometry,* **19** (1998), 461–469.

[35]   D. R. Woodall, Thrackles and deadlock, in: *Combinatorial Mathematics and its Applications* (Proc. Conf., Oxford, 1969), Academic Press, London, 1971, 335–347.

János Pach

City College
CUNY and Courant Institute of
Mathematical Sciences
New York University
New York, NY
U.S.A.

pach@cims.nyu.edu

Radoš Radoičić

Department of Mathematics
Massachusetts Institute of Technology
Cambridge, MA
U.S.A.

rados@math.mit.edu

Géza Tóth

Rényi Institute
Hungarian Academy of Sciences
H-1364 Budapest
P.O.B. 127
Hungary

geza@renyi.hu

BOLYAI SOCIETY
MATHEMATICAL STUDIES, 15

Conference on Finite
and Infinite Sets
Budapest, pp. 301–315.

Notes on CNS Polynomials and Integral Interpolation

A. PETHŐ*

Dedicated to the 70th birthday of Professors V. T. Sós and A. Hajnal

1. Introduction

1.1. CNS polynomials

Let $P(x) = p_d x^d + \cdots + p_0 \in \mathbb{Z}[x]$, with $p_d=1$. It is called a CNS polynomial if every element of the factor ring $R = \mathbb{Z}[x]/P(x)\mathbb{Z}(x)$ has a unique representative of form

$$(1) \qquad \sum_{i=0}^{\ell} a_i x^i, \qquad 0 \le a_i < |p_0|, \quad 0 \le i \le \ell.$$

This definition is equivalent with the following one: for any $A(x) \in \mathbb{Z}[x]$ there exist uniquely integers ℓ, $0 \le a_i < |p_0|$, $0 \le i \le \ell$, $p_\ell \ne 0$ if $\ell > 0$, such that

$$(2) \qquad A(x) \equiv \sum_{i=0}^{\ell} a_i x^i \quad (\text{mod } P).$$

In the sequel the polynomial staying on the right hand side of (2) will be called the CNS representation of $A(x)$.

*Research partially supported by Hungarian National Foundation for Scientific Research Grant Nos 29330 and 38225.

It is clear that there exist for any $A(x) \in \mathbb{Z}[x]$ unique integers, $A_0, \ldots, A_d$, such that

$$(3) \qquad A(x) \equiv \sum_{i=0}^{d-1} A_i x^i \quad (\mathrm{mod}\ P).$$

One gets $\sum_{i=0}^{d-1} A_i x^i$ from $A(x)$ by dividing it by $P(x)$. In contrast one obtains $\sum_{i=0}^{\ell} a_i x^i$ from $A(x)$ by a "backward" division process with $P(x)$. More precisely, let $A(x) = \sum_{i=0}^{\infty} A_i x^i \in \mathbb{Z}[x]$, $A_i = 0$ for all but finitely many indices, and let

$$T(A) = \sum_{i=0}^{\infty} \left(A_{i+1} - p_{i+1} \left\lfloor \frac{A_0}{p_0} \right\rfloor \right) x^i,$$

where $\lfloor x \rfloor$ denotes the integer part of x. Then

$$A(x) = A_0 - p_0 \left\lfloor \frac{A_0}{p_0} \right\rfloor + x T(A).$$

To obtain the CNS representation of A one has to compute $T(A), T^2(A), \ldots$ until $T^{\ell}(A) = 0$ for some $\ell \geq 0$.

Unfortunately this "backward" division process often does not terminate; it can become divergent (e.g. -1 for $P(x) = x^2 + 4x + 2$) or ultimately periodic (e.g. -1 for $P(x) = x - 2$). Therefore the characterization of CNS polynomials is not a trivial problem. In the sequel the set of CNS polynomials will be denoted by $\mathcal{C}$.

The concept of CNS polynomials was introduced in [14] as a generalization of canonical number systems [7], [6], [8] or radix representations [5] in algebraic number fields. Generalizing a result of [9] *I* gave the following algorithmic characterization of square-free CNS polynomials [14].

Theorem AR. *Let $P(x) \in \mathbb{Z}[x]$ be square-free. Let $\alpha_1, \ldots, \alpha_d$ and D denote its zeroes and discriminant respectively. Then $P(x) \in \mathcal{C}$ if and only if*

(i) *$|\alpha_i| > 1$ and $\alpha_i < -1$, if $\alpha_i \in \mathbb{R}$, $1 \leq i \leq d$,*

(ii) *every* $A(x) = \sum_{i=0}^{d-1} A_i x^i \in \mathbb{Z}[x]$ *with*

$$|A_i| \le |D|^{-2} \left(d h(P)^{d-1} \right)^{d/2} \max_{1 \le j \le d} \frac{|\alpha_j|}{|\alpha_j| - 1}, \qquad i = 0, \ldots, d-1$$

has a CNS representation. Here $h(P)$ *denotes the maximum of the absolute values of the coefficients of* $P(x)$.

This result was generalized for not necessarily square-free polynomials, but without giving an explicit upper bound for the $A_i's$ in [3].

By changing the basis $1, x, \ldots, x^{d-1}$ of R to $w_1, w_2, \ldots, w_d$ where $w_i = \sum_{j=0}^{i} p_{d-j} x^j$, $i = 1, \ldots, d$, H. Brunotte [4] realized that the CNS property of polynomials can be characterized by properties of the mapping $\tau : \mathbb{Z}^d \to \mathbb{Z}^d$,

$$(4) \qquad \tau \left((a_1, \ldots, a_d)^T \right) = \left(- \left\lfloor \frac{a_1 p_1 + \cdots + a_d p_d}{p_0} \right\rfloor, a_1, \ldots, a_{d-1} \right)^T .$$

His original algorithm was simplified in [3]. We give here the version appeared in [1, Lemma 2.6].

Lemma B. *Assume that* $E \subseteq \mathbb{Z}^d$ *has the following properties*

(i) $(1, 0, \ldots, 0) \in E$,

(ii) $-E \subseteq E$,

(iii) $\tau(E) \subseteq E$,

(iv) *for every* $\mathbf{e} \in E$ *there exists some* $k \in \mathbb{N}$ *with* $\tau^k(\mathbf{e}) = 0$.

Then $P \in \mathcal{C}$.

Akiyama and Rao [3] pointed out, that if all zeroes of $P(x)$ are lying outside the closed unit disc then τ is ultimately contractive, i.e. there exists a finite set $E \subseteq \mathbb{Z}^d$ satisfying (i)–(iii) of Lemma B. Their argument is based on Theorem AR and the connection between the mappings T and τ. Here we will give a direct proof of the following theorem.

Theorem 1. *Let $P(x)$ be such that all of its zeroes are lying outside the closed unit disc. Then there exists a constant c depending only on d and $H(P)$ such that if*

$$E = \left\{ (a_1, \ldots, a_d)^T \in \mathbb{Z}^d : \max \left\{ |a_i|, \ 1 \leq i \leq d \right\} \leq c \right\}$$

then there exist for any $\mathbf{b} \in \mathbb{Z}^d$ a positive integer k with $\tau^k(\mathbf{b}) \in E$.

In the case when $P(x)$ is square-free, i.e. $P(x)$ has no multiple roots, we are able to give a nice explicit form for c.

Theorem 2. *Let $P(x)$ be square-free and denote $\alpha_1, \ldots, \alpha_d$ its zeros. Assume that $|\alpha_i| > 1$, $1 \leq i \leq d$. Then we may take*

$$c = \left\lfloor p_0 \sum_{i=1}^{d} \frac{|\alpha_i|^{d-1}}{\left(|\alpha_i| - 1 \right) \left| P'(\alpha_i) \right|} \right\rfloor,$$

where $P'(x)$ denotes the derivative of $P(x)$.

1.2. Integral interpolation

Let $m_1, \ldots, m_d \in \mathbb{Z}$ be pairwise distinct and $a_1, \ldots, a_d \in \mathbb{Z}$. We call $\mathbf{a} = (a_1, \ldots, a_d)^T \in \mathbb{Z}^d$ *integral interpolable* by $\mathbf{m} = (m_1, \ldots, m_d)^T$ if there exists an $I(x) \in \mathbb{Z}(x)$ such that

(5) $$I(m_i) = a_i, \quad i = 1, \ldots, d.$$

It is well known that there exists always an $I(x) \in \mathbb{Q}[x]$ satisfying (5), but very often $I(x)$ has non-integer coefficients. The Chinese remainder theorem (see e.g. Mignotte and Ştefănescu [12]) gives necessary and sufficient condition for the solvability of the integral interpolation problem, but it is usually very complicated. We are intend to give here an other condition based on CNS polynomials.

The vector $\mathbf{a} \in \mathbb{Z}^d$ is called *simultaneously representable* by $\mathbf{m} \in \mathbb{Z}^d$ if there exist integers $0 \leq q_0, \ldots, q_\ell < M = |m_1 \ldots m_d|$ such that

(6) $$a_i = \sum_{j=0}^{\ell} q_j m_i^j, \quad i = 1, \ldots, d.$$

This concept was introduced by Indlekofer, Kátai and Racskó [11].

We start with a simple observation.

Proposition 1. *If* $\mathbf{a} \in \mathbb{Z}^d$ *is simultaneously representable by* $\mathbf{m}$ *then it is integral interpolable by* $\mathbf{m}$.

Proof. If (6) holds then take $I(x) = \sum_{j=0}^{\ell} q_j x^j \in \mathbb{Z}[x]$, which satisfies obviously (5). ∎

The converse of Proposition 1 is not true. Take for example $m_1 = 1$, $m_2 = 2$, then for any pair $(a_1, a_2)^T \in \mathbb{Z}^2$ the polynomial $I(x) = (a_2 - a_1)x + (2a_1 - a_2)$ satisfies (5). On the other hand, if a_1 or a_2 is negative, then (6) can never hold, because its right hand side is always non-negative.

The next theorem connects integral interpolation, simultaneous representation and CNS polynomials.

Theorem 3. *Let* $P(x) = \prod_{i=1}^{d}(x - m_i) \in C$. *Then* $\mathbf{a} \in \mathbb{Z}^d$ *is simultaneously representable by* $\mathbf{m}$ *if and only if it is integral interpolable by* $\mathbf{m}$.

Proof. By Proposition 1 it is enough to consider the case, when $\mathbf{a}$ is integral interpolable by $\mathbf{m}$. Then there exists an $I(x) \in \mathbb{Z}[x]$ satisfying (5). As $P(x)$ is a CNS polynomial, whose constant term is $(-1)^d m_1 \ldots m_d$, there exist integers $0 \leq q_0, \ldots, q_\ell < M = |m_1 \ldots m_d|$ such that

$$I(x) \equiv \sum_{j=0}^{\ell} q_j x^j \quad (\mathrm{mod}\ P(x)),$$

which means

$$I(x) = \sum_{j=0}^{\ell} q_j x^j + Q(x)P(x)$$

with a $Q(x) \in \mathbb{Z}[x]$. Substituting here $x = m_1, \ldots, m_d$ and using (5) and $P(m_i) = 0$, $i = 1, \ldots, d$ we obtain (6). ∎

Remark that if $P(x) = \prod_{i=1}^{d}(x - m_i) \notin C$ then there exist infinitely many $\mathbf{a} \in \mathbb{Z}^d$ which is integral interpolable by $\mathbf{m}$, but not simultaneously representable. Indeed, as $P \notin C$ there exists $I(x) \in \mathbb{Z}[x]$ of degree less then d, which does not have a CNS representation with respect to $P(x)$. Choosing $a_i = I(m_i)$, $i = 1, \ldots, d$ the vector $\mathbf{a} = (a_1, \ldots, a_d)^T$ is integral interpolable by **m**. If **a** would be simultaneously representable by **m** then there would exist integers $0 \le q_1, \ldots, q_\ell < M$ such that

$$a_i = \sum_{j=0}^{\ell} q_j m_i^j, \qquad i = 1, \ldots, d.$$

Taking $J(x) = \sum_{j=0}^{\ell} q_j x^j$ then $I(x) \equiv J(x) \pmod{x - m_i}$, $i = 1, \ldots, d$ hold and as the polynomials $x - m_i$ are pairwise relatively primes $I(x) \equiv J(x)$ (mod $P(x)$) by the Chinese remainder theorem. Hence $J(x)$ would be the CNS representation of $I(x)$ with respect to $P(x)$ which is a contradiction. Hence **a** is not integral representable by **m**.

Let $Q(x) = \sum_{i=0}^{h} q_i x^i \in \mathbb{Z}[x]$ with $0 \le q_i < M$. Then $I_Q(x) = Q(x) + x^{h+1} I(x)$ does not have a CNS representation because $T^{h+1}(I_Q(x)) = I(x)$. The set

$$S = \left\{ (a_{Q_1}, \ldots, a_{Q_d}) : a_{Q_i} = I_Q(m_i), \; Q \text{ as above} \right\}$$

is obviously infinite. The elements of S are not simultaneously representable by **m**.

In the sequel we will prove under some assumptions that $\prod_{i=1}^{d} (x - m_i)$ belongs to $\mathcal{C}$.

Theorem 4. *Let* $m_1, \ldots, m_d \le -2$ *be such that*

$$\sum_{i=1}^{d} \frac{1}{|m_i|} \le 1.$$

Then $P(x) = \prod_{i=1}^{d} (x - m_i) \in \mathcal{C}$.

Proof. Let $\prod_{i=1}^{d} (x - m_i) = x^d + p_{d-1} x^{d-1} + \cdots + p_0$. By the assumptions all coefficients of $P(x)$ are positive. We have $p_j = \sum_{1 \le i_1 < \cdots < i_{d-j} \le d} |m|_{i_1} \cdots |m|_{i_{d-j}}$, $j = 0, \ldots, d - 1$. Further

(7) $$\left(\frac{p_j}{\binom{d}{j}} \right)^2 \ge \frac{p_{j+1}}{\binom{d}{j+1}} \cdot \frac{p_{j-1}}{\binom{d}{j-1}}, \qquad j = 0, \ldots, d - 2$$

hold by Newton's inequality between symmetric means. (See [10, p. 52] or the original work of I. Newton [13].)

We have

$$\frac{p_1}{p_0} = \sum_{i=1}^{d} \frac{1}{|m_i|} \leq 1,$$

i.e. $p_0 \geq p_1$. Assume that we proved already $p_0 \geq p_1 \geq \cdots \geq p_j$ for some $1 \leq j < d - 2$. Applying (7) we obtain

$$p_{j+1} \leq \binom{d}{j-1} \cdot \binom{d}{j+1} \cdot \binom{d}{j}^{-2} \cdot \frac{p_j}{p_{j-1}} \cdot p_j < p_j.$$

The theorem is proved. ∎

In the next theorem we show that the same assertion is true if d is small.

Theorem 5. *Let $d \leq 4$ and $m_1, \ldots, m_d \leq -2$ be pairwise distinct integers. Then $P(x) = \prod_{j=1}^{d} (x - m_j) \in \mathcal{C}$.*

To prove Theorem 5 we need a lemma.

Lemma 1. *Let $P(x) = x^4 + p_3 x^3 + p_2 x^2 + p_1 x + p_0 \in \mathbb{Z}[x]$ be such that*

(i) $p_1 > p_0 > p_1 - p_2/2 + p_3$,

(ii) $p_1 \leq 2p_2 \geq 8p_3$, $p_3 \leq 3$.

Then $P(x) \in \mathcal{C}$.

Proof. Consider the following set of quadruples

$$E_0 = \{ (0,0,0,0), (0,0,0,1), (0,0,1,-2), (0,0,1,-1), (0,0,1,0),$$
$$(0,1,-2,1), (0,1,-2,2), (0,1,-1,0), (0,1,-1,1), (0,1,-1,2), (0,1,0,-1),$$
$$(1,-2,2,-2), (1,-2,2,-1), (1,-2,1,0), (1,-1,0,0), (1,-1,0,1),$$
$$(1,-1,1,-2), (1,-1,1,-1), (1,-1,1,0), (1,-1,2,-2), (1,-1,2,-1),$$
$$(1,0,-1,0), (1,0,-1,1), (1,0,-1,2), (1,0,0,-1), (1,0,0,0), (2,-2,1,0),$$
$$(2,-2,2,-1), (2,-1,0,0), (2,-1,0,1), (2,-2,2,-2)\}$$

and put $E = E_0 \cup -E_0$. Notice that if $(a_1, a_2, a_3, a_4)^T \in E$, then $a_i a_{(i+1) \bmod 4} \leq 0$ and if $a_i = \pm 2$, then $a_{i+1} \neq 0$. It is easy to see that

E satisfies the conditions (i) and (ii) of Lemma B. It remains to show (iii) and (iv), which will be done by considering several cases. If $\tau(\mathbf{e}_1) = \mathbf{e}_2$, then we will write $\mathbf{e}_1 \to \mathbf{e}_2$. Further $\mathbf{e} \xrightarrow{A}$ means that $\tau(\mathbf{e})$ belongs to case A considered earlier.

I $(0, 1, a_3, a_4) \longrightarrow (0, 0, 1, a_3) \longrightarrow (0, 0, 0, 1) \longrightarrow 0$

II $a_2 < 0$, $(1, a_2, a_3, a_4) \longrightarrow (0, 1, a_2, a_3) \xrightarrow{I}$

III $a_1 > 0$, $(-1, a_2, a_3, a_4) \longrightarrow (1, -1, a_2, a_3) \xrightarrow{II}$

IV $(1, 0, a_3, a_4) \longrightarrow \begin{cases} (-1, 1, 0, a_3) \xrightarrow{III} & \text{if } p_1 + a_3 p_3 + a_4 > p_0 \\ (0, 1, 0, a_3) \xrightarrow{I} & \text{otherwise} \end{cases}$

V $(2, a_2, a_3, a_4) \longrightarrow (-1, 2, a_2, a_3) \xrightarrow{III}$

VI $(0, -1, a_3, a_4) \longrightarrow (1, 0, -1, a_3) \xrightarrow{IV}$

VII $(0, 0, -1, a_4) \longrightarrow (1, 0, 0, -1) \xrightarrow{IV}$

VIII $(0, 0, 0, -1) \longrightarrow (1, 0, 0, 0) \xrightarrow{IV}$

IX $(-2, a_2, a_3, a_4) \longrightarrow (2, -2, a_2, a_3) \xrightarrow{V}$

This shows that (iii) and (iv) hold for our set E, hence $P(x) \in \mathcal{C}$. ∎

Proof of Theorem 5. If $\sum_{i=1}^{d} \frac{1}{|m_i|} \leq 1$ then the assertion follows from Theorem 4.

If $d \leq 3$ then $\sum_{i=1}^{d} \frac{1}{|m_i|} \leq 1$ holds except when $(m_1, m_2, m_3) = (-2, -3, -4), (-2, -3, -5)$. The corresponding polynomials $x^3 + 9x^2 + 26x + 24$ and $x^3 + 10x^2 + 31x + 30$ belong to $\mathcal{C}$ by Proposition 3.12 (3) and (1) of [2].

Let $d = 4$. Then there are three infinite families

$$(m_1, m_2, m_3, m_4) = (-2, -3, -4, -m),$$

$$(-2, -3, -5, -m) \quad \text{and} \quad (-2, -3, -6, -m)$$

with $m \geq 5, 6$ and 7 respectively and 64 further values

$$
\begin{aligned}
(-2, -3, -7, -m), &\qquad 8 \leq m \leq 41, \\
(-2, -3, -8, -m), &\qquad 9 \leq m \leq 23, \\
(-2, -3, -9, -m), &\qquad 10 \leq m \leq 17, \\
(-2, -3, -10, -m), &\qquad 11 \leq m \leq 14 \quad \text{and} \\
(-2, -3, -11, -m), &\qquad 12 \leq m \leq 13
\end{aligned}
$$

for which $\sum_{i=1}^{4} \frac{1}{|m_i|} > 1$.

It is easy to see that the corresponding polynomials satisfy the assumptions of Lemma 1, thus they belong to $\mathcal{C}$. ∎

Using the method of the proof of Theorem 5 one could probably prove the same assertion for $d = 5$ too. Unfortunately the number of cases, which must be handled separately is much larger, one has three two parametric and 64 one parametric families and a lot of sporadic cases. Hence we need new ideals.

On the other hand, in light of Theorems 4 and 5 we do not see any reason not to formulate the following conjecture.

Conjecture 1. *Let* $m_1, \ldots, m_d \leq -2$ *be pairwise distinct integers. Then*
$$
P(x) = \prod_{i=1}^{d} (x - m_i) \in \mathcal{C}.
$$

2. PROOF OF THEOREMS 1 AND 2

To prove Theorems 1 and 2 we need some preparation from linear algebra and from linear recurring sequences. More precisely we have to analyze the mapping τ defined by equation (4). For $\mathbf{a} \in \mathbb{Z}^d$ let us define

$$
\tau^k(\mathbf{a}) = \begin{cases} \mathbf{a}, & \text{if } k = 0, \\ \tau\left(\tau^{k-1}(\mathbf{a})\right), & \text{if } k > 0. \end{cases}
$$

Further let $\mathbf{P} \in \mathbb{Z}^{d \times d}$ be the matrix

$$
\mathbf{P} = \begin{pmatrix} -p_1/p_0 & \cdots & -p_{d-1}/p_0 & -p_d/p_0 \\ 1 & \cdots & 0 & 0 \\ \vdots & \ddots & \vdots & \vdots \\ 0 & \cdots & 1 & 0 \end{pmatrix}.
$$

With these definitions we have the following assertion

Lemma 2. *Let* $\mathbf{a} = (a_1, \ldots, a_d)^T \in \mathbb{Z}^d$ *and* $1 \le k \in \mathbb{Z}$. *Then* $\tau^k(\mathbf{a}) \in \mathbb{Z}^d$ *and there exist* $-1 < \delta_1, \ldots, \delta_k \le 0$ *such that*

$$
\tau^k(\mathbf{a}) = \mathbf{P}^k \mathbf{a} + \sum_{j=1}^{k} \mathbf{P}^{k-j} \boldsymbol{\delta}_j
$$

holds, where $\boldsymbol{\delta}_j = (\delta_j, 0, \ldots, 0)^T \in \mathbb{R}^d$.

Proof. Let $k = 1$. Then $\tau(\mathbf{a}) \in \mathbb{Z}^d$, which can be written in the form

$$
\tau(\mathbf{a}) = \left(-\frac{a_1 p_1 + \cdots + a_d p_d}{p_0} + \delta_1, a_1, \ldots, a_{d-1} \right)^T = \mathbf{P}\mathbf{a} + \boldsymbol{\delta}_1,
$$

where $\boldsymbol{\delta}_1 = (\delta_1, 0, \ldots, 0)^T \in \mathbb{R}^d$ with $-1 < \delta_1 \le 0$.

Assume that the assertion is true for $k - 1 \ge 0$. Then $\tau^{k-1}(\mathbf{a}) \in \mathbb{Z}^d$, thus $\tau^k(\mathbf{a}) \in \mathbb{Z}^d$ is true by (4) too. Let $\tau^{k-1}(\mathbf{a}) = \left(a_1^{(k-1)}, \ldots, a_d^{(k-1)}\right)^T \in \mathbb{Z}^d$ then

$$
\tau^k(\mathbf{a}) = \left(-\left\lfloor \frac{a_1^{(k-1)} p_1 + \cdots + a_d^{(k-1)} p_d}{p_0} \right\rfloor, a_1^{(k-1)}, \ldots, a_{d-1}^{(k-1)} \right)^T
$$

$$
= \left(-\frac{a_1^{(k-1)} p_1 + \cdots + a_d^{(k-1)} p_d}{p_0} + \delta_k, a_1^{(k-1)}, \ldots, a_{d-1}^{(k-1)} \right)^T,
$$

where δ_k satisfies the inequalities $-1 < \delta_k \le 0$ by the definition of the integer part function. Thus

$$
\tau^k(\mathbf{a}) = \mathbf{P}\tau^{k-1}(\mathbf{a}) + (\delta_k, 0, \ldots, 0)^T
$$

and by the induction hypothesis the assertion of the lemma follows. $\blacksquare$

Let $\{G_n\}_{n=0}^{\infty}$ be the linear recurring sequence defined by the initial terms $G_0 = \ldots G_{d-2} = 0$, $G_{d-1} = 1$ and by the difference equation

(8)
$$G_{n+d} = -\frac{p_1}{p_0} G_{n+d-1} - \cdots - \frac{p_d}{p_0} G_n.$$

Let further $\mathbf{G}_n = (G_{n+d-1}, \ldots, G_n)^T$ and for $n \geq d-1$ denote by $\mathcal{G}_n$ the $d \times d$ matrix, whose columns are $\mathbf{G}_n, \ldots, \mathbf{G}_{n-d+1}$. Then we have obviously

$$\mathcal{G}_n = \mathbf{P}\mathcal{G}_{n-1} \quad \text{for} \quad n = d, d+1, \ldots.$$

This implies

(9)
$$\mathcal{G}_{n+d-1} = \mathbf{P}^n \, \mathcal{G}_{d-1} \qquad \text{for} \quad n \geq 0.$$

As $\mathcal{G}_{d-1}$ is a non singular matrix we obtain

(10)
$$\mathbf{P}^n = \mathcal{G}_{n+d-1}\mathcal{G}_{d-1}^{-1}.$$

On the other hand if $\beta_1, \ldots, \beta_h$ denote the distinct zeroes of the polynomial

$$P^*(x) = x^d + \frac{p_1}{p_0} x^{d-1} + \cdots + \frac{p_d}{p_0} = x^d P\left(\frac{1}{x}\right)$$

with multiplicity $e_1, \ldots, e_h \geq 1$, then

(11)
$$G_n = g_1(n)\beta_1^n + \cdots + g_h(n)\beta_h^n$$

hold for any $n \geq 0$, where $g_i(x)$, $1 \leq i \leq h$ denote polynomials with coefficients of the field $\mathbb{Q}(\beta_1, \ldots, \beta_h)$ of degree at most $e_i - 1$. (See [15].)

Denoting by $p_{ij}^{(n)}$, $1 \leq i, j \leq d$, $n \geq 0$ the entries of $\mathbf{P}^n$ then (10) and (11) imply

(12)
$$p_{ij}^{(n)} = \sum_{\ell=1}^{h} g_{ij\ell}(n)\beta_\ell^n.$$

Proof of Theorem 1. As we explained in the introduction it is enough to consider polynomials of degree at most $d-1$, say

$$A(x) = \sum_{i=0}^{d-1} a_i x^i, \qquad a_i \in \mathbb{Z}.$$

Let $\mathbf{a} = (a_0, \ldots, a_{d-1})^T \in \mathbb{Z}^d$. We have to prove that there exists a $k \geq 0$ such that $\tau^k(\mathbf{a}) \in E$. From the proof it will be clear how to choose the constant c.

Let $k \geq 1$. Then there exist by Lemma 2 $-1 < \delta_1, \ldots, \delta_k \leq 0$ such that

$$(13) \qquad \tau^k(\mathbf{a}) = \mathbf{P}^k \mathbf{a} + \sum_{j=1}^{k} \mathbf{P}^{k-j} \delta_j.$$

In the sequel let $\tau^k(\mathbf{a})_i$ denote the i-th coordinate of $\tau^k(\mathbf{a})$. Then (13) implies

$$\tau^k(\mathbf{a})_i = \sum_{j=1}^{d} p_{ij}^{(k)} a_j + \sum_{j=1}^{k} p_{i1}^{(k-j)} \delta_j.$$

Observe that by (9) the first column of $\mathbf{P}^n$ is exactly $\mathbf{G}_n$. Using this, (11), (12) and the last equation we obtain

$$\tau^k(\mathbf{a})_i = \sum_{j=1}^{d} a_j \sum_{\ell=1}^{h} g_{ij\ell}(k) \beta_\ell^k + \sum_{\ell=1}^{h} \sum_{j=0}^{k-1} \delta_{k-j} g_\ell(j) \beta_\ell^i.$$

Thus

$$(14) \qquad \left| \tau^k(\mathbf{a})_i \right| \leq \sum_{j=1}^{d} |a_j| \sum_{\ell=1}^{h} |g_{ij\ell}(k)| \, |\beta_\ell|^k + \sum_{\ell=1}^{h} \sum_{j=0}^{k-1} |\delta_{k-j}| \, |g_\ell(j)| \, |\beta_\ell|^j.$$

The roots of $P^*(x)$ are the reciprocal of the roots of $P(x)$, hence $|\beta_\ell| < 1$, $1 \leq \ell \leq h$. If k is large enough then the first summand of (14) is less than 1. Similarly, there exists a j_0 such that if $j > j_0$ then $|g_\ell(j)| |\beta_\ell|^j < |\beta_\ell|^{j/2}$. As $|\delta_j| \leq 1$, $j \geq 1$ we can estimate the second summand of (14) as follows

$$\sum_{\ell=1}^{h} \sum_{j=0}^{k-1} |\delta_{k-j}| \, |g_\ell(j)| \, |\beta_\ell|^j \leq \sum_{\ell=1}^{h} \sum_{j=0}^{j_0} |g_\ell(j)| \, |\beta_\ell|^j + \sum_{\ell=1}^{h} \sum_{j=j_0+1}^{j_0} |\beta_\ell|^{j/2}$$

$$\leq \sum_{\ell=1}^{h} \sum_{j=j_0+1}^{\infty} |g_\ell(j)| \, |\beta_\ell|^j + \sum_{\ell=1}^{h} \frac{|\beta_\ell|^{(j_0+1)/2}}{1 - |\beta_\ell|^{1/2}}.$$

Hence, taking

$$c = 1 + \sum_{\ell=1}^{h} \sum_{j=0}^{j_0} |g_\ell(j)| \, |\beta_\ell|^j + \sum_{\ell=1}^{h} \frac{|\beta_\ell|^{(j_0+1)/2}}{1 - |\beta_\ell|^{1/2}}$$

then

$$\left| \tau^k(\mathbf{a})_i \right| \leq c$$

and c depends only on the height and degree of $P(x)$. ∎

To prove Theorem 2 we need one more lemma

Lemma 3. Let the linear recurring sequence $\{G_n\}_{n=0}^{\infty}$ be defined by (8). Assume that $P(x)$ is square-free and denote $\alpha_1, \ldots, \alpha_d$ its roots. Then

$$G_n = -p_0 \sum_{h=1}^{d} \frac{\alpha_h^{d-n-2}}{P'(\alpha_h)}.$$

Proof. By a result of M. Ward [16] we have

$$G_n = \sum_{h=1}^{d} \frac{\beta_h^n}{P^{*\prime}(\beta_h)}.$$

As $\beta_h = \frac{1}{\alpha_h}$, $h = 1, \ldots, d$ and

$$P^{*\prime}(\beta_h) = \prod_{\substack{j=1 \\ j \neq h}}^{d} (\beta_h - \beta_j) = \prod_{\substack{j=1 \\ j \neq h}}^{d} \left(\frac{1}{\alpha_h} - \frac{1}{\alpha_j} \right) = \frac{(-1)^{2d-1}}{p_0 \alpha_h^{d-2}} \prod_{\substack{j=1 \\ j \neq h}}^{d} (\alpha_h - \alpha_j).$$ ∎

Proof of Theorem 2. We are using the notations introduced in the proof of Theorem 1. As the roots of $P(x)$ are simple the polynomials $g_{ij\ell}(x)$ are constants. Further $\beta_h = 1/\alpha_h$, $h = 1, \ldots, d$. After these preparations, using also Lemma 3 inequality (14) can be rewritten as

$$\left| \tau^k(\mathbf{a})_i \right| \leq \sum_{j=1}^{d} |a_j| \sum_{\ell=1}^{d} |g_{ij\ell}| \left| \frac{1}{\alpha_\ell} \right|^k + \sum_{\ell=1}^{d} \sum_{j=0}^{k-1} |\delta_{k-j} p_0| \frac{|\alpha_\ell|^{d-2}}{|P'(\alpha_\ell)|} |\alpha_\ell|^{-j}.$$

Now there exists for any $\varepsilon > 0$ a k_0 such that if $k > k_0$ then the first summand is less than ε. For the second summand we obtain

$$\sum_{\ell=1}^{d} |p_0| \frac{|\alpha_\ell|^{d-2}}{|P'(\alpha_\ell)|} \sum_{j=0}^{k-1} |\delta_{k-j}| |\alpha_\ell|^{-j} \leq \sum_{\ell=1}^{d} |p_0| \frac{|\alpha_\ell|^{d-2}}{|P'(\alpha_\ell)|} \cdot \frac{1}{1 - \frac{1}{|\alpha_\ell|}}$$

$$= |p_0| \sum_{\ell=1}^{d} \frac{|\alpha_\ell|^{d-1}}{(|\alpha_\ell| - 1) |P'(\alpha_\ell)|}.$$

Hence

$$\left| \tau^k(\mathbf{a})_i \right| \le \varepsilon + |p_0| \sum_{\ell=1}^{d} \frac{|\alpha_\ell|^{d-1}}{\left(|\alpha_\ell| - 1\right)\left|P'(\alpha_\ell)\right|}.$$

As $\tau^k(\mathbf{a})_i \in \mathbb{Z}$ and ε can be chosen arbitrary small we obtain

$$\left| \tau^k(\mathbf{a})_i \right| \le \left\lfloor |p_0| \sum_{\ell=1}^{d} \frac{|\alpha_\ell|^{d-1}}{\left(|\alpha_\ell| - 1\right)\left|P'(\alpha_\ell)\right|} \right\rfloor. \qquad \blacksquare$$

REFERENCES

[1] Sh. Akiyama, H. Brunotte and A. Pethő, Cubic CNS polynomials, notes on a conjecture of W. J. Gilbert, *J. Math. Anal. and Appl.,* **281** (2003), 402–415.

[2] Sh. Akiyama and A. Pethő, On canonical number systems, *Theoret. Comp. Sci.,* **270** (2002), 921–933.

[3] Sh. Akiyama and H. Rao, *New criteria for canonical number systems, Acta Arith.,* **111** (2004), 5–25.

[4] H. Brunotte, On trinomial bases of radix representations of algebraic integers, *Acta Sci. Math. (Szeged),* **67** (2001), 407–413.

[5] W. J. Gilbert, Radix representations of quadratic fields, *J. Math. Anal. and Appl.,* **83** (1981), 264–274.

[6] I. Kátai and B. Kovács, Kanonische Zahlsysteme in der Theorie der quadratischen Zahlen, *Acta Sci. Math.,* **42** (1980), 99–107.

[7] I. Kátai and J. Szabó, Canonical number-systems for complex integers, *Acta Sci. Math.,* **37** (1975), 255–260.

[8] B. Kovács, Canonical number systems in algebraic number fields, *Acta Math. Hung.,* **37** (1981), 405–407.

[9] B. Kovács and A. Pethő, Number systems in integral domains especially in orders of algebraic number fields, *Acta Sci. Math. Szeged,* **55** (1991), 287–299.

[10] G. Hardy, J. E. Littlewood and G. Pólya, *Inequalities,* Cambridge the University Press, 2nd ed. 1952.

[11] K.-H. Indlekofer, I. Kátai and P. Racskó, Number systems and fractal geometry, Probability theory and applications, Essays to the Mem. of J. Mogyoródi, *Math. Appl.,* **80** (1992), 319–334.

[12] M. Mignotte and D. Ştefănescu, *Polynomials: an algorithmic approach,* Singapore: Springer, 1999.

[13] I. Newton, *Arithmetica universalis: sive de compositione et resolutione arithemetic liber,* 1701.

[14] A. Pethő, On a polynomial transformation and its application to the construction of a public key cryptosystem, in: *Computational Number Theory*, Proc., Walter de Gruyter Publ. Comp., Eds.: A. Pethő, M. Pohst, H. G. Zimmer and H. C. Williams (1991), 31–44.

[15] T. N. Shorey and R. Tijdeman, *Exponential Diophantine Equations,* Cambridge Univ. Press 1986.

[16] M. Ward, The maximal prime divisors of linear recurrences, *Can. J. Math.,* **6** (1954), 455–462.

Attila Pethő

Department of Computer Science
University of Debrecen
H-4010 Debrecen
P.O. Box 12, Hungary

pethoe@math.klte.hu

BOLYAI SOCIETY
MATHEMATICAL STUDIES, 15

Conference on Finite
and Infinite Sets
Budapest, pp. 317–328.

THE EVOLUTION OF AN IDEA – GALLAI'S ALGORITHM

A. RECSKI and D. SZESZLÉR

Vera T. Sós is probably the single most influential person for orienting the research interest of many of the participants of this conference towards discrete mathematics. It is appropriate to recall that the single most influential person for orienting *his* interest towards discrete mathematics was his secondary school math teacher, Tibor Gallai who achieved outstanding results in several areas of graph theory. In this note the first forty years of the influence on VLSI design of a classic result of Gallai about the perfectness of interval graphs is described.

1. INTRODUCTION

The first classic result in the topic of VLSI (Very Large Scale Integrated) routing is probably Gallai's linear time algorithm. From a graph-theoretical viewpoint it is nothing else but an alternative proof of the fact that interval graphs are perfect.

The design of VLSI circuits is a broad area, it covers a wide range of substantially different problems that arise during the design process. One of these problems is *detailed routing* which can be formulated in the following way. Assume that the devices of the electric equipment have already been placed on the four boundaries of a rectangular circuit board. Our task is to interconnect certain given subsets (or *nets*) of the pins (or *terminals*) of these devices by wires. Wires belonging to different nets must not intersect or get closer to each other than a given distance. To this end, it is mostly assumed that the wires must go on a given 3-dimensional rectangular grid consisting of a number of planar layers, each of them parallel with the circuit board. (In the 1-layer, that is, 2-dimensional case the problem is unsolvable in most cases.) Wires can leave a layer for a consecutive one at any gridpoint. To sum it up from a graph-theoretical viewpoint, the detailed routing problem consists of finding vertex-disjoint Steiner-trees (trees with a given terminal vertex set) in a 3-dimensional rectangular grid graph. In this context, the given vertex sets of the trees are the nets and the Steiner-trees themselves are called *wires*.

Traditionally, detailed routing was considered a 2-dimensional problem because the number of layers was very small compared to the length and the width of the board. (Originally, in the ancient times of printed circuit technology there were only two layers: the two sides of the board. Later the number of layers was gradually extended to $3, 4, \ldots$) Since recent technology permits more and more layers (6, 8 or even more) a 'real' 3-dimensional approach becomes reasonable. In this paper we first give a brief survey of 2-dimensional results with a special emphasis on those that use Gallai's algorithm or an idea similar to it. Then we turn our attention to 3-dimensional routing and we survey a few related results.

2. 2-DIMENSIONAL ROUTING

2.1. Single Row Routing

Within detailed routing, the easiest special case is *single row routing*. In a single row routing problem all the terminals of each net are situated on one boundary (say, the upper boundary) of the circuit board. Hence the specification of such a problem only fixes the number of columns of the grid (the *length* of the problem). Therefore the usual formulation of single row routing is to fix the number of layers and ask for the minimum *width* routing, that is, a routing that occupies the minimum number of rows in the grid.

A solution of a (not necessarily single row) routing problem is said to belong to the *Manhattan model* if consecutive layers contain wire segments of different directions only. That is, layers with horizontal (parallel with the upper boundary) and vertical (perpendicular to the upper boundary) wire segments alternate. This notion is motivated by the fact that for certain technologies it is advantageous not to have long parallel wire segments on two consecutive layers. Therefore there are many results that provide routings in the Manhattan model. If a solution does not belong to the Manhattan model, it is said to be in the *unconstrained model*.

Gallai's algorithm solves the single row routing problem with optimal width in the 2-layer Manhattan model. Such a routing problem together with a possible solution is shown in Fig. 1. In Fig. 1 solid dots denote the terminals and sets of terminals marked with a common number form the nets. Wire segments of the two layers are denoted by continuous and dashed lines, respectively. Empty dots denote the edges of the wires that join adjacent vertices of the two layers (these are called *vias*).

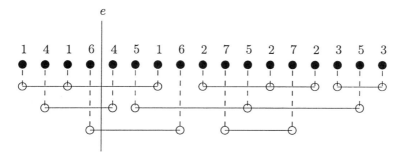

Fig. 1

For every vertical line e that cuts the grid into two we define its *congestion* $c(e)$: it is the number of nets that are divided into two by e (that is, the number of nets that have terminals both left to e and right to e). For example, the congestion of the line e in Fig. 1 is $c(e) = 3$. The maximum congestion of all vertical lines that cut the grid into two is called the *density* of the problem. It is straightforward that the density is a lower bound on the width of any routing (again, in the 2-layer Manhattan model).

Theorem 1 (T. Gallai) [3]. *The minimum width of a solution of a single row routing problem in the 2-layer Manhattan model is equal to the density of the problem.*

The proof involves a linear time algorithm (the 'left edge algorithm'). The connection between the above result and the perfectness of interval graphs is almost straightforward. A horizontal interval is associated with every net, stretching from its leftmost terminal to its rightmost terminal. The density is equal to the clique number of the corresponding interval graph. A colouring with an equal number of colours can easily be transformed into an optimal width routing: nets belonging to a common colour class can be routed in a common row. In Fig. 2 the interval graph corresponding to the routing problem of Fig. 1 is coloured using three colours; the solution of the routing problem obtained from this colouring is shown in Fig. 1.

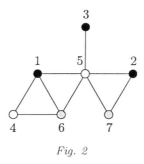

Fig. 2

We mention that no polynomial time algorithm is known to find an optimum width solution for a single row routing problem in the unconstrained 2-layer model.

2.2. Channel Routing

By the *channel routing problem* we mean the special case of detailed routing in which all the terminals of each net are situated on two opposite boundaries of the grid (say, the upper and lower boundaries). Again, the usual setting of the problem is to fix the number of layers and ask for the minimum width routing, if at all a routing exists. However, channel routing is much more complicated than single row routing as it is shown by the following theorem.

Theorem 2 (T. G. Szymanski, 1985) [12]. *It is NP-complete to decide whether a channel routing problem is solvable in the 2-layer Manhattan model with width at most w (where w is part of the input).*

Therefore it is worthwhile to look at this problem under less strict conditions: either in the 2-layer unconstrained model, or in the multilayer Manhattan model.

It is true that every channel routing problem is solvable in the 2-layer unconstrained model in polynomial time (with a sufficiently large width). This was first proved by M. Marek-Sadowska and E. Kuh [6]. Later A. Recski and F. Strzyzewski found a linear time algorithm which also uses Gallai's algorithm as a 'subroutine'.

Theorem 3 (A. Recski and F. Strzyzewski, 1990) [8]. *Every channel routing problem can be solved in linear time in the 2-layer unconstrained model.*

Their algorithm does not give an optimal width solution. The complexity of the naturally arising question of finding a minimum width routing is not known, but according to the widely accepted conjecture of D. S. Johnson [5] it is NP-hard.

It is also true that every channel routing problem is solvable in the k-layer Manhattan model for every $k \geq 3$. This again can be proved by a simple modification of Gallai's algorithm. The complexity of finding a minimum width routing is in this case known to be NP-hard [7].

2.3. Switchbox Routing

In the *switchbox routing problem* terminals may occur on all four boundaries of the circuit board. Since the specification of such a problem fixes both the length and the width of the board, the number of layers is to be optimized. We suppose that the corners of the board are not occupied by a terminal and routings must not use them either. We also suppose that the wires can access the terminals on any layer.

We have seen that in case of single row and channel routing two layers were always sufficient to solve any problem (and if we restrict ourselves to the Manhattan model, three layers were needed in case of channel routing). This, however, is not true for switchbox routing. Moreover, no fixed number of layers suffice, which is shown by the following theorem.

Theorem 4 (S. E. Hambrusch, 1985) [4]. *For every positive integer k there exists a switchbox routing problem that cannot be solved on k layers in the unconstrained model.*

Proof. Consider the switchbox routing problem of Fig. 3. The congestion of the line e is $n + w$, that is, each of the $n + w$ nets have terminals on both sides of e. Therefore the existence of a routing on k layers implies $n + w \leq kw$ since there are w rows on every layer. From this we have $\frac{n}{w} + 1 \leq k$. The value of n and w can be chosen such that this inequality does not hold, which proves the theorem. ∎

Obviously, the background of the phenomenon involved in the above theorem is the fact that in case of switchbox routing both the length n and the width w are given by the specification. Denote the ratio $\max\left(\frac{n}{w}, \frac{w}{n}\right)$ by

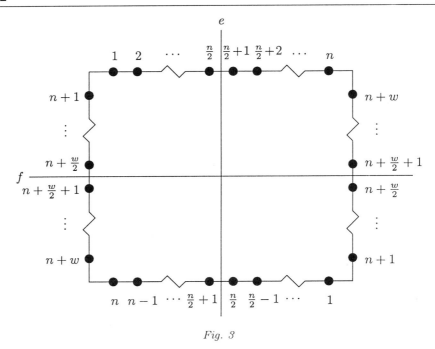

Fig. 3

m. The proof of the above theorem also includes the following statement: $\lceil m \rceil + 1$ is a lower bound on the minimum number of layers of a solution in the worst case. A slight modification of the proof shows that if we restrict ourselves to the Manhattan model then at least $2\lceil m \rceil + 1$ layers are needed in the worst case if $m > 1$ and 4 layers are needed if $m = 1$.

It is a natural question whether there is also an upper bound for the necessary number of layers as a function of m? The following theorem answers this question in the affirmative.

Theorem 5 (E. Boros, A. Recski and F. Wettl, 1995) [2]. *Any switchbox routing problem can be solved in linear time on at most 18 layers if $m \le 2$ and on at most $2m + 14$ layers if $m > 2$ in the unconstrained model.*

Later the bounds given in the above theorem were improved. The construction of the following result can also be regarded as a generalization of Gallai's method.

Theorem 6 (D. Szeszlér, 1997) [11]. *Any switchbox routing problem can be solved in linear time on at most $2\lceil m \rceil + 4$ layers in the Manhattan model.*

3. 3-DIMENSIONAL ROUTING

Although the solution of a switchbox problem can require arbitrarily many layers, switchbox routing can still be regarded as a 2-dimensional problem: the input consists of four sequences (the terminals on the four boundaries) and the output consists of a fixed number of planar layers (provided that the value of m defined in the previous section is fixed).

Due to the quick improvement of routing technology, research has recently turned towards 'real' 3-dimensional routing. In the *single active layer routing problem* (or *SALRP* for short) the terminals to be interconnected are situated on a rectangular planar grid of size $w \times n$ and the routing should be realized in a cubic grid of height h above the original grid that contains the terminals. Evidently, the height h is to be optimized. Henceforth we will use the term 'vertical direction' to refer to the direction of h (that is, the direction perpendicular to the $w \times n$ rectangle) and not for the direction of w.

One can easily see even in small instances like 4×1 or 2×2 that a routing is usually impossible unless either the length n or the width w may be extended by introducing extra rows or columns between rows and columns of the original grid.

By a *spacing of s_n in direction n* we are going to mean that we introduce $s_n - 1$ pieces of extra columns between every two consecutive columns (and also to the right hand side of the rightmost column) of the original grid. A *spacing of s_w in direction w* is defined analogously. This way the length of the grid is extended to $n' = s_n \cdot n$ and the width is extended to $w' = s_w \cdot w$.

A very similar argument to that of Theorem 4 provides a lower bound on the height h in the worst case.

Lemma 1. *For any given n and s_w there exists a routing problem that cannot be solved with height h smaller than $\frac{n}{2s_w}$.*

Proof. Let, for simplicity, the width and the length be even, let $w = 2a$ and $n = 2b$. Consider the following example. Suppose that each net consists of two terminals in central-symmetric position as shown in Fig. 4.

The number of nets is an. Since each net is cut into two by the central vertical line e, any routing with width $w' = s_w \cdot w$ and height h must satisfy $w'h \geq an$. Therefore $h \geq (w/2w')n$, hence $h \geq \frac{n}{2s_w}$. ∎

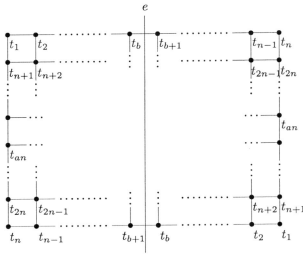

Fig. 4

The following lemma, on the other hand, provides a trivial upper bound for the height.

Lemma 2. *If $s_w \geq 2$ and $s_n \geq 2$ then every routing problem can be solved with height $h \leq \frac{wn}{2}$.*

Proof. We assign a separate layer to each net. For every terminal we introduce a vertical (parallel with the height) wire segment to connect the terminal with the layer of its net. The interconnection of the terminals of each net can now be performed trivially on its layer using the extra rows and columns guaranteed by the spacing in both directions.

Since 1-terminal nets can be disregarded, the number of nets is at most $\frac{1}{2}nw$ thus $h \leq \frac{wn}{2}$ follows immediately. ∎

The above lemma is a partial explanation for the phenomenon that the nature of single active layer routing seems to depend fundamentally on whether only one of the quantities s_w and s_n is at least 2 or both of them.

3.1. The $s_n = 1$ case

An alternative interpretation of Lemma 2 is that if we fix w then there is a routing of height $h = O(n)$, provided that $s_w, s_n \geq 2$. The truth of the same statement is not at all obvious in the $s_n = 1$ case. However, the following result shows that such a statement is true if $s_w \geq 8$ holds.

Theorem 7 [9]. *If $s_w \geq 8$ then for any fixed value of w and for any n a single active layer routing problem can always be solved in time $t = O(n)$ and with height $h = O(n)$ such that the length n is preserved or increased by at most one. Both linear bounds are best possible.*

Our algorithm gives $t = O(w^3 n)$ and $h = O(wn)$. (The straightforward lower bound for the time is the length of the input, that is, $t = \Omega(wn)$.)

The proof of the above theorem is somewhat lengthy and highly technical. Nevertheless, the basic idea is again a 3-dimensional modification of the Gallai algorithm. This is only illustrated by the following remarks.

Suppose at first that $w = 1$. Then what we have is essentially a single row routing problem with density d. Each net determines an interval of length at most n and these intervals can be packed in a vertex-disjoint way into d parallel lines, usually called *tracks*, using the Gallai algorithm. Using the classical 2-layer Manhattan model, we can arrange the tracks in a horizontal plane, as shown in the top of Fig. 5, thus realizing a routing with $w' = d$ and $h = 2$. However, alternatively these tracks can occupy either a vertical plane, leading to $w' = 2$ and $h = d$, or two vertical planes, leading to $w' = 3$ and $h = \lceil d/2 \rceil$, see the middle and the bottom drawing of Fig. 5, respectively. (Theoretically one can pack the tracks to more vertical planes and thus ensure $h = \lceil 3d/(2w') \rceil$ for larger values of w' as well but it does not seem to be interesting.) Throughout in Figures 5 and 6 continuous lines denote wires while dotted lines are for the indication of coplanarity only.

Similarly, if $w = 2$ then we have a channel routing problem with density d and using the same linear time algorithm we can always realize a routing with $w' = d + 1$ and $h = 3$ or with $w' = 3$ and $h = d + 1$, see Fig. 6.

Actually, it is the right hand side of Fig. 6 that shows the essential idea of the proof of Theorem 7.

3.2. The $s_n, s_w \geq 2$ case

We have seen in Lemma 2 that every SALRP problem can trivially be solved with height $h = \frac{wn}{2}$ if $s_w, s_n \geq 2$. This provides an upper bound of $h = O(n^2)$ in the $n = w$ case. However, in 2000 Aggarwal et al. [1] proved the following theorem using elaborate probabilistic methods.

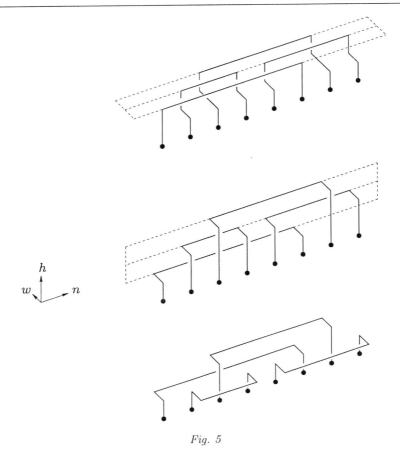

Fig. 5

Theorem 8 (A. Aggarwal, J. Kleinberg and D. P. Williamson, 2000) [1]. *If each net consists of two terminals only then the nets of an $n \times n$ SALRP can be partitioned into $O(n \log^2 n)$ classes such that each class of nets can be routed on a copy of the grid (of size $n \times n$).*

An easy corollary of this theorem is that if $s_w = s_n = 2$ and each net consists of two terminals only than every SALRP can be solved with height $h = O(n \log^2 n)$. The following result shows that actually $h = O(n)$ also suffices, even if multiterminal nets are also allowed.

Theorem 9 [10]. *Any SALRP can be solved with $s_n \geq \lceil \frac{w}{2n} \rceil + 1$, $s_w = 2$ and height $h = 6n$. Furthermore, if each net consists of two terminals only then $h = 3n$ also suffices.*

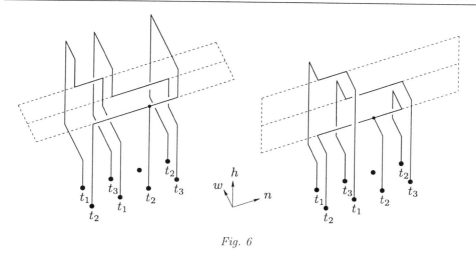

Fig. 6

Since the role of n and w is symmetric, $n \geq w$ can be assumed without loss of generality and thus we obtain the following corollary.

Corollary 1 [10]. *Any SALRP can be solved with $s_w = s_n = 2$ and height $h = 6 \max(n, w)$. If each net consists of two terminals only then $h = 3 \max(n, w)$ also suffices.*

The constructions of the above results can be performed by a polynomial algorithm. (If $w = \Theta(n)$ is assumed then the algorithms work in $O\left(A^{\frac{3}{2}}\right)$ time, where $A = w \cdot n$ is the size of the input.)

Acknowledgement. Research partially supported by the Hungarian National Science Foundation (Grant Numbers OTKA 42559 and 44733).

REFERENCES

[1] A. Aggarwal, J. Kleinberg and D. P. Williamson, Node-disjoint paths on the mesh and a new trade-off in VLSI layout, *SIAM J. Computing*, **29** (2000), 1321–1333.

[2] E. Boros, A. Recski and F. Wettl, Unconstrained multilayer switchbox routing, *Annals of Operations Research*, **58** (1995), 481–491.

[3] T. Gallai, His unpublished results were announced in A. Hajnal and J. Surányi, Über die Auflösung von Graphen in vollständige Teilgraphen, *Annales Univ. Sci. Budapest. Eötvös Sect. Math.*, **1** (1958), 115–123.

[4] S. E. Hambrusch, Channel routing in overlap models, *IEEE Trans. Computer-Aided Design of Integrated Circ. Syst.*, **CAD-4** (1985), 23–30.

[5] D. S. Johnson, The NP-completeness column: An ongoing guide, *J. Algorithms,* **5** (1984), 147–160.

[6] M. Marek-Sadowska and E. Kuh, General channel-routing algorithm, *Proc. IEE (GB),* **130** (1983), 83–88.

[7] A. Recski, Channel routing in the dogleg-free multilayer Manhattan model, *Proc. ECCTD'97* (1997), 39–43.

[8] A. Recski and F. Strzyzewski, Vertex-disjoint channel routing on two layers, *Integer programming and combinatorial optimization* (Ravi Kannan and W. R. Pulleyblank, ed.), University of Waterloo Press (1990), 397–405.

[9] A. Recski and D. Szeszlér, 3-dimensional single active layer routing, *Discrete and Computational Geometry*, Lecture Notes in Computer Science **2098**, 318–329, Springer, Berlin (2001).

[10] A. Recski and D. Szeszlér, Routing Vertex Disjoint Steiner-Trees in a Cubic Grid – an Application in VLSI, *submitted*.

[11] D. Szeszlér, Switchbox routing in the multilayer Manhattan model, *Annales Univ. Sci. Budapest. Eötvös Sect. Math.*, **40** (1997), 155–164.

[12] T. G. Szymanski, Dogleg channel routing is NP-complete, *IEEE Trans. Computer-Aided Design of Integrated Circ. Syst.*, **CAD-4** (1985), 31–41.

András Recski

Department of Computer Science and Information Theory
Budapest University of Technology and Economics
H-1521 Budapest
Hungary

recski@cs.bme.hu

Dávid Szeszlér

Department of Computer Science and Information Theory
Budapest University of Technology and Economics
H-1521 Budapest
Hungary

szeszler@cs.bme.hu

BOLYAI SOCIETY
MATHEMATICAL STUDIES, 15

Conference on Finite
and Infinite Sets
Budapest, pp. 329–339.

ON THE NUMBER OF ADDITIVE REPRESENTATIONS
OF INTEGERS

A. SÁRKÖZY[*]

*Dedicated to András Hajnal and Vera T. Sós
on the occasion of their 70th birthday*

1. INTRODUCTION

The set of the integers, nonnegative integers, resp. positive integers is denoted by $\mathbb{Z}, \mathbb{N}_o$ and $\mathbb{N}$. $\mathcal{A}, \mathcal{B}, \ldots$ denote (finite or infinite) subsets of $\mathbb{N}_o$, and their counting functions are denoted by $A(n), B(n), \ldots$ so that, e.g.,

$$A(n) = \big| \{a : 0 < a \leq n, \ a \in \mathcal{A}\} \big|.$$

The asymptotic density $d(\mathcal{A})$ of the set $\mathcal{A} \subset \mathbb{N}_o$ is defined by

$$d(\mathcal{A}) = \lim_{n \to +\infty} \frac{A(n)}{n}$$

if this limit exists. $\mathcal{A}_1 + \mathcal{A}_2 + \cdots + \mathcal{A}_k$ denotes the set of the integers that can be represented in the form $a_1 + a_2 + \cdots + a_k$ with $a_1 \in \mathcal{A}, a_2 \in \mathcal{A}, \ldots, a_k \in \mathcal{A}_k$; in the special case $\mathcal{A}_1 = \mathcal{A}_2 = \cdots = \mathcal{A}_k = \mathcal{A}$ we write

$$\mathcal{A}_1 + \mathcal{A}_2 + \cdots + \mathcal{A}_k = \mathcal{A} + \mathcal{A} + \cdots + \mathcal{A} = k\mathcal{A}.$$

[*]Partially supported by Hungarian National Foundation for Scientific Research, Grant
No. T 029759.

For $\mathcal{A} \subset \mathbb{N}_o, k \in \mathbb{N}$ the number of solutions of the equations

$$a_1 + a_2 + \cdots + a_k = n, \quad a_1, a_2, \ldots, a_k \in \mathcal{A},$$

$$a_1 + a_2 + \cdots + a_k = n, \quad a_1 \leq a_2 \leq \cdots \leq a_k, \quad a_1, a_2, \ldots, a_k \in \mathcal{A},$$

$$a_1 + a_2 + \cdots + a_k = n, \quad a_1 < a_2 < \cdots < a_k, \quad a_1, a_2, \ldots, a_k \in \mathcal{A}$$

is denoted by $r_1(\mathcal{A}, n, k)$, $r_2(\mathcal{A}, n, k)$, resp. $r_3(\mathcal{A}, n, k)$, and in the special case $k = 2$ we write $r_i(n) = r_i(\mathcal{A}, n) = r_i(\mathcal{A}, n, 2)$ for $i = 1, 2, 3$. For $k, g \in \mathbb{N}$, $B_k[g]$ denotes the class of all (finite or infinite) sets $\mathcal{A} \subset \mathbb{N}_o$ such that for all $n \in \mathbb{N}$ we have $r_2(\mathcal{A}, n, k) \leq g$, i.e., the equation

$$a_1 + a_2 + \cdots + a_k = n, \quad a_1 \leq a_2 \leq \cdots \leq a_k, \quad a_1, a_2, \ldots, a_k \in \mathcal{A}$$

has at most g solutions.

If $F(n) = O\big(G(n)\big)$, then we write $F(n) \ll G(n)$. $c_1, c_2, \ldots$ denote positive absolute constants.

In [7], [8], [9] and [11] Paul Erdős, Vera T. Sós and I, and in [17] Vera T. Sós and I studied the irregularity properties and the range of the additive representation functions $r_i(\mathcal{A}, n, k)$. In this paper my goal is to give a survey of these 5 papers and the most important related results. (We also studied difference sets, Sidon sets and Ramsey type additive problems involving general sequences in [10], [11], [12], [13] and [17]; these results have been surveyed or will be surveyed elsewhere.)

2. The early days

As an answer to a question of S. Sidon, in 1956 Erdős [3] proved the following result:

Theorem 1. *There is an infinite set $\mathcal{A} \subset \mathbb{N}$ such that*

$$(1) \qquad c_1 \log n < r_1(\mathcal{A}, n) < c_2 \log n \quad \text{for} \quad n > n_o.$$

In two papers Erdős and I extended the problem by estimating

$$\big| r_1(\mathcal{A}, n) - F(n) \big|$$

for "nice" functions $F(n)$. First we proved [5]:

Theorem 2. *If $F(n)$ is an arithmetic function satisfying*

$$(2) \qquad\qquad F(n) \to +\infty,$$

$$(3) \qquad\qquad F(n+1) \geq F(n) \quad \text{for} \quad n \geq n_o$$

and

$$(4) \qquad\qquad F(n) = o\left(\frac{n}{(\log n)^2}\right),$$

then

$$(5) \qquad\qquad \max_{n \leq N}\left|r_1(\mathcal{A}, n) - F(n)\right| = o\left(\left(F(N)\right)^{1/2}\right)$$

cannot hold.

Indeed, we proved this in the sharper form that (5) cannot hold in mean square sense.

Later we proved [6] that if $F(n)$ is a "nice" function, then there is an $\mathcal{A}$ with

$$\left|r_1(\mathcal{A}, n) - F(n)\right| \ll \left(F(n)\log n\right)^{1/2}:$$

Theorem 3. *If $F(n)$ is an arithmetic function satisfying*

$$F(n) > 36 \log n \quad \text{for} \quad n > n_o,$$

and there exist a real function $g(x)$, defined for $0 < x < +\infty$, and real numbers x_o, n_1, such that

(i) $g'(x)$ *exists and it is continuous for* $0 < x < +\infty$,

(ii) $g'(x) \leq 0$ *for* $x \geq x_o$,

(iii) $0 < g(x) < 1$ *for* $x \geq x_o$,

(iv) $\left|F(n) - 2\int_0^{n/2} g(x)g(n-x)\,dx\right| < \left(F(n)\log n\right)^{1/2}$ *for $n > n_1$, then there exists a sequence $\mathcal{A}$ such that*

$$\left|r_1(\mathcal{A}, n) - F(n)\right| < 8\left(F(n)\log n\right)^{1/2} \quad \text{for} \quad n > n_2.$$

In particular, it follows from this theorem that

(i) there is an $\mathcal{A}$ satisfying (1);

(ii) if $\omega(n)$ is a "smooth" increasing function with $\omega(n) \to +\infty$ (say, $\omega(n) = \log \log n$), then there is an $\mathcal{A}$ with

$$r_1(\mathcal{A}, n) \sim \omega(n) \log n;$$

(iii) for all $0 < \alpha < 1$, there is an $\mathcal{A}$ with

$$\left| r_1(\mathcal{A}, n) - n^\alpha \right| \ll n^{\alpha/2} (\log n)^{1/2}.$$

3. Vera joins us (and a cure for an incurable disease)

In the January of 1984, not much after completing our papers [5], [6] with Erdős, the two of us and Vera T. Sós travelled to India for a few weeks. It was a long flight (18 hours or so), we had to transfer twice. Thus to spend the time, soon we started to discuss mathematics. Vera asked Erdős and me on our most recent results. We told her about our results quoted as Theorems 2 and 3 above. After learning our results, she asked a few more exciting new questions on the irregularities of the additive representation functions $r_i(\mathcal{A}, n)$. Unfortunately, after 1 or 2 hours I was forced to quit the discussion since I did not feel well. Soon I realized that I had a bad case of flu, my fever approaching 39C °.

Nearly one day later we arrived. When our plane landed in Madras, a heavy monsoon rain had just endeed, the temperature was around 36–37C °, and the tarmac of the airport was steaming. It was quite unpleasant, however, soon I was feeling better, and in one hour my flu was completely over (and it never returned again): the "steam bath" cured it! So that not only we had a very pleasant and, as you will see, fruitful time in India, but as a byproduct of our trip I also discovered the cure of the incurable disease of the influenza. (Unfortunately, this discovery is of not much use: it is just too costly, lengthy and complicated to travel to India for a cure, and while you work on it, you have a good chance to recover spontaneously.)

Anyway, in India we continued the discussion, and we completed the work after our return to Hungary. We ended up with 3 triple papers. First in [7] we studied the following problem: what condition is needed to ensure

$$(6) \qquad \limsup_{n \to +\infty} \left| r_1(\mathcal{A}, n + 1) - r_1(\mathcal{A}, n) \right| = +\infty?$$

We gave such a condition in terms of the function

$$B(\mathcal{A}, N) = \left| \{n : n \leq N, \ n \in \mathcal{A}, \ n - 1 \notin \mathcal{A}\} \right| :$$

Theorem 4. *If*

$$\lim_{N \to +\infty} B(\mathcal{A}, N) N^{-1/2} = +\infty,$$

then (6) holds.

We also showed that this result is nearly sharp:

Theorem 5. *For all $\varepsilon > 0$, there exists an infinite sequence $\mathcal{A}$ such that*

$$B(\mathcal{A}, N) \gg N^{1/2-\varepsilon}$$

and $r_1(\mathcal{A}, N)$ is bounded (so that $\left| r_1(\mathcal{A}, N+1) - r_1(\mathcal{A}, N) \right|$ is also bounded).

In [8] and [9] we studied the monotonicity properties of the three representation functions $r_1(n)$, $r_2(n)$, $r_3(n)$. Let $\mathcal{A}$ be a set of positive integers which can be obtained from $\mathbb{N}$ by dropping finitely many integers (i.e., $\mathbb{N} \setminus \mathcal{A}$ is finite). Then clearly, each of the three functions $r_1(\mathcal{A}, n), r_2(\mathcal{A}, n), r_3(\mathcal{A}, n)$ is monotone increasing from a certain point on. So the question is: are there any other "non-trivial" sets $\mathcal{A}$ (i.e., sets $\mathcal{A}$ for which $\mathbb{N} \setminus \mathcal{A}$ is infinite) so that the function $r_i(\mathcal{A}, n)$ $(i = 1, 2, 3)$ is monotone increasing from a certain point on? Somewhat unexpectedly, the answer depends very much on that which of the three functions $r_1(\mathcal{A}, n)$, $r_2(\mathcal{A}, n)$, $r_3(\mathcal{A}, n)$ is considered. First in [8] we proved:

Theorem 6. *The function $r_1(\mathcal{A}, n)$ is monotone increasing from a certain point on for an infinite set $\mathcal{A} \subset \mathbb{N}$ if and only if $\mathbb{N} \setminus \mathcal{A}$ is finite.*

Theorem 7. *There is an infinite set $\mathcal{A} \subset \mathbb{N}$ such that*

$$A(n) < n - c_3 \, n^{1/3} \quad \text{for} \quad n > n_o$$

and $r_3(\mathcal{A}, n)$ is monotone increasing for $n > n_1$.

Theorem 8. *If $\mathcal{A} \subset \mathbb{N}$ is an infinite set with*

$$A(n) = o \left(\frac{n}{\log n} \right),$$

then the functions $r_2(\mathcal{A}, n)$ and $r_3(\mathcal{A}, n)$ cannot be monotone increasing from a certain point on.

Later in [9] we improved on Theorem 8 in case of the function $r_2(\mathcal{A}, n)$:

Theorem 9. *If* $\mathcal{A} \subset \mathbb{N}$ *is an infinite set such that*

$$(7) \qquad \lim_{n \to \infty} \frac{n - A(n)}{\log n} = +\infty,$$

then the function $r_2(\mathcal{A}, n)$ *cannot be monotone increasing from a certain point on.*

Note that in [1] Balasubramanian (who learned on our results Theorems 6, 7 and 8 during our Indian visit) also proved Theorem 9 independently. Interestingly enough, although his proof is different from ours, he has exactly the same condition (7) in his theorem.

In [11] Erdős, Vera T. Sós and I studied the range of the additive representation functions $r_i(\mathcal{A}, n)$. For $i = 1, 2, 3$ let $\mathcal{R}_i(\mathcal{A})$ denote the range of the function $r_i(\mathcal{A}, n)$, i.e., $\mathcal{R}_i(\mathcal{A})$ denotes the set of the integers m such there is a number $n \in \mathbb{N}$ with

$$(8) \qquad r_i(\mathcal{A}, n) = m,$$

and let $\mathcal{R}_i^\infty(\mathcal{A})$ denote the set of the integers m such that there are infinitely many integers n satisfying (8). We proved:

Theorem 10. *For a set* $\mathcal{B} \subset \mathbb{N}_o$, *there is a set* $\mathcal{A}$ *with*

$$\mathcal{R}_i(\mathcal{A}) = \mathcal{B}, \quad \mathcal{A} \subset \mathbb{N}$$

if and only if either $\mathcal{B} = \{0, 1\}$ *or* $\{0, 1, 2\} \subset \mathcal{B}$.

(The cases $i = 2, 3$ could be handled similarly.)

Theorem 11. *For each* $i = 1, 2, 3$ *and for all* $\mathcal{B} \subset \mathbb{N}_o$, *the equation*

$$\mathcal{R}_i^\infty(\mathcal{A}) = \mathcal{B}$$

can be solved.

In [17], one of the problems studied by Vera T. Sós and me was a conjecture of Erdős and Freud [4]. They conjectured:

Conjecture 1. *If* $\mathcal{A} \subset \mathbb{N}$ *is an infinite set such that* $r_2(\mathcal{A}, n)$ *is bounded, then* $r_2(\mathcal{A}, n)$ *must assume the value 1 infinitely often, i.e., there are infinitely many integers* $n \in 2\mathcal{A}$ *whose representation in the form*

$$(9) \qquad a + a' = n, \quad a, a' \in \mathcal{A}, \quad a \le a'$$

is unique.

Moreover, they wrote: "Probably there are "more" integers n with a unique representation of the form (9) than integers n with more than one representation". We showed in [17] that this is not so, at least for $\mathcal{A} \in B_2[g]$, $g \geq 3$:

Theorem 12. *For every $g \in \mathbb{N}, g \geq 2$ there is an infinite set $\mathcal{A} \subset \mathbb{N}_o$ such that $\mathcal{A} \in B_2[g]$ and for $\varepsilon > 0$, $n > n_o$ we have*

$$\left|\{n : n \leq N, \ r_2(\mathcal{A}, n) = 1\}\right| < (1+\varepsilon)\frac{2}{2g-3}\left|\{n : n \leq N, \ r_2(\mathcal{A}, n) > 1\}\right|.$$

On the other hand, we conjectured that the following sharpening of Conjecture 1 is true:

Conjecture 2. *If $r_2(\mathcal{A}, n)$ is bounded then we have*

$$(10) \qquad \limsup_{N \to \infty} \frac{\left|\{n : n \leq N, \ r_2(\mathcal{A}, n) = 1\}\right|}{\left|\{n : n \leq N, \ r_2(\mathcal{A}, n) \geq 1\}\right|} > 0.$$

In [17] we also showed that for any fixed finite set $\mathcal{U}$, there is an infinite set $\mathcal{A} \subset \mathbb{N}$ such that $r_2(\mathcal{A}, n)$, apart from a "thin" set of integers n, assumes only values from $\mathcal{U}$ with about the same frequency. For $\mathcal{A} \subset \mathbb{N}, u \in \mathbb{N}$, denote the set of the integers $n \in \mathbb{N}$ with

$$r_2(\mathcal{A}, n) = u$$

by $\mathcal{S}_u(\mathcal{A})$ (so that $\cup_{u=1}^{+\infty} \mathcal{S}_u(\mathcal{A}) = 2\mathcal{A}$).

Theorem 13. *Let $k \in \mathbb{N}$ and let $u_1 < u_2 < \cdots < u_k$ be positive integers. Then there is an infinite set $\mathcal{A} \subset \mathbb{N}_o$ such that writing*

$$\mathcal{B} = \mathbb{N} \setminus \left(\cup_{i=1}^{k} \mathcal{S}_{u_i}(\mathcal{A}) \right)$$

we have

$$\mathcal{S}_{u_i}(\mathcal{A}, N) = \frac{N}{k} + O(N^\alpha)$$

and

$$B(N) = O(N^\alpha)$$

with $\alpha = \frac{\log 3}{\log 4}$.

(Here $\mathcal{S}_{u_i}(\mathcal{A}, N)$ denotes the counting function of $\mathcal{S}_{u_i}(\mathcal{A})$.)

4. Recent developments and unsolved problems

Cs. Sándor [15] disproved Conjectures 1 and 2 above and, indeed, he constructed an infinite set $\mathcal{A} \subset \mathbb{N}$ such that $r_2(\mathcal{A}, n) \leq 3$ for all n but $r_2(\mathcal{A}, n) \neq 1$ for $n > n_o$. However, it is still possible that Conjecture 1 is true if the upper bound for $r_2(\mathcal{A}, n)$ is 2, i.e., $\mathcal{A} \in B_2[2]$:

Problem 1. Is it true that if $\mathcal{A} \in B_2[2]$, then $r_2(\mathcal{A}, n)$ must assume the value 1 infinitely often?

G. Horváth [14] extended Theorem 2 from sums $a + a'$ of 2 terms to sums of k terms. Note that Theorem 3 has no similar extension to the k term case (with $k > 2$) yet. Namely, Theorem 3 (and all the other results of similar nature) are proved by a probablisitic argument, and this approach usually fails for $k > 2$ (because of the lack of independence of the events involved).

Problem 2. Prove a result of type Theorem 3 for sums of $k > 2$ terms, i.e., a similar result with $r_1(\mathcal{A}, n, k)$ in place of $r_1(\mathcal{A}, n) \big(= r_1(\mathcal{A}, n, 2) \big)$.

G. Dombi [2] *constructed* sequences $\mathcal{A}$ of density $1/2$ for which $r_1(\mathcal{A}, n, k)$ is monotone for large n if $k > 4$. The point of his result is that he gave constructions (using the Rudin–Shapiro and Thue–Morse sequences), while in all the other known results of similar nature existence proofs are given (using probability theory).

Finally, I will present a few further related unsolved problems selected from [11], [16] and [17], and also a couple of new ones.

First a problem related to Theorems 2 and 3:

Problem 3. Does there exist an arithmetic function $F(n)$ satisfying (2), (3) and (4) and a set $\mathcal{A} \subset \mathbb{N}$ such that

$$\big| r_1(\mathcal{A}, n) - F(n) \big| = o\big((F(n))^{1/2} \big)$$

holds on a sequence of integers n whose density is 1?

The next problem is the extension of (6) to the case of more than 2 summands:

Problem 4. For $k \in \mathbb{N}$, $k > 2$, what condition is needed to ensure

$$\limsup_{n \to +\infty} \big| r_1(\mathcal{A}, n + 1, k) - r_1(\mathcal{A}, n, k) \big| = +\infty?$$

Some problems related to Theorems 6, 7, 8 and 9:

Problem 5. Does there exist a set $A \subset \mathbb{N}$ such that $\mathbb{N} \setminus A$ is infinite and

$$r_1(A, n+1) \geq r_1(A, n)$$

holds on a sequence of integers n whose density is 1?

Problem 6. Does there exist an infinite set A such that $\mathbb{N} \setminus A$ is infinite and $r_2(A, n)$ is increasing from a certain point on?

Problem 7. Does there exist an infinite set $A \subset \mathbb{N}$ such that its lower (or even upper) asymptotic density is less than 1, and $r_3(A, n)$ is monotone increasing from a certain point on?

Problem 8. What condition (on A) is needed to ensure that the function $r_i(A, n)$ $(i = 1, 2, 3)$ assumes infinitely many "locally small", resp. "locally large" values, i.e.,

$$r_i(A, n) < \min \left\{ r_i(A, n-1), r_i(A, n+1) \right\},$$

resp.

$$r_i(A, n) > \max \left\{ r_i(A, n-1), r_i(A, n+1) \right\}?$$

Problem 9. What can one say on the monotonicity of the functions $r_i(A, n, k)$ $(i = 1, 2, 3)$ in the case $k > 2$? In particular, I conjecture:

Conjecture.

(i) If $k \geq 2$ and $r_i(A, n, k)$ $(i = 1, 2, 3)$ is increasing (in n) from a certain point on, then

$$A(n) = O(n^{2/k - \varepsilon})$$

cannot hold.

(ii) If $k \geq 2$, then there is a set $A \subset \mathbb{N}$ such that

$$A(n) = O(n^{2/k + \varepsilon})$$

and $r_i(A, n, k)$ $(i = 1, 2, 3)$ is increasing.

Problem 10. When and how the results on the monotonicity of $r_i(A, n)$ $(i = 1, 2, 3)$ can be extended from sums $a_1 + a_2 + \cdots + a_k$ to linear forms $b_1 a_1 + b_2 a_2 + \cdots + b_k a_k$ where $b_1, b_2, \ldots, b_k$ are fixed positive integers?

REFERENCES

[1] R. Balasubramanian, A note on a result of Erdős, Sárközy and Sós, *Acta Arith.*, **49** (1987), 45–53.

[2] G. Dombi, Additive properties of certain sets, *Acta Arith.*, to appear.

[3] P. Erdős, Problems and results in additive number theory, *Colloque sur la Thèorie des Numbers (CBRM)* (Bruxelles, 1956), 127–137.

[4] P. Erdős and R. Freud, On Sidon-sequences and related problems, *Mat. Lapok,* **1** (1991), 1–44 (in Hungarian).

[5] P. Erdős and A. Sárközy, Problems and results on additive properties of general sequences, I, *Pacific J.,* **118** (1985), 347–357.

[6] P. Erdős and A. Sárközy, Problems and results on additive properties of general sequences, II, *Acta Math. Acad. Sci. Hungar.,* **48** (1986), 201–211.

[7] P. Erdős, A. Sárközy and V. T. Sós, Problems and results on additive properties of general sequences, III, *Studia Sci. Math. Hungar.,* **22** (1987), 53–63.

[8] P. Erdős, A. Sárközy and V. T. Sós, Problems and results on additive properties of general sequences, IV, in: *Number Theory, Proceedings,* Ootacamund, India, 1984, Lecture Notes in Mathematics 1122, Springer-Verlag, 1985; 85–104.

[9] P. Erdős, A. Sárközy and V. T. Sós, Problems and results on additive properties of general sequences, V, *Monatshefte Math.,* **102** (1986), 183–197.

[10] P. Erdős, A. Sárközy and V. T. Sós, On a conjecture of Roth and some related problems, I., in: *Irregularities of Partitions,* eds. G. Halász and V. T. Sós, Algorithms and Combinatorics **8**, Springer-Verlag, Berlin–Heidelberg–New York, 1989; 47–59.

[11] P. Erdős, A. Sárközy and V. T. Sós, On additive properties of general sequences, *Discrete Math.,* **136** (1994), 75–99.

[12] P. Erdős, A. Sárközy and V. T. Sós, On sum sets of Sidon sets, I., *J. Number Theory,* **47** (1994), 329–347.

[13] P. Erdős, A. Sárközy and V. T. Sós, On sum sets of Sidon sets, II., *Israel J. Math.,* **90** (1995), 221–233.

[14] G. Horváth, On an additive property of sequences of nonnegative integers, *Periodica Math. Hungar.,* to appear.

[15] Cs. Sándor, Range of bounded additive representation functions, *Periodica Math. Hungar.,* **42** (2001), 169–177.

[16] A. Sárközy, Unsolved problems in number theory, *Periodica Math. Hungar.,* **42** (2001), 17–35.

[17] A. Sárközy and V. T. Sós, On additive representation functions, in: *The Mathematics of Paul Erdős,* vol. I, eds. R. L. Graham and J. Nešetřil, Algorithms and Combinatorics **13**, Springer-Verlag, 1997, 129–150.

A. Sárközy

Eötvös Loránd University
Department of Algebra and Number
Theory
H-1117 Budapest
Pázmány Péter Sétány 1/C.
Hungary

BOLYAI SOCIETY
MATHEMATICAL STUDIES, 15

Conference on Finite
and Infinite Sets
Budapest, pp. 341–358.

A Lifting Theorem on Forcing LCS Spaces

L. SOUKUP[*]

Denote by $\mathcal{THIN}(\alpha)$ the statement that there is an LCS space of height α and width ω. We prove, for each regular cardinal κ, that if there is a "natural" c.c.c poset P such that $\mathcal{THIN}(\kappa)$ holds in V^P then there is a "natural" c.c.c poset Q as well such that $\mathcal{THIN}(\delta)$ holds in V^Q for each $\delta < \kappa^+$.

1. Introduction

A topological space X is called *scattered* if its every non-empty subspace has an isolated point. Denoting by $I(Y)$ the isolated points of a subspace $Y \subset X$ for each ordinal α define the αth *Cantor–Bendixson level* of the space X, $I_\alpha(X)$, as follows:

$$I_\alpha(X) = I\left(X \setminus \cup\{ I_\beta(X) : \beta < \alpha\}\right).$$

The minimal α with $I_\alpha(X) = \emptyset$ is called the height of X and denoted by $\mathrm{ht}(X)$. Define the width of X, $wd(X)$, as follows: $wd(X) = \sup\{|I_\alpha(X)| : \alpha < \mathrm{ht}(X)\}$. The *cardinal sequence* of X, $\mathrm{CS}(X)$, is the sequence of the cardinalities of its Candor–Bendixson levels, i.e.

$$\mathrm{CS}(X) = \langle|I_\alpha(X)| : \alpha < \mathrm{ht}(X)\rangle.$$

The following problem was first posed by R. Telgarsky in 1968 (unpublished): *Does there exist a locally compact, scattered (in short: LCS) space*

---

[*]The author was partially supported by Hungarian Foundation for Scientific Research, grant No. 37758 and the Bolyai Scholarship of the Hungarian Academy of Sciences.

with height ω_1 and width ω? After some consistency results Rajagopalan, in [8], constructed such a space in ZFC.

To simplify the formulation of the forthcoming results denote by $\mathcal{THIN}(\alpha)$ the statement that there is an LCS space of height α and width ω. (A scattered space is called *thin* iff it has width ω.)

In [4] I. Juhász and W. Weiss showed $\mathcal{THIN}(\alpha)$ for each $\alpha < \omega_2$. W. Just proved, in [5], that this result is sharp in the following sense. Add ω_2 Cohen reals to a ZFC model satisfying CH. Then, in the generic extension, $2^\omega = \omega_2$ and $\mathcal{THIN}(\omega_2)$ fails. So you can not prove $\mathcal{THIN}(\alpha)$ for each $\alpha < (2^\omega)^+$ in ZFC.

Just's result was improved in [3] by I. Juhász, S. Shelah, L. Soukup and Z. Szentmiklóssy: if we add Cohen reals to a model of set theory satisfying CH, then, in the new model, every LCS space has at most ω_1 many countable levels.

The notion of Δ-function (see definition 1.1 below) was introduced in [2]. In that paper Baumgartner and Shelah proved that (a) the existence of a Δ-function is consistent with ZFC, (b) if there is a Δ-function then $\mathcal{THIN}(\omega_2)$ holds in a *natural* c.c.c forcing extension. We will explain later, in Section 3, what we mean under "natural poset". Roughly speaking, "natural" means that the elements of the posets are just finite approximations of the locally compact right-separating neighbourhoods of the points of the desired space. Building on their method, but using much more involved combinatorics, Martinez [6] proved that if there is a strong Δ-function, then for each $\delta < \omega_3$ there is a c.c.c poset P_δ such that $\mathcal{THIN}(\delta)$ holds in V^{P_δ}. These results naturally raised the following problem.

Problem 1. Does $\mathcal{THIN}(\omega_2)$ imply $\mathcal{THIN}(\delta)$ for each $\delta < \omega_3$?

Although this question remains still open we prove a "lifting theorem" claiming that if there is a natural poset P_{ω_2} such that $\mathcal{THIN}(\omega_2)$ holds in $V^{P_{\omega_2}}$ then for each $\delta < \omega_3$ there is a natural poset P_δ such that $\mathcal{THIN}(\delta)$ holds in V^{P_δ}: the posets used by Martinez can be constructed directly from the poset applied by Baumgartner and Shelah without even mentioning the Δ-function. Moreover, our lifting theorem works for each cardinal κ, not only for ω_2! Since there is no Δ-function on ω_3 you can not expect to apply the method of Baumgartner and Shelah to prove $\mathcal{THIN}(\omega_3)$. However, if anybody can construct a "natural" c.c.c poset P such that $\mathcal{THIN}(\omega_3)$ holds in V^P then our theorem gives immediately the consistency of $\mathcal{THIN}(\alpha)$ for each $\alpha < \omega_4$.

To formulate this statement more precisely we introduce some notation, so we postpone the formulation of our main result till Theorem 3.15.

First we recall some definition and results.

Definition 1.1. Let $f : [\omega_2]^2 \to [\omega_2]^{\leq\omega}$ be a function with $f\{\alpha,\beta\} \subset \alpha\cap\beta$ for $\{\alpha,\beta\} \in [\omega_2]^2$.

(1) We say that two finite subsets x and y of ω_2 are *good for* f provided that for $\alpha \in x\cap y$, $\beta \in x\setminus y$ and $\gamma \in y\setminus x$ we always have

(a) $\alpha < \beta, \gamma \Longrightarrow \alpha \in f\{\beta,\gamma\}$,

(b) $\alpha < \beta \Longrightarrow f\{\alpha,\gamma\} \subset f\{\beta,\gamma\}$,

(c) $\beta < \gamma \Longrightarrow f\{\alpha,\beta\} \subset f\{\alpha,\gamma\}$.

(2) The function f is a Δ-*function* if every uncountable family of finite subsets of ω_2 contains two elements x and y which are are good for f.

(3) The function f is a *strong* Δ-*function* if every uncountable family $\mathcal{A}$ of finite subsets of ω_2 contains an uncountable subfamily $\mathcal{B}$ such that any two sets x and y from $\mathcal{B}$ are good for f.

Theorem (Velickovic). *If $\square_{\omega_1}$ holds then there is a strong Δ-function.*

For the proof see [1].

2. A METHOD TO FORCE THIN LCS SPACES WITH PRESCRIBED CARDINAL SEQUENCE

Recall that given a topological space $\langle X, \tau_X \rangle$ a function $f : X \to \tau_X$ is called *neighbourhood assignment* iff $x \in f(x)$ for each $x \in X$.

Assume that X is an LCS space. Define the function $\mathrm{ht} : X \to \mathrm{ht}(X)$ by the formula $x \in I_{\mathrm{ht}(x)}(X)$. Since LCS spaces are 0-dimensional, we can fix a neighbourhood-assignment $U : X \to \tau_X$ such that $U(x)$ is a compact-open neighbourhood of x with

$$U(x)\setminus\{x\} \subset I_{<\mathrm{ht}(x)}(X) = \bigcup\{I_\alpha(X) : \alpha < \mathrm{ht}(x)\}.$$

The family $\{U(x), X\setminus U(x) : x \in X\}$ is a subbase of X.

The space is *coherent* iff we can choose U in such a way that $x \in$ U (y) implies U $(x) \subset$ U (y). Such a U is a called *coherent neighbourhood assignment*.

If U is coherent then we can define a partial order $\lhd_U$ on X by taking $x \lhd_U y$ iff $x \in$ U (y). Since clearly U $(x) = \{y \in X : y \lhd_U x\}$ we have that $\lhd_U$ determines the neighbourhood assignment U.

If $\lhd$ is an arbitrary partial order on X then define the topology $\tau_\lhd$ on X generated by the family $\{ U_\lhd(x), X \setminus U_\lhd(x) : x \in X \}$ as a subbase, where $U_\lhd(x) = \{y \in X : y \lhd x\}$. As we have seen if U witnesses that $\langle X, \tau \rangle$ is coherent then $\tau_{\lhd_U} = \tau$.

So the topologies of coherent LCS spaces are determined by partial orderings. We would like to determine certain properties of a partial ordering in such a way that if some partial order $\langle X, \lhd \rangle$ has those properties then $\langle X, \tau_\lhd \rangle$ is an LCS-space with prescribed Cantor–Bendixson levels.

To formulate these properties we investigate some covering properties of the family $\{ U(x) : x \in X \}$, where U is a coherent neighbourhood assignment on some LCS-space X.

If $x \notin$ U (y) and $y \notin$ U (x) then

$$U(x) \cap U(y) \subset \bigcup \{ U(z) : z \in U(x) \cap U(y) \}.$$

Since U $(x) \cap$ U (y) is compact there is a finite set $i\{x, y\} \in [U(x) \cap U(y)]^{<\omega}$ such that

$$U(x) \cap U(y) \subset \bigcup \{ U(z) : z \in i\{x, y\} \}.$$

We will enumerate some properties of $\lhd$ and the function i. Let $\delta = $ ht (X) and for $\alpha < \delta$ write $X_\alpha = I_\alpha(X)$.

(I) if $x \in X_\alpha$, $y \in X_\beta$ and $x \lhd y$ then either $x = y$ or $\alpha < \beta$,

(II) $\forall \{x, y\} \in [X]^2 \ \forall z \in X \big((z \lhd x \wedge z \lhd y) \text{ iff } \exists t \in i\{x, y\} z \lhd t \big).$

(III) if $x \in X_\alpha$ and $\beta < \alpha$ then the set $\{y \in X_\beta : y \lhd x\}$ is infinite.

Proposition 2.1. *Assume that $\{X_\alpha : \alpha < \delta\}$ is a partition of a given set X, $\lhd$ is a partial order on X and $i : [X]^2 \to [X]^{<\omega}$ is a function satisfying (I)–(III). Then $\mathcal{X} = \langle X, \tau_\lhd \rangle$ is a (coherent) LCS space with $I_\alpha(\mathcal{X}) = X_\alpha$ for $\alpha < \delta$.*

Proof. X is right-separated, i.e. scattered, witnessed by any well-ordering extending the well-founded partial ordering $\lhd$ because of (I).

For each $x \in X$, the family

$$\mathbb{U}(x) = \left\{ U_\lhd(x) \setminus \bigcup_{y \in F} U_\lhd(y) \ : \ F \in \left[U_\lhd(x) \setminus \{x\} \right]^{<\omega} \right\}$$

is a neighbourhood base of x. Indeed, if $x \neq y$ then $U_\lhd(x) \cap U_\lhd(y) = U_\lhd(x)$ provided $x \in U_\lhd(y)$ and

$$U_\lhd(x) \setminus U_\lhd(y) = U_\lhd(x) \setminus \bigcup_{z \in i\{x,y\}} U_\lhd(z)$$

provided $x \notin U_\lhd(y)$, where $i\{x, y\} \in \left[U_\lhd(x) \setminus \{x\} \right]^{<\omega}$.

Lemma 2.2. $I_\alpha(\mathcal{X}) = X_\alpha$.

Proof. First we show by induction on α that if $x \in X_\alpha$, $U \in \mathbb{U}(x)$ and $\beta \leq \alpha$ then $U \cap X_\beta \neq \emptyset$. For $\beta = \alpha$ we have $x \in U \cap X_\beta$ so we can assume $\beta < \alpha$. Assume that $U = U_\lhd(x) \setminus \bigcup \left\{ U_\lhd(z) \ : \ z \in F \right\}$, where $F \in \left[U_\lhd(x) \setminus \{x\} \right]^{<\omega}$. Let $\mu = \max \left\{ \nu \ : \ F \cap X_\nu \neq \emptyset \right\}$ and $\gamma = \max \{ \mu, \beta \}$. Since $\gamma < \alpha$ by (III) we can pick $t \in \left(X_\gamma \cap U_\lhd(x) \right) \setminus F$. Then $U_\lhd(t) \setminus \bigcup \left\{ U_\lhd(z) \ : \ z \in F \right\} \subset U$ is a neighbourhood of t which intersects X_β by the inductive hypothesis because $t \in X_\gamma$ and $\beta \leq \gamma < \alpha$.

Now prove the statement of the lemma by induction on α. Let $Y = X \setminus \bigcup_{\beta < \alpha} I_\beta(X) = X \setminus \bigcup_{\beta < \alpha} X_\beta$. If $x \in X_\alpha$ then $U(x) \cap Y = \{x\}$, so $X_\alpha \subset I(Y)$. If $x \in X_\gamma$ for some $\gamma > \alpha$ then for any neighbourhood of U we have $U \cap X_\alpha \neq \emptyset$, i.e. $U \cap Y \neq \{x\}$, and so $x \notin I(Y)$. Thus $I(Y) = X_\alpha$ which was to be proved. ∎

Lemma 2.3. $U_\lhd(x)$ is compact in $\mathcal{X}$.

Proof. We prove this statement by induction on $\mathrm{ht}\,(x)$. By Alexander's subbase lemma it suffices to show that any cover $\mathcal{V}$ of $U_\lhd(x)$ by members of $\left\{ U_\lhd(y) \ : \ y \in X \right\}$ and their complements has a finite subcover. Let $V \in \mathcal{V}$ be such that $x \in V$. If $V = U_\lhd(y)$ then $U_\lhd(x) \subset U_\lhd(y)$ so we have a one element covering. So we can assume that $V = X \setminus U_\lhd(y)$. Then

$$U_\lhd(x) \setminus V = U_\lhd(x) \cap U_\lhd(y) = \bigcup \left\{ U_\lhd(z) \ : \ z \in i\{x,y\} \right\}.$$

For each $z \in i\{x, y\}$ we have $\mathrm{ht}\,(z) < \mathrm{ht}\,(x)$ and so $U_\lhd(z)$ is compact, and so $U_\lhd(x) \setminus V$ is compact as well. Thus there is a finite $\mathcal{W} \subset \mathcal{V}$ with $U_\lhd(x) \setminus V \subset \bigcup \mathcal{W}$. Hence $\mathcal{W} \cup \{V\}$ is a finite cover of $U_\lhd(x)$. ∎

This completes the proof of Proposition 2.1. ■$_{2.1}$

We say that $\lhd$ is an *LCS-order on* X iff $\langle X, \tau_\lhd \rangle$ is an LCS-space.

So our strategy to force an LCS space with a prescribed cardinal sequence $\langle \kappa_\alpha : \alpha < \delta \rangle$ is the following. Let $X_\alpha = \{\alpha\} \times \kappa_\alpha$ for $\alpha < \delta$ and put $X = \bigcup\{X_\alpha : \alpha < \delta\}$. Now we try to add generically a partial ordering $\lhd$ on X and a function $i : [X]^2 \to [X]^{<\omega}$ satisfying (I)–(III) using finite approximations. That is, a typical forcing condition is a triple $\langle a, \leq, i \rangle$, where a is a finite subset of X, $\leq$ is a partial order on a, and i is a function on $[a]^2$ such that $\langle a, \leq, i \rangle$ satisfies (I) and (II). (III) would be guaranteed by some density argument. This type of forcing was introduced by Judy Roitmann to get thin superatomic Boolean algebras (LCS spaces).

The main problem is that the poset of all the possible finite approximations may not satisfy c.c.c. That is the point where the Δ-function came into the picture. Baumgartner and Shelah, and later Martinez, applied this function to select a suitable subfamily of the conditions which satisfies c.c.c. Our strategy will be different: we show that if there is a suitable poset which introduces $\mathcal{THIN}(\kappa)$ then for each $\delta < \kappa^+$ there is a a suitable poset which introduces $\mathcal{THIN}(\delta)$.

This strategy will be carried out in the next section in a special situation.

3. LIFTING THEOREM

Fix a cardinal $\kappa \geq \omega$ and let $\pi : \kappa^+ \times \omega \to \kappa^+$ be the natural projection: $\pi(\langle \alpha, n \rangle) = \alpha$.

Define the poset $P^0 = \langle P^0, \prec \rangle$ as follows. The underlying set P^0 consists of triples $\langle a, \leq, i \rangle$ satisfying the following requirements:

(i) $a \in [\kappa^+ \times \omega]^{<\omega}$,

(ii) $\leq$ is a partial ordering on a,

(iii) $\forall\{x, y\} \in [a]^2$ if $x \leq y$ the $\pi(x) < \pi(y)$,

(iv) $i : [a]^2 \to \mathcal{P}(a)$ is a function,

(v) $\forall\{x, y\} \in [a]^2$ if $\pi(x) = \pi(y)$ then $i\{x, y\} = \emptyset$,

(vi) $\forall\{x, y\} \in [a]^2$ if $x \leq y$ then $i\{x, y\} = \{x\}$.

Write $p = \langle a^p, \leq^p, i^p \rangle$ for $p \in P^0$. Define the function $h^p : a^p \to \mathcal{P}(a^p)$ by the formula $h^p(x) = \{y \in a^p : y \leq^p x\}$. For $b \subset a^p$ write $h^p[b] = \bigcup\{h^p(x) : x \in b\}$.

Let $p \prec q$ iff $a^q \subset a^p$,

$$\leq^q = \leq^p \cap (a^q \times a^q),$$

$$i^q \subset i^p.$$

Clearly $\prec$ is a partial ordering on P^0.

Let

$$P^* = \{\langle a, \leq, i \rangle \in P^0 : \forall\{x, y\} \in [a]^2 \; \forall z \in a$$

$$(z \leq x \wedge z \leq y) \text{ iff } \exists t \in i\{x, y\} \; z \leq t\}.$$

Fact 3.1. *For $p \in P^0$,*

$$p \in P^* \text{ iff } \forall\{x, y\} \in [a^p]^2 h^p(x) \cap h^p(y) = h^p[i^p\{x, y\}].$$

The elements of P^* can be considered as the natural finite approximations of an LCS-order on $\kappa^+ \times \omega$ and the witnessing function i.

Definition 3.2. Two condition $p, q \in P^0$ are *twins* iff (i)–(ii) below hold, where $a = a^p \cap a^q$:

(i) $\leq^p {\restriction} a = \leq^q {\restriction} a$,

(ii) $i^p \restriction [a]^2 = i^q \restriction [a]^2$.

Definition 3.3. Let $p, q \in P^0$ be twins. A condition $r \in P^0$ is an *amalgamation of p and q* iff

(a) $a^r = a^p \cup a^q$

(b) $\leq^r$ is the partial ordering on a^r generated by $\leq^p \cup \leq^q$,

(c) $i^r \supset i^p \cup i^q$.

Let
$$\mathfrak{amalg}\,(p, q) = \{r : r \text{ is an amalgamation of } p \text{ and } q\}.$$

When we speak about amalgamations of two conditions we will always assume that these conditions are twins.

Fact 3.4. *If $r \in P^0$ is an amalgamation of p and q, then*

(1) $\leq^r \restriction a^p = \leq^p$,

(2) $r \prec p$ and $r \prec q$,

(3) *If $x \in a^p$ and $y \in a^q$ then $x \leq^r y$ iff there is $z \in a^p \cap a^q$ such that $x \leq^p z \leq^q y$.*

Fact 3.5. *If $r \in P^0$ is an amalgamation of p and q, moreover $p, q \in P^*$ then*

$$\forall \{x, y\} \in [a^p]^2 \cup [a^q]^2 \ h^r(x) \cap h^r(y) = h^r \left[i\{x, y\} \right].$$

Proof. Assume that $\{x, y\} \in [a^p]^2$ and let $z \in \left(h^r(x) \cap h^r(y) \right) \cap a^q$. Then there are $u, v \in a^p \cap a^q$ with $z \leq^q u \leq^p x$ and $z \leq^q v \leq^p y$. Since $q \in P^*$ there is $w \in i^q \{u, v\}$ with $z \leq^q w$. Since $i^p \{u, v\} = i^q \{u, v\}$ we have $w \in a^p \cap a^q$. Thus $w \in h^p(x) \cap h^p(y)$. Since $q \in P^*$, there is $t \in i^p \{x, y\}$ with $w \leq^p t$. Thus $z \leq^q w \leq^p t \in i^p \{x, y\} = i^r \{x, y\}$ and hence $z \in h^r \left[i^r \{x, y\} \right]$.   ∎ 3.5

For $A \subset \kappa^+$ let

$$P_A^* = \{p \in P^* : a^p \subset A \times \omega\}.$$

Next we introduce three properties, (K^+), D_1^A and D_2^A, of posets $\langle P, \prec \rangle$, where $P \subset P_A^*$ for some $A \subset \kappa^+$. The first one is a strong version of property (K), the two others are density requirements.

Definition 3.6. Let $P \subset P^*$. The poset $\mathcal{P} = \langle P, \prec \rangle$ has *property (K^+)* iff

$$\forall S \in [P]^{\omega_1} \exists T \in [S]^{\omega_1} \forall \{p, q\} \in [T]^2$$

p and q have an amalgamation in P.

Definition 3.7. For a condition $p \in P^0$ and $x \in (\kappa^+ \times \omega) \setminus a^p$ define $q = p \uplus \{x\} \in P^0$ as follows:

- $a^q = a^p \cup \{x\}$,

- $\leq^q = \leq^p \cup \left\{ \langle x, x \rangle \right\}$,

- $i^p \subset i^q$,

- $i^q \{x, y\} = \emptyset$ for $y \in a^p$.

Fact 3.8. $p \uplus \{x\} \in P_A^*$ for each $p \in P_A^*$ and $x \in (A \times \omega) \setminus a^p$.

Definition 3.9. Let $P \subset P_A^*$. The poset $\mathcal{P} = \langle P, \prec \rangle$ has property D_1^A iff

$$p \uplus \{x\} \in P \text{ for each } p \in P \text{ and } x \in (A \times \omega) \setminus a^p.$$

Definition 3.10. For $p \in P_A$, $x \in a^p$, $y_0, y_1, \ldots, y_{n-1} \in (A \times \omega) \setminus a^p$ with $\pi(y_0) < \pi(y_1) < \ldots \pi(y_{n-1}) < \pi(x)$ define the condition $q = p \uplus_x \langle y_0, \ldots, y_{n-1} \rangle \in P^0$ as follows:

- $a^q = a^p \cup \{y_0, \ldots, y_{n-1}\}$,

- $\leq^q = \leq^p \cup \{ \langle y_i, y_j \rangle : i \leq j < n \rangle \} \cup \{ \langle y_i, z \rangle : z \in a^p, x \leq^p z \}$,

- $i^p \subset i^q$,

- $i^q\{y_i, y_j\} = y_{\min(i,j)}$,

- $i^q\{y_i, z\} = \begin{cases} y_i & \text{if } x \leq^p z \\ \emptyset & \text{otherwise} \end{cases}$ for $z \in a^p$.

Fact 3.11. If $p \in P_A^*$, $x \in a^p$, $y_0, y_1, \ldots, y_{n-1} \in (A \times \omega) \setminus a^p$ with $\pi(y_0) < \pi(y_1) < \ldots \pi(y_{n-1}) < \pi(x)$, then $q = p \uplus_x \langle x_0, \ldots, x_{n-1} \rangle \in P_A^*$.

Definition 3.12. Let $P \subset P_A^*$. The poset $\mathcal{P} = \langle P, \prec \rangle$ has property D_2^A iff

$\forall \{\alpha, \beta\} \in A$, $\alpha < \beta$, there is a finite set of ordinals $L^P(\alpha, \beta) = \{\alpha_0, \ldots, \alpha_{n-1}\} \in [A]^{<\omega}$ such that $\alpha = \alpha_0 < \alpha_1 < \ldots \alpha_{n-1} < \beta$ and if $p \in P$, $x \in a^p$ with $\pi(x) = \beta$ and $x_i \in (A \times \omega) \setminus a^p$ with $\pi(x_i) = \alpha_i$ for $i < n$, then $p \uplus_x \langle y_0, \ldots y_{n-1} \rangle \in P$.

Definition 3.13. Let $A \subset \kappa^+$ and $P \subset P^*$. The poset $\mathcal{P} = \langle P, \prec \rangle$ is A-*nice* iff $P \subset P_A^*$ and P has properties (K^+), (D_1^A) and (D_2^A). For $\delta < \kappa^+$ let $\mathcal{NAT}(\delta)$ be the statement that there is δ-nice poset P_δ.

Proposition 3.14. *If a poset P is δ-nice then P has property (K) and $\mathcal{THIN}(\delta)$ holds in V^P.*

Proof. By Fact 3.4(2) property (K^+) implies property (K). Let $G \subset P$ be a generic filter. Put $A = \bigcup\{a^p : p \in G\}$, $i = \bigcup\{i^p : p \in G\}$ and $\prec = \bigcup\{\prec^p : p \in G\}$. Then $A = \delta \times \omega$ by (D_1^δ). The partial ordering $\prec$ satisfies (I) because every $p \in P$ satisfies (iii). The function $i : [\delta \times \omega]^2 \rightarrow [\delta \times \omega]^{<\omega}$ satisfies (II) because every element of P is in P^*. Finally (III) holds because (D_2^δ) can be applied in a suitable density argument. Thus $\langle \delta \times \omega, \tau_\prec \rangle$ is an LCS space with levels $\{\alpha\} \times \omega$ for $\alpha < \delta$. ∎ 3.14

After this preparation we are able to formulate the main lifting theorem.

Theorem 3.15. $\mathcal{NAT}(\kappa)$ *implies* $\mathcal{NAT}(\delta)$ *for each cardinal* κ *and ordinal* $\delta < \kappa^+$.

First, in Lemma 3.16 below, we show that our lifting theorem works downwards. Although $\mathcal{THIN}(\kappa)$ clearly implies $\mathcal{THIN}(\delta)$ for $\delta < \kappa$ we should prove $\mathcal{NAT}(\delta)$ for $\delta < \kappa$ as well, because we will use the posets witnessing this to prove $\mathcal{NAT}(\gamma)$ for $\gamma \geq \kappa$.

If $p \in P^0$ and $I \subset \kappa^+$ let

$$p \restriction I = \left\langle a^p \cap (I \times \omega), \leq^p \restriction (I \times \omega), i^p \restriction [I \times \omega]^2 \right\rangle.$$

Observe that

- $p \restriction I \in P^0$ iff $i^p\{x,y\} \subset I \times \omega$ for each $\{x,y\} \in \left[a^p \cap (I \times \omega)\right]^2$,

- if $p \in P^*$ and $p \restriction I \in P^0$ then $p \restriction I \in P^*$.

Lemma 3.16. $\mathcal{NAT}(\kappa)$ *implies* $\mathcal{NAT}(\delta)$ *for* $\delta < \kappa$.

Proof. Fix $P_\kappa \subset P_\kappa^*$ such that $\mathcal{P}_\kappa = \langle P_\kappa, \prec \rangle$ has properties (K^+), (D_1^κ) and (D_2^κ). Let $\mathcal{P}_\delta = \langle P_\delta, \prec \rangle$, where $P_\delta = \{p \restriction \delta : p \in P_\kappa\}$.

We should check that $\mathcal{P}_\delta$ also has has properties (K^+), (D_1^δ) and (D_2^δ).

(K^+): Let $\{p_\nu \restriction \delta : \nu < \omega_1\} \in [P_\delta]^{\omega_1}$. We can assume that for each $\{\nu, \mu\} \in [\omega_1]^2$ p_ν and p_μ have an amalgamation $r_{\nu,\mu} \in P_\kappa$. Hence $r_{\nu,\mu} \restriction \delta \in P_\delta$ is an amalgamation of $p_\nu \restriction \delta$ and $p_\mu \restriction \delta$.

(D_1^δ) is easy: $(p \restriction \delta) \uplus \{x\} = (p \uplus \{x\}) \restriction \delta$ for $\pi(x) < \delta$.

(D_2^δ) is also easy: $(p \restriction \delta) \uplus_x \langle y_0, \ldots, y_{n-1} \rangle = \left(p \uplus_x \langle y_0, \ldots, y_{n-1} \rangle\right) \restriction \delta$ for $\pi(x) < \delta$. ∎ 3.16

Proof of Theorem 3.15. Since we know the statement for $\delta < \kappa$ we prove the theorem by induction on $\delta \geq \kappa$. When we constructed P_δ we will also have $P_A \subset P_A^*$ for each $A \subset \kappa^+$ with order type δ such that $\mathcal{P}_A = \langle P_A, \prec \rangle$ has properties (K^+), (D_1^A) and (D_2^A).

We will write $L^A(\alpha, \beta)$ for $L^{P_A}(\alpha, \beta)$. Let $L^A(\alpha, \alpha) = \emptyset$.

Successor step:

Assume that $\mathcal{P}_\delta$ is constructed. Then we can get $\mathcal{P}_{\delta+1}$ as follows.

A p is in $P_{\delta+1}$ iff

(i) $p \in P_{\delta+1}^*$,

(ii) $p \upharpoonright \delta \in P_\delta$,

(iii) $\forall \{x, y\} \in [a^p]^2$ if $\pi(x) < \delta$ and $\pi(y) = \delta$ then either $i\{x, y\} = x$ (i.e. $x \leq^p y$) or $i\{x, y\} = \emptyset$ (i.e. $h^p(x) \cap h^p(y) = \emptyset$).

We show that $\mathcal{P}_{\delta+1} = \langle P_\delta, \prec \rangle$ works, i.e. it satisfies properties (K^+), $(D_1^{\delta+1})$ and $(D_2^{\delta+1})$.

Lemma 3.17. $\mathcal{P}_{\delta+1}$ satisfies (K^+).

Proof. Let $\{p_\nu : \nu < \omega_1\} \in [\mathcal{P}_{\delta+1}]^{\omega_1}$, $p_\nu = \langle a_\nu, \leq_\nu, i_\nu \rangle$, $h_\nu = h^{p_\nu}$. Without loss of generality

(a) $\forall \{\nu, \mu\} \in [\omega_1]^2 \exists r_{\nu,\mu} \in P_\delta r_{\nu,\mu}$ is an amalgamation of p_ν and p_μ.

(b) $\exists t a_\nu \cap (\{\delta\} \times \omega) = t$.

(c) $\{a_\nu : \nu < \omega_1\}$ forms a Δ-system with kernel a.

(d) $\leq_\nu \upharpoonright a = \leq_\mu \upharpoonright a$ for each $\{\nu, \mu\} \in [\omega_1]^2$

(iii) and (d) together imply that

(e) $\forall \{\nu, \mu\} \in [\omega_1]^2 \forall x \in t \forall y \in (a \setminus t) i_\nu \{x, y\} = i_\mu \{x, y\}$.

Now for each $\{\nu, \mu\} \in [\omega_1]^2$ the conditions p_ν and p_μ are twins and we can define $r \in P^0$ as follows:

- r is an amalgamation of p_ν and p_μ

- $r \upharpoonright \delta = r_{\nu,\mu}$.

If $\{x, y\} \in [a^r]^2 \setminus ([a^p]^2 \cup [a^q]^2)$ then $\{x, y\} \in [a^{r_{\nu,\mu}}]^2$. Hence $r_{\nu,\mu} \in P_\delta^*$ and Fact 3.5 imply that $r \in P_{\delta+1}^*$. Thus $r \in P_{\delta+1}$. ∎ 3.17

Lemma 3.18. $\mathcal{P}_{\delta+1}$ satisfies $(D_1^{\delta+1})$.

Straightforward.

Lemma 3.19. $\mathcal{P}_{\delta+1}$ satisfies $(D_2^{\delta+1})$.

Proof. For $\alpha < \beta < \delta$ let $L_{\alpha,\beta}^{\delta+1} = L_{\alpha,\beta}^\delta$. For $\alpha < \delta$ let $L_{\alpha,\delta}^{\delta+1} = \{\alpha\}$. ∎ 3.19

The successor step is done.

Limit step:

Assume that δ is limit ordinal, and $\mathcal{P}_A$ is constructed for each $A \subset \kappa^+$ with order type $< \delta$.

Fix a club $C \subset \delta$, $C = \{\gamma_\zeta : \zeta < \mathrm{cf}(\delta)\}$. Let $I_\zeta = [\gamma_\zeta, \gamma_{\zeta+1})$ for $\zeta < \mathrm{cf}(\delta)$.

Let $\rho : \delta \to \mathrm{cf}(\delta)$ s.t. $\rho(\alpha) = \zeta$ iff $\alpha \in I_\zeta$.

Let $p \in P_\delta$ iff

(δ1) $p \in P_\delta^*$,

(δ2) $p \restriction C \in P_C$,

(δ3) $p \restriction I_\zeta \in P_{I_\zeta}$ for each $\zeta < \mathrm{cf}(\delta)$,

(δ4) $\forall x, y \in a^p$ if $x \leq^p y$, $\gamma_\zeta < \pi(x) < \gamma_{\zeta+1} \leq \pi(y)$ then $\exists u \in a^p x \leq^p u \leq^p y$ and $\pi(u) = \gamma_{\zeta+1}$

(δ5) $\forall x, y \in a^p$ if $x \leq^p y$, $\pi(x) < \gamma_\xi \leq \pi(y) < \gamma_{\xi+1}$ then $\exists v \in a^p x \leq^p v \leq^p y$ and $\pi(v) = \gamma_\xi$

(δ6) $\forall x, y \in a^p$, $\gamma_\zeta \leq \pi(x) < \gamma_{\zeta+1} \leq \gamma_\xi \leq \pi(y) < \gamma_{\xi+1}$, $x \not\leq^p y$ then

$$ i^p\{x, y\} \subset \bigcup \left\{ i^p\{u, v\} : u \leq^p x, v \leq^p y, \pi(u) = \gamma_\zeta, \pi(v) = \gamma_\xi \right\}. $$

We show that $\mathcal{P}_\delta = \langle P_\delta, \prec \rangle$ works, i.e. it satisfies properties (K^+), (D_1^δ) and (D_2^δ).

Lemma 3.20. $\mathcal{P}_\delta$ satisfies (K^+).

Proof. Let $\{p_\nu : \nu < \omega_1\} \in [P_\delta]^{\omega_1}$, $p_\nu = \langle a_\nu, \leq_\nu, i_\nu \rangle$, $h_\nu = h^{p_\nu}$. Let $c_\nu = \{\eta < \mathrm{cf}(\delta) : a_\nu \cap I_\eta \neq \emptyset\}$. By thinning out the sequence $\{p_\nu : \nu < \omega_1\}$ we can assume that

(a) $\{a_\nu : \nu \in \omega_1\}$ forms a Δ-system with kernel d,

(b)   there is a partial ordering $\leq^d$ on d such that $\leq_\nu \restriction d = \leq^d$ for each $\nu \in \omega_1$,

(c) $\{c_\nu : \nu < \omega_1\}$ forms a Δ-system with kernel c,

(d) $\forall \eta \in c \exists e_\eta \forall \nu \in \omega_1 a_\nu \cap \left(\{\gamma_\eta, \gamma_\eta + 1\} \times \omega \right) = e_\eta$,

(e) $\forall \eta \in c\forall\{\nu, \mu\} \in [\omega_1]^2$ the conditions $p_\nu \upharpoonright I_\eta$ and $p_\mu \upharpoonright I_\eta$ have an amalgamation $r^\eta_{\nu,\mu} = \langle a^\eta_{\nu,\mu}, \leq^\eta_{\nu,\mu}, i^\eta_{\nu,\mu}\rangle$ in P_{I_η}.

(f) $\forall\{\nu, \mu\} \in [\omega_1]^2$ the conditions $p_\nu \upharpoonright C$ and $p_\mu \upharpoonright C$ have an amalgamation $r^C_{\nu,\mu} = \langle a^C_{\nu,\mu}, \leq^C_{\nu,\mu}, i^C_{\nu,\mu}\rangle$ in P_C.

(g) $i_\nu\{x, y\} = i_\mu\{x, y\}$ for each $\{x, y\} \in [d]^2$ and $\{\nu, \mu\} \in [\omega_1]^2$.

To ensure (g) fix $\{x, y\} \in [d]^2$. If $\rho(x) = \rho(y) = \eta$ then (g) holds by (e): $i_\nu\{x, y\} = i_\mu\{x, y\}$. If $\{\pi(x), \pi(y)\} \in [C]^2$ then $i_\nu\{x, y\} = i_\mu\{x, y\} \overset{\text{def}}{=}$ $i^C\{x, y\}$ by (f). If $\eta = \rho(x) \neq \rho(y) = \sigma$ then by ($\delta 6$) we have

$$i_\nu\{x, y\} \subset \bigcup\{i_\nu\{u, v\} : u \leq_\nu x, v \leq_\nu y, \pi(u) = \gamma_{\rho(x)}, \pi(v) = \gamma_{\rho(y)}\} \subset$$

$$\bigcup\{i^C\{u, v\} : u \in e_{\rho(x)}, v \in e_{\rho(y)}\},$$

i.e. $i_\nu\{x, y\}$ is a subset of a fixed finite set for each $\nu \in \omega_1$. So, by thinning out our sequence we can guarantee that (g) holds.

Claim 3.20.1. p_ν and p_μ are twins for each $\{\nu, \mu\} \in [\omega_1]^2$.

Fix $\{\nu, \mu\} \in [\omega_1]^2$. Define $r = \langle a, \leq, i\rangle \in P^0$ as follows:

(r1) $a = a_\nu \cup a_\mu$,

(r2) $\leq$ is the partial ordering on a generated by $\leq_\nu \cup \leq_\mu$,

(r3)

$$i\{x, y\} = \begin{cases} i_\nu\{x, y\} & \text{if } \{x, y\} \in [a_\nu]^2, \\ i_\mu\{x, y\} & \text{if } \{x, y\} \in [a_\mu]^2, \\ i^C_{\nu,\mu}\{x, y\} & \text{if } \{x, y\} \in [C]^2, \\ i^\eta_{\nu,\mu}\{x, y\} & \text{if } \{x, y\} \in [I_\eta]^2, \\ M(x, y) & \text{otherwise,} \end{cases}$$

where

$$M(x, y) = \bigcup\{i\{u, v\} : \{u, v\} \in [a]^2,\ u \leq x,\ v \leq y,$$

$$\pi(u) = \gamma_{\rho(x)},\ \pi(v) = \gamma_{\rho(y)}\}.$$

Claim 3.20.2. r is an amalgamation of p_ν and p_μ.

Claim 3.20.3. $\leq \restriction C \times \omega \; = \; \leq^C_{\nu,\mu}$.

Proof. Let $x, y \in a \cap (C \times \omega)$, $x \leq y$. We can assume that $x \in a_\nu$ and $y \in a_\mu$, and $\rho(x) < \rho(y)$. Then, by Fact 3.4(3), there is $z \in d$ with $x \leq_\nu z \leq_\mu y$. Then, applying ($\delta 5$) for x and z in p_ν there is $v \in a_\nu$ such that $x \leq_\nu v \leq_\nu z$ and $\pi(v) = \gamma_{\rho(z)}$. Since $z \in a$ we have $\rho(z) \in c$ and so $v \in e_{\rho(z)} \subset d$. Thus $x \leq^C_{\nu,\mu} y$ because $x \leq_\nu v \leq_\mu y$ and $v \in d \cap (C \times \omega)$. $\blacksquare$ 3.20.3

Claim 3.20.4. r satisfies ($\delta 2$) and ($\delta 3$).

Proof. $r \restriction I_\eta = r^\eta_{\nu,\mu} \in P_{I_\eta}$ is clear for each $\eta < \mathrm{cf}(\delta)$ and $r \restriction C = r^C_{\nu,\mu} \in P_C$ follows from Claim 3.20.3. $\blacksquare$ 3.20.4

Claim 3.20.5. r satisfies ($\delta 4$).

Proof. Assume that $\{x, y\} \in [a]^2$, $x \leq y$, $\gamma_\eta < \pi(x) < \gamma_{\eta+1} \leq \pi(y)$. We can assume that $x \in a_\nu \setminus a_\mu$ and $y \in a_\mu \setminus a_\nu$. Pick $z \in d$ such that $x \leq_\nu z \leq_\mu y$.

If $\gamma_\eta < \pi(z) < \gamma_{\eta+1}$ then applying ($\delta 4$) for the pair $\{z, y\}$ in p_μ we obtain $u \in a_\mu$ such that $z \leq_\mu u \leq_\mu y$ and $\pi(u) = \gamma_{\eta+1}$. Then this u works for $\{x, y\}$.

If $\gamma_{\eta+1} \leq \pi(z)$ then applying ($\delta 4$) for the pair $\{x, z\}$ in p_ν we obtain $v \in a_\nu$ such that $x \leq_\nu v \leq_\nu z$ and $\pi(v) = \gamma_{\eta+1}$. $\blacksquare$ 3.20.5

Claim 3.20.6. r satisfies ($\delta 5$).

Proof. Assume that $\{x, y\} \in [a]^2$, $x \leq y$, $\pi(x) < \gamma_\eta \leq \pi(y) < \gamma_{\eta+1}$. We can assume that $x \in a_\nu \setminus a_\mu$ and $y \in a_\mu \setminus a_\nu$. Pick $z \in d$ such that $x \leq_\nu z \leq_\mu y$.

If $\gamma_\eta \leq \pi(z) < \gamma_{\eta+1}$ then applying ($\delta 5$) for the pair $\{x, z\}$ in p_ν we obtain an $u \in a_\nu$ such that $x \leq_\nu u \leq_\nu z$ and $\pi(u) = \gamma_\eta$. Then this u works for $\{x, y\}$.

If $\gamma_{\eta+1} < \pi(z)$ then applying ($\delta 5$) for the pair $\{z, y\}$ in p_μ we obtain a $v \in a_\mu$ such that $z \leq_\mu v \leq_\mu y$ and $\pi(v) = \gamma_\eta$. $\blacksquare$ 3.20.6

Claim 3.20.7. r satisfies ($\delta 6$).

Straightforward from the construction of i.

Claim 3.20.8. r satisfies ($\delta 1$): $r \in P^*_\delta$.

Proof. Write $h = h^r$. Let $\{x, y\} \in [a]^2$ be $\leq$-incomparable elements. By Fact 3.5 we can assume that $x \in a_\nu \setminus a_\mu$ and $y \in a_\mu \setminus a_\nu$. Let $z \in h(x) \cap h(y)$.

Case 1. $\rho(x) = \rho(y)$.

Let $\eta = \rho(x)$. Since $r \upharpoonright I_\eta = r^\eta_{\nu,\mu} \in P^*_{I_\eta}$ we can assume that $\pi(z) < \gamma_\eta$. Applying ($\delta 5$) for the pairs $\{z, x\}$ and $\{z, y\}$ we obtain u and v, respectively, such that $\pi(u) = \pi(v) = \gamma_\eta$, $z \leq u \leq x$, $z \leq v \leq y$. Since $\eta \in c_\nu \cap c_\mu = c$ we have $\{u, v\} \subset d$. Since $z \in h(u) \cap h(v)$ we have $u = v$ by Fact 3.5. Hence there is $t \in i^\eta_{\nu,\mu}\{x, y\} = i\{x, y\}$ with $u \leq^\eta_{\nu,\mu} t$. Thus $z \leq t \in i\{x, y\}$.

Case 2. $\rho(z) = \rho(x) < \rho(y)$, $\pi(z) = \gamma_{\rho(x)}$.

Applying ($\delta 5$) for the pair $\{z, y\}$ there is $u \in a$ such that $z \leq u \leq y$ and $\pi(u) = \gamma_{\rho(y)}$. Then $i\{z, u\} = \{z\}$ and $i\{z, u\} \subset i\{x, y\}$.

Case 3. $\rho(z) = \rho(x) < \rho(y)$, $\pi(z) > \gamma_{\rho(x)}$.

Applying ($\delta 4$) for the pair $\{z, y\}$ there is $u \in a$ such that $z \leq u \leq y$ and $\pi(u) = \gamma_{\rho(x)+1}$. If $u \in a_\mu$ then there is $w \in d \cap (I_{\rho(X)} \times \omega)$ such that either $z \leq_\nu w \leq_\mu u$ or $z \leq_\mu w \leq_\nu x$ by Fact 3.4(3). Hence $\rho(x) \in c$ and so $u \in e_{\rho(x)} \subset d \subset a_\nu$. Thus $u \in a_\nu$. Thus $z \in h_\nu(x) \cap h_\nu(u)$, hence by Fact 3.5 and by ($\delta 6$) we have $x \leq_\nu u$. Hence $x \leq y$, contradiction, this case is not possible.

Case 4. $\rho(z) < \rho(x) < \rho(y)$.

Applying ($\delta 4$) for the pairs $\{z, x\}$ and $\{z, y\}$ we obtain u and v, respectively, such that $\pi(u) = \pi(v) = \sigma \in C$, $z \leq u \leq x$, $z \leq v \leq y$. Then $z \in h(u) \cap h(v)$ so, by case 1, we have $u = v$. Applying ($\delta 5$) for the pairs $\{u, x\}$ and $\{u, y\}$ we obtain t and w such that $u \leq t \leq x$, $u \leq w \leq y$, $\pi(v) = \gamma_{\rho(x)}$, $\pi(w) = \gamma_{\rho(y)}$. Since $r \upharpoonright C = r^C_{\nu,\mu} \in P^*_C$ there is $s \in i\{t, w\}$ with $u \leq s$. Then $z \leq s$ and $s \in i\{x, y\}$. $\blacksquare$ 3.20.8

Hence P_δ satisfies (K^+). $\blacksquare$ 3.20

Lemma 3.21. P_δ satisfies (D^δ_1).

Proof. Assume that $p \in P_\delta$ and $z \in (\delta \times \omega) \setminus a^p$. Let $q = p \uplus \{x\}$. We need to show that $q \in P_\delta$, i.e., q satisfies ($\delta 1$)–($\delta 6$).

($\delta 1$) follows from Fact 3.8.

If $z \notin C \times \omega$ then $q \upharpoonright C = p \upharpoonright C \in P_C$ because $p \in P_\delta$. If $z \in C \times \omega$ then $q \upharpoonright C = (p \upharpoonright C) \uplus \{z\} \in P_C$ because $p \upharpoonright C \in P_C$ and P_C satisfies (D^C_1). Hence ($\delta 2$) holds. Similar arguments work for ($\delta 3$).

As for ($\delta 4$), let $\{x, y\} \in [a^q]^2$ with $x \leq^q y$. Then $\{x, y\} \in [a^p]^2$ because z and the elements of a^p are $\leq^q$-incomparable. So we can apply property

($\delta 4$) for $\{x, y\}$ in p to get a suitable $u \in a^p \subset a^q$. Similar arguments work for ($\delta 5$).

As for ($\delta 6$), let $x, y \in [a^q]^2$. If $z \in \{x, y\}$ then $i^q\{x, y\} = \emptyset$ so the required inclusion holds trivially. Otherwise $\{x, y\} \in [a^p]^2$ so we can use property ($\delta 6$) for p to get the required inclusion. $\blacksquare_{3.21}$

Lemma 3.22. P_δ satisfies (D_2^δ).

Proof. If $\{\alpha, \beta\} \in [I_\eta]^2$ for some η then let $L^\delta(\alpha, \beta) = L^{I_\eta}(\alpha, \beta)$. Otherwise, if $\alpha \in I_\eta$, $\beta \in I_\sigma$, $\eta < \sigma$, then let $\alpha^+ = \min(C \setminus \alpha + 1)$ and put

$$L^\delta(\alpha, \beta) = \{\alpha, \alpha^+\} \cup L^C(\alpha^+, \gamma_\sigma) \cup L^{I_\sigma}(\gamma_\sigma, \beta).$$

Enumerate $L^P(\alpha, \beta)$ as $\alpha = \alpha_0 < \alpha_1 < \cdots < \alpha_{n-1} < \beta$. Let $p \in P_\delta$, $z \in a^p$ with $\pi(z) = \beta$ and $z_i \in (\delta \times \omega) \setminus a^p$ with $\pi(z_i) = \alpha_i$ for $i < n$. Let $q = p \uplus_z \langle z_0, \ldots z_{n-1} \rangle$.

We should show that $q \in P_\delta$, i.e. q satisfies ($\delta 1$)–($\delta 6$).

We will consider only the harder case, i.e. when $\alpha \in I_\eta$, $\beta \in I_\sigma$, $\eta < \sigma$. Fix $1 \le m < n$ such that $L^C(\alpha^+, \gamma_\sigma) = \{\alpha_1, \ldots, \alpha_{k-1}\}$ and $L^{I_\sigma}(\gamma_\sigma, \beta) = \{\alpha_k, \ldots, \alpha_{m-1}\}$, i.e.

$$\alpha_0 = \alpha < \alpha_1 = \alpha^+ = \gamma_{\eta+1} < \cdots < \alpha_m = \gamma_\sigma < \cdots < \alpha_{n-1} < \beta.$$

($\delta 1$) follows from Fact 3.11.

($\delta 2$): If $z \notin C \times \omega$ then

$$q \upharpoonright C = ((p \upharpoonright C) \uplus \{z_k\}) \uplus \{z_{k-1}\} \cdots \uplus \{z_\ell\} \in P_C,$$

where $\ell = 1$ if $\alpha_0 \notin C$ and $\ell = 0$ if $\alpha_0 \in C$, because P_C satisfies D_1^C.

If $z \in C \times \omega$ then

$$q \upharpoonright C = (p \upharpoonright p) \uplus_z \langle z_\ell, \ldots, z_k \rangle \in P_C,$$

where $\ell = 1$ if $\alpha_0 \notin C$ and $\ell = 0$ if $\alpha_0 \in C$, because P_C satisfies D_2^C.

($\delta 3$): Let $\zeta < \mathrm{cf}\, \delta$. If $\zeta = \sigma$ then

$$q \upharpoonright I_\sigma = (p \upharpoonright I_\sigma) \uplus_z \langle z_k, \ldots, z_{n-1} \rangle \in P_{I_\sigma}$$

because P_{I_σ} satisfies $D_2^{I_\sigma}$. If $\gamma_\zeta = \alpha_i$ for some $i \in \{0, \ldots k - 1\}$ then

$$q \upharpoonright I_\zeta = (p \upharpoonright I_\zeta) \uplus \{z_i\} \in P_{I_\zeta}$$

because P_{I_ζ} satisfies $D_1^{I_\zeta}$.

Otherwise $q \restriction I_\zeta = p \restriction I_\zeta \in P_{I_\zeta}$.

($\delta 4$): Let $\{x, y\} \in [a^q]^2$ with $x \leq^q y$ and $\gamma_\zeta < \pi(x) < \gamma_{\zeta+1} \leq \pi(y)$. If $x \in a^p$ then $y \in a^p$ so we can apply ($\delta 4$) in p the get a suitable u. So we can assume that $x \in \{z_0, \ldots, z_{n-1}\}$. Since $\gamma_\zeta < \pi(x) < \gamma_{\zeta+1} \leq \pi(y)$ we have $x = z_0$ or $z \in \{z_{k+1}, \ldots, z_{n-1}\}$. If $x = z_0$ then $u = z_1$ works. If $x = z_i$ for some $k < i < n$ then $\xi = \sigma$ so $\gamma_{\sigma+1} \leq \pi(y)$ implies $y \in a^p$. Hence $\gamma_\sigma < \pi(z) < \gamma_{\sigma+1} \leq \pi(y)$ and so applying ($\delta 4$) in p for the pair $\{z, y\}$ we get $u \in a^p$ with $z \leq^p u \leq^y$ and $\pi(u) = \gamma_{\sigma+1}$. Thus this u works for $\{x, y\}$ in q.

($\delta 5$): Let $\{x, y\} \in [a^q]^2$ with $x \leq^q y$ and $\pi(x) < \gamma_\xi \leq \pi(y) < \gamma_{\xi+1}$. If $x \in a^p$ then $y \in a^q$ so we can apply $\delta 4$ in p the get a suitable v. So we can assume that $x \in \{z_0, \ldots, z_{n-1}\}$.

If $\xi = \sigma$ then $v = z_k$ works.

If $\xi > \sigma$ then $z \leq^p y$ and $\pi(z) = \gamma_\sigma < \gamma_\xi \leq \pi(y) < \gamma_{\zeta+1}$ so we can apply ($\delta 5$) in p for the pair $\{z, y\}$ to get a suitable v.

If $\xi < \sigma$ then $y \in \{z_1, \ldots z_{k-1}\}$ so $v = y$ works.

($\delta 6$): Let $\{x, y\} \in [a^p]^2$ If $\{x, y\} \in [a^p]^2$ then we can apply ($\delta 6$) for p to get the required inclusion. We can assume that $x \in \{z_0, \ldots, z_{n-1}\}$. Then $i^q\{z, y\} = \emptyset$ by the construction of $q = p \uplus_z \langle z_0, \ldots, z_{n-1}\rangle$ because x and y are incomparable and so $z \not\leq^p y$. $\blacksquare_{3.22}$

Thus the limit step is done as well, which completes the inductive construction, so Theorem 3.15 is proved. $\blacksquare_{3.15}$

We conclude the paper with the result we quoted in the abstract.

Theorem 3.23. *If there is a κ-nice poset P for some regular cardinal κ then there is a c.c.c poset Q such that $\mathcal{THIN}(\delta)$ holds in V^Q for each $\delta < \kappa^+$.*

Proof. Using Theorem 3.15 we fix, for each $\delta < \kappa^+$, a δ-nice poset P_δ. Let Q be the finite-support product of $\{P_\delta : \delta < \kappa^+\}$. Since every P_δ has property (K), so has Q.

Let $\mathcal{G}$ be a Q-generic filter and let $\delta < \kappa^+$ be arbitrary. Then $\mathcal{G}_\delta = \{p(\delta) : p \in \mathcal{G} \wedge \delta \in \operatorname{dom} p\}$ is a P_δ-generic filter, hence $\mathcal{THIN}(\delta)$ holds in $V[\mathcal{G}_\delta]$ winessed by some space X_δ by Proposition 3.14. Since $V[\mathcal{G}_\delta] \subset V[\mathcal{G}]$ the space X_δ witnesses $\mathcal{THIN}(\delta)$ in $V[\mathcal{G}]$. $\blacksquare$

REFERENCES

[1] M. Bekkali, *Topics in set theory. Lebesgue measurability, large cardinals, forcing axioms, rho-functions. Notes on lectures by Stevo Todorčević.* Lecture Notes in Mathematics, 1476. Springer-Verlag, Berlin, 1991.

[2] J. E. Baumgartner and S. Shelah, Remarks on superatomic Boolean algebras, *Ann. Pure Appl. Logic*, **33**, no. 2 (1987), 109–129.

[3] I. Juhász, S. Shelah, L. Soukup and Z. Szentmiklóssy, Cardinal sequences and Cohen real extensions, submitted to *Fund. Math.*

[4] I. Juhász and W. Weiss, On thin-tall scattered spaces, *Colloquium Mathematicum*, vol. XL (1978) 63–68.

[5] W. Just, Two consistency results concerning thin-tall Boolean algebras, *Algebra Universalis*, **20**, no. 2 (1985), 135–142.

[6] J. C. Martínez, A forcing construction of thin-tall Boolean algebras, *Fundamenta Mathematicae*, **159**, no. 2 (1999), 99-113.

[7] Judy Roitman, Height and width of superatomic Boolean algebras, *Proc. Amer. Math. Soc.*, **94**, no. 1 (1985), 9–14.

[8] M. Rajagopalan, A chain compact space which is not strongly scattered, *Israel J. Math.*, **23** (1976), 117–125.

Lajos Soukup

Alfréd Rényi Institute of Mathematics

`soukup@renyi.hu`

BOLYAI SOCIETY
MATHEMATICAL STUDIES, 15

Conference on Finite
and Infinite Sets
Budapest, pp. 359–380.

EXTREMAL FUNCTIONS FOR GRAPH MINORS

A. THOMASON

The extremal problem for graph minors is to determine, given a fixed graph H, how many edges a graph G can have if it does not have H as a minor. It turns out that the extremal graphs are pseudo-random; the sense of this has best been expressed by Vera T. Sós in a question answered by Joseph Myers.

   This survey describes what is known about the extremal function and discusses some related matters.

1. INTRODUCTION

We say that the graph H is a *minor* or *subcontraction* of the graph G, written $G \succ H$, if H can be obtained from G by deleting some vertices and edges and by contracting some other edges. This is equivalent to the statement that $V(G)$ contains disjoint subsets W_u, $u \in V(H)$, such that the subgraph $G[W_u]$ induced by W_u is connected for each $u \in V(H)$ and there is an edge in G between W_u and W_v whenever $uv \in E(H)$.

   This survey describes what is currently known about the fundamental extremal question regarding graph minors, namely, how many edges are needed in G to ensure that $G \succ H$? It is now possible to give a fairly full answer to this question. In the first place, it turns out that there is a close connection with the theory of random graphs and with the theory of pseudo-random graphs. This connection is expressed best by a question of Vera T. Sós; her question, and the answer subsequently given by Joseph Myers, are discussed in §5. Secondly, the variation of the extremal function with H can be described in terms of a structural property of H, reminiscent of the way in which, in classical extremal graph theory, the extremal function depends on the chromatic number. In the present case, the relevant structural property is again a kind of partition of H, by means of weights, that is defined in §1.2 and discussed in detail in §6.

We also describe briefly (in §8 and §9) some other extremal problems for minors, such as what connectivity or girth forces a graph to have a given minor. This area has enjoyed some substantial recent advances, but there remain significant open questions about which little, as yet, is known.

1.1. Background

The source of the basic extremal problem for minors is, arguably, the remarkable paper of Wagner [38], in which he proved that the Four Colour Theorem is equivalent to the assertion that $G \succ K_5$ for every graph G that needs five colours to colour it. Hadwiger [10] in 1943 famously conjectured that $G \succ K_t$ for every graph G that needs t colours to colour it. This assertion is trivial for $t \leq 3$, and Hadwiger proved it for $t = 4$. Much more recently, Robertson, Seymour and Thomas [31] have proved the conjecture for $t = 6$ by showing that it follows from the Four Colour Theorem. For a good survey of Hadwiger's conjecture see Toft [37].

In 1964 Wagner [39] proved that $G \succ K_t$ provided the chromatic number of G is sufficiently large (2^{t-3} will do). Mader [22] then developed the idea that the chromatic number might not be the significant parameter; he showed that $G \succ K_t$ provided merely that the *average degree* of G is sufficently large. He therefore introduced the function

$$c(t) = \min \big\{ c \, : \, e(G) \geq c|G| \text{ implies } G \succ K_t \big\},$$

proving that $c(t) \leq 2^{t-3}$ (see Lemma 2.1) and later [23] that $c(t) \leq 8\lceil t \log_2 t \rceil$. Thus we are led to the extremal problem for complete graph minors.

In fact, for small t, much more precise information is available. Write $F + G$ for the join of two graphs F and G, meaning their disjoint union with all edges added between. Observe that the graph $K_{t-2} + \overline{K}_{n-t+2}$ does not have a K_t minor, and neither does the graph $K_{t-5} + P$ if P is a maximal planar graph. These graphs all have $(t-2)|G| - \binom{t-1}{2}$ edges. Dirac [7] demonstrated that if $t \leq 5$ then this is the exact maximum number of edges in G if $G \not\succ K_t$, and Mader [23] extended this to $t \leq 7$. But the seductive pattern stops here; as Mader pointed out, the complete 5-partite graph with two vertices in each class has $40 = 6|G| - 20$ edges and no K_8 minor. (Jørgensen [12] later proved that this is the maximum size of graphs with no K_8 minor, and characterized the extremal graphs. He could thereby (see [11]) extend to $t \leq 8$ the cases in which the following conjecture is known to hold: that if G has a partition into $V_1, \ldots, V_t$ such that $G[V_i \cup V_j]$ is connected for $i \neq j$, then $G \succ K_t$. This conjecture is one of several, related to Hadwiger's conjecture, made by Las Vergnas and Meyneil [21].)

For larger values of t the divergence of the extremal function from the simple pattern just described is much greater. Random graphs provide examples showing that $c(t)$ is of order at least $t\sqrt{\log t}$. This was noticed by several people at about the same time (for example Kostochka [15, 16], and also Fernandez de la Vega [9] based on Bollobás, Catlin and Erdős [2]). Kostochka [15, 16] proved that the correct order of growth for $c(t)$ is indeed $t\sqrt{\log t}$ (see also [32]).

1.2. Recent developments

Recently, the asymptotic value of $c(t)$ was determined.

Theorem 1.1 ([34]). *There exists a constant* $\alpha = 0.3190863\ldots$ *such that*

$$c(t) = \bigl(\alpha + o(1)\bigr)\, t\sqrt{\log t}.$$

The constant α can be explicitly described (see §3); it is simply the best constant that can be obtained from randomly generated lower bounds (note that logarithms are natural unless stated otherwise).

It is evident from Theorem 1.1 that there is a connection between random graphs and extremal functions for minors, though the connection is still closer than first appears. The extremal graphs must be pseudo-random graphs of specified order and density, or else a more-or-less disjoint union of such graphs ([34, 27]). The connection has been captured best by Vera T. Sós in a question which, loosely speaking, is this: if a graph of positive density has no minor bigger than what might be found in a random graph of the same density, must the graph itself be pseudo-random? Myers [26] has given a positive answer to this question. We explain this question more precisely, together with its answer, in §5.

Even more recently, the asymptotic value of the average degree that implies a general H minor has been determined, and the strong connection with pseudo-random graphs persists. Let

$$c(H) = \min\{c : e(G) \geq c|G| \text{ implies } G \succ H\},$$

so that $c(t) = c(K_t)$. The results about $c(H)$ are expressed in terms of a parameter $\gamma(H)$ of the graph H, defined as the minimum average vertex weight amongst weightings satisfying a certain condition.

Definition 1.2. Let H be a graph of order t. We define

$$\gamma(H) = \min_w \frac{1}{t} \sum_{u \in H} w(u) \qquad \text{such that} \qquad \sum_{uv \in E(H)} t^{-w(u)w(v)} \leq t,$$

where the minimum is over all assignments $w : V(H) \to \mathbf{R}^+$ of non-negative weights to the vertices of H.

A uniform weighting w shows that $0 \leq \gamma(H) \leq 1$ for all H and, more generally, $\gamma(H) \leq \sqrt{\tau}$ if H has at most $|H|^{1+\tau}$ edges. In §6.2 we shall describe ways of estimating $\gamma(H)$ fairly precisely, but it is worth pointing out here that, amongst H with $|H|^{1+\tau}$ edges, almost all H and all regular H satisfy $\gamma(H) \approx \sqrt{\tau}$; indeed, $\gamma(H)$ will not be significantly smaller than this unless H has some very restrictive structure.

The extremal result for H, if H has t vertices, is then this.

Theorem 1.3 ([28]). *There exists a constant* $\alpha = 0.3190863\ldots$ *such that*

$$c(H) = \big(\gamma(H)\alpha + o(1)\big)\, t\sqrt{\log t}$$

for every graph H of order t, where the $o(1)$ term is a term tending to zero as $t \to \infty$.

1.3. Contents of this article

We begin in §2 with some preliminary remarks about the extremal function; in particular, it is seen why only dense graphs are of importance in the study of the extremal problem. There follows in §3 a discussion of minors of random graphs and in §4 an explanation of what lies behind Theorem 1.1. The discussion of Sós's question in §5 should nevertheless be comprehensible without first reading the earlier parts.

After that, we go on in §6 to consider the general extremal problem for contractions to a fixed graph H (not necessarily complete). In §7 we comment on an application of the extremal problem to linking in graphs. We finish with some remarks about other conditions on a graph that imply it has large minors; in §8 it is seen how large girth can replace large minimal degree as such a condition, and lastly in §9 we look at how large connectivity might do the same.

2. INITIAL OBSERVATIONS

Here is a simple lemma that implies the existence of the function $c(t)$.

Lemma 2.1. *Let d be an integer and let G be minimal, with respect to taking minors, in the class*

$$\left\{ G \,:\, e(G) \geq d|G| \right\}.$$

Then every edge of G is in at least d triangles; in particular, if H is the neighbourhood subgraph of some vertex, then $e(H) \geq \frac{d}{2}|H|$.

Proof. If G is minimal then G is non-empty and, for every edge uv, the graph G/uv obtained by contracting uv satisfies $e(G/uv) < d(|G| - 1)$. Thus more than d edges are lost by contracting uv, meaning that uv is in at least d triangles. So, if H is the neighbourhood graph of u, then $\delta(H) \geq d$. ∎

The bound $c(t) \leq 2^{t-3}$ follows at once from Lemma 2.1 by induction on t, because a graph G with $e(G) \geq 2^{t-3}|G|$ contracts to a graph containing a vertex u joined to a graph H with $H \succ K_{t-1}$.

Now if G is minimal in $\left\{ G \,:\, e(G) \geq d|G| \right\}$ then $e(G) = d|G|$ (else just remove an edge), so if u is a vertex of minimal degree then $|H| = \delta(G) \leq 2d$. Thus, if we can find a large complete minor in any graph H with $\delta(H) \geq |H|/2$, we can find a large complete minor in any graph at all. In fact, the function $c(t)$ is completely determined by minors of dense graphs, as we explain in §4.

The simple idea of Lemma 2.1 can be exploited further by considering graphs minimal in the class $\left\{ G \,:\, e(G) \geq f(|G|),\, |G| \geq m \right\}$ where $f(n)$ is an integer-valued function chosen so that $f(m) > \binom{m}{2}$ for some m. Then the class contains no graph of order m so a minimal graph must, by the argument above, satisfy $G \succ H$ where $|H| \leq 2f(|G|)$ and $\delta(H) \geq f(|G|) - f(|G|-1)$. A couple of choices that are helpful in different contexts, both essentially due to Mader [23], are these.

First, let $f(|G|) = d|G| - kd$. Provided $k \leq d/2$ we can take $m = d$. This choice gives the same conclusion as Lemma 2.1 but with the extra property that $\kappa(G) \geq k + 1$, as can easily be shown. This choice is useful when determining the extremal function $c(t)$.

Secondly, with the choice $f(G) = \lceil \beta d |G| (1 + \log(|G|/\beta d))/2 \rceil$, where β satisfies $1 = \beta(1 + \log(2/\beta))$, we can take $m = \lceil \beta d \rceil$. The function is chosen both so that $f(|G|) - f(|G| - 1)$ is large for $|G| \leq 2d$ and also so that the graph H from Lemma 2.1, with $|H| \leq 2d$ and $\delta(H) \geq d$, lies in the class. Applying the above arguments to this H produces, after a little calculation, the following result.

Lemma 2.2. *Let* $\beta = 0.37\ldots$ *be as above. Let* G *be a graph with* $e(G) \geq d|G|$. *Then* $G \succ H$, *where* $|H| \leq d + 2$ *and* $2\delta(H) \geq |H| + \lfloor \beta d \rfloor - 1$.

The main point of this lemma is that the minimum degree is bounded below away from $|H|/2$. This has useful consequences, as we describe in §7.

3. RANDOM GRAPHS

Let $G(n, p)$ denote a graph of order n whose edges are chosen independently and at random with probability p.

Theorem 3.1. *Given* $\varepsilon > 0$ *there exists* $T = T(\varepsilon)$ *with the following property. Let* $t > T$, *let* $\varepsilon < p < 1 - \varepsilon$, *let* $q = 1 - p$ *and let* $n = \lfloor (1 - \varepsilon) t \sqrt{\log_{1/q} t} \rfloor$. *Then* $G(n, p) \succ K_t$ *with probability less than* ε.

By choosing $q = \lambda$ where $\lambda = 0.284668\ldots$ is the root of the equation $1 - \lambda + 2\lambda \log \lambda = 0$, we obtain from Theorem 3.1 graphs that have no K_t minor and that have average degree $pn \sim \alpha t \sqrt{\log t}$ where $\alpha = (1 - \lambda)/2\sqrt{\log(1/\lambda)}$. This straightaway gives half of Theorem 1.1, namely $c(t) \geq (\alpha + o(1)) t \sqrt{\log t}$.

Theorem 3.1 is best possible, as shown by Bollobás, Catlin and Erdős [2], in the sense that if $n = (1 + \varepsilon) t \sqrt{\log_{1/q} t}$ then $G(n, p)$ almost surely has a

K_t minor, but this follows in any case from the stronger Theorem 4.1 in §4. For our purposes, random graphs are needed only as a supply of graphs without H minors, for any specified H.

It is worth seeing what determines whether or not $G(n,p) \succ H$ with high probability. Let the vertices of $G(n,p)$ be partitioned into sets W_u, $u \in V(H)$. We need $G[W_u]$ to be connected and we need an edge between W_u and W_v whenever $uv \in E(H)$. The first of these is, in practice, easily arranged — it is the second condition that is the harder to satisfy. The probability that it is satisfied for a particular partition is

$$\prod_{uv \in E(H)} \left(1 - q^{|W_u||W_v|}\right) \approx \exp\left\{ - \sum_{uv \in E(H)} q^{|W_u||W_v|}\right\}.$$

So the partitions most likely to work are those where $\sum_{uv \in E(H)} q^{|W_u||W_v|}$ is minimized, and it is the way in which this sum minimizes, for a particular H, that decides which random graphs have H minors and so, in turn, decides the value of $c(H)$.

By far the most common case is that where, in the minimizing choice, all $|W_u|$ are equal; that is, $|W_u| = n/t$ where $t = |H|$. The expected number of successful partitions is then around $t^n \exp\left\{ - e(H)q^{n^2/t^2}\right\}$, there being about t^n possible partitions. For a graph with $e(H) = t^{1+\tau}$ edges this expected value is small or large according to whether n is less than, or greater than, $\sqrt{\tau}\, t\sqrt{\log_{1/q} t}$, so this is the threshold value of n at which H minors appear.

For general H, put $w(u) = |W_u|/\sqrt{\log_{1/q} t}$, and write $\overline{w} = n/t\sqrt{\log_{1/q} t}$ for the average value of w. Choosing $|W_u|$ to minimize the sum above is the same as choosing w to minimize $\sum_{uv \in E(H)} t^{-w(u)w(v)}$. Writing M for this minimum value, the expectation becomes $t^n \exp(-M)$; since $n = \overline{w}\, t\sqrt{\log_{1/q} t}$, the threshold region for n is when M is approximately t. It can now be seen that the quantity $\overline{w}$ determining this threshold is precisely the parameter $\gamma(H)$ defined in §1.2.

4. COMPLETE MINORS OF DENSE GRAPHS

The main theorem relevant to the extremal properties of complete minors is the following one, a slightly weakened version of that appearing in [34].

Theorem 4.1 ([34]). *Given $\varepsilon > 0$ there exists $T = T(\varepsilon)$ with the following property. Let $t > T$, let $\varepsilon < p < 1 - \varepsilon$, let $q = 1 - p$ and let $n = \left\lceil (1 + \varepsilon)\, t \sqrt{\log_{1/q} t} \right\rceil$. Then every graph G of order n and connectivity $\kappa(G) \geq n(\log\log\log n)/(\log\log n)$ has a K_t minor.*

Thus, every graph of positive density (except those which are nearly disconnected) has complete minors at least as large as those in random graphs of the same density. Some kind of connectivity requirement is obviously required since, for example, the minors of a union of two disjoint graphs of order $n/2$ and density $1/2$ are the minors in the individual components, and they would not be expected to correspond to the minors in a typical graph of order n and density $1/4$.

To prove Theorem 4.1 we must find a partition of $V(G)$ into sets W_u, $u \in V(K_t)$, such that each $G[W_u]$ is connected and such that there is an edge between W_u and W_v whenever $uv \in E(K_t)$. Just as in §3, the first requirement can be arranged fairly straightforwardly, and it is the second that needs care. A natural approach would be to take a random partition of the n vertices into t parts of size n/t each, in the hope that, even if not all the required edges materialize, at most $o(t)$ of them fail, and by dropping any vertex of K_t that is incident with one of these failed edges, we are still left with a complete minor on $t - o(t)$ vertices, which is good enough.

The reason this approach does not succeed directly is because the degrees in the graph G may vary wildly. In order for the argument to work it is necessary that a randomly chosen part of size $l = \sqrt{\log t}$ be joined to all but not much more than nq^l vertices; a second random part would then fail to have an edge to the first random part with probability around $q^{l \times l}$, so behaving much as if the graph were itself random. However, the expected number of vertices not joined to our first random part is $\sum_{x \in G} q(x)^l$, where x has $q(x)n$ non-neighbours, and this expected value can be much larger than nq^l if the degrees differ.

It transpires that two properties of a randomly chosen part are needed to make things work: both the part itself, and its set of non-neighbours, must be spread uniformly throughout the vertices of different degrees; that is, these sets must contain their fair share of the vertices of each degree, in a sense that can be made precise. All but $o(t)$ of the parts, which can be discarded, have both these two properties, and between the remaining parts, all but $o(t)$ of the desired edges materialize, and so we can proceed according to our initial strategy. (In the proof given in [34], the parts are in fact chosen at random only from those that are spread uniformly through the vertices, and so only the spread of the non-neighbours is an issue. On the other hand, in the proof given in [28] of Theorem 6.2 below, which extends Theorem 4.1 to general H, the parts are chosen entirely at random.)

4.1. The extremal function $c(t)$

The remaining half of Theorem 1.1, that is, the upper bound on $c(t)$, can be derived from Theorem 4.1 in this way. Writing $d = \alpha t \sqrt{\log t}$, it is enough to show that if G is minimal in the class $\{G : e(G) \geq d|G|\}$ then $G \succ K_t$. This minimal graph G is either small and dense, or sparse but large. In the first case, Theorem 4.1 implies straightaway that $G \succ K_t$. In the second case, we can assume by the arguments of §2 that G is reasonably well connected and that each edge is in at least d triangles. A few judicious applications of Theorem 4.1 then produce a large number of small minors that can be combined to form a K_t minor. In fact, a minor much larger than K_t can be formed, and from this it follows that extremal graphs arise only from the first case, and they are therefore essentially disjoint unions of small dense pieces.

4.2. Directed graphs

All the above arguments can be made to work for directed graphs, where the minor being sought is DK_t, the complete directed graph of order t with an edge in each direction between each pair of vertices. The extremal digraphs turn out just to be those obtained from the undirected case by replacing each edge by a double edge — details are in [34].

5. PSEUDO-RANDOMNESS AND SÓS'S QUESTION

As indicated in the §4.1, the extremal graphs for the function $c(t)$ are formed by first taking random-like graphs of the appropriate order and density, and then forming as large a graph as desired by taking (almost) disjoint unions of the random-like pieces. Thus extremal graphs must be looked for in the class of pseudo-random, or quasi-random, graphs as discussed by Chung, Graham and Wilson [4] or in [33].

Now it is not true that all pseudo-random graphs behave as well as random graphs in terms of not having large minors. In fact, in [35] it is shown that most of the standard examples of pseudo-random graphs with n vertices have complete minors with $\Theta(n)$ vertices, compared with only $\Theta(n/\sqrt{\log n})$ for random graphs. Indeed, Mader's request [25] for an explicit graph whose largest complete minor has $o(n)$ vertices remains unanswered; in general it seems hard to find a graph G whose largest minor has $o(\delta(G))$ vertices. Alon [1] has nevertheless shown that random Cayley graphs have minors no larger than $\Theta(n/\sqrt{\log n})$.

Sós has expressed the connection between the extremal theorems and quasi-randomness in the most succinct way. Although quasi-randomness does not preclude the presence of large minors, she asked whether quasi-randomness is necessary for the absence of large minors. To be precise, she asked whether a graph of density p and order $t\sqrt{\log_{1/q} t}$, and having no K_t minor, must necessarily be quasi-random.

The standard arguments about quasi-random graphs, even when properly quantified, are not quite strong enough to answer Sós's question. The issue has been settled by Myers [26] in the following way (at the same time giving a more precise description of the extremal graphs for the function $c(t)$.)

To understand Myers' theorem, consider a graph G whose vertex set is partitioned into two sets, X and Y, and define the three densities

$$p_X = \frac{e(X)}{\binom{|X|}{2}}, \qquad p_{XY} = \frac{e(X,Y)}{|X||Y|}, \qquad p_Y = \frac{e(Y)}{\binom{|Y|}{2}}$$

where $e(X)$, $e(Y)$ and $e(X,Y)$ are the numbers of edges of G spanned by X, spanned by Y and joining X to Y. Likewise define $q_X = 1 - p_X$, $q_{XY} = 1 - p_{XY}$ and $q_Y = 1 - p_Y$. It is the principal feature of quasi-random graphs that G is quasi-random if and only if $p_{X'}$ differs little from p_X for every X' with $|X'| = |X|$, which of course implies that each of p_X, p_{XY} and p_Y are close to p, the density of G. Note that, whether or not G is quasi-random, the density of G satisfies

$$q = x^2 q_X + 2x(1-x)q_{XY} + (1-x)^2 q_Y$$

if G is large, where $q = 1 - p$ and $x = |X|/|G|$.

Consider now a randomly generated graph $G(n, x, p_X, p_{XY}, p_Y)$, having n vertices partitioned into two sets X and Y, where $|X| = xn$; the edges are chosen independently, with probability p_X inside X, p_{XY} between X and Y and p_Y inside Y. The proof of Theorem 3.1 is readily modified to show that the threshold value of n at which a K_t minor almost surely appears in $G(n, x, p_X, p_{XY}, p_Y)$ is

$$n = \left(1 + o(1)\right) t \sqrt{\log_{1/q^*} t} \qquad \text{where} \qquad q^* = q_X^{x^2} \, q_{XY}^{2x(1-x)} \, q_Y^{y^2}.$$

By taking logarithms and applying Jensen's inequality it can be seen that

$$q \geq q^*$$

with equality if and only if $q_X = q_{XY} = q_Y = q$.

Thus, so far as the sizes of complete minors are concerned, the constrained random graph $G(n, x, p_X, p_{XY}, p_Y)$ of density $1 - q$ behaves like the ordinary but denser random graph $G(n, 1 - q^*)$.

We can now state Myers' generalization of Theorem 4.1.

Theorem 5.1 (Myers [26]). *Given $\varepsilon > 0$ there exists $T = T(\varepsilon)$ with the following property. Let $t > T$, let $\varepsilon < p < 1 - \varepsilon$, let $q = 1 - p$ and let $n = \left\lceil (1 + \varepsilon) t \sqrt{\log_{1/q} t} \right\rceil$. Let G be a graph of order n and connectivity $\kappa(G) \geq n(\log \log \log n)/(\log \log n)$, having a vertex partition into X and Y as described above, where $\varepsilon < q_X, q_{XY}, q_Y \leq 1$ and $q^* < 1 - \varepsilon$. Then $G \succ K_s$ where*

$$s = \left\lceil \sqrt{\frac{\log(1/q^*)}{\log(1/q)}} \, t \right\rceil .$$

In other words, a graph G with a partition as described will have complete minors at least as large as those found in $G(n, 1 - q^*)$. It follows immediately that if a graph as described in Theorem 4.1 has no minor significantly larger than K_t then q_X is approximately equal to q for every subset X of size xn, implying that G is quasi-random.

The proof of Theorem 5.1 is similar to that of Theorem 4.1, except that the vertices of X and Y are ordered separately, and the parts W_u are chosen so that each is sure to contain a representative sample of both X and Y. The principal difficulty is that the ordering of X, say, must respect the number of neighbours a vertex has both in X and in Y; however, by ordering with respect to a certain subtle parameter, a suitable linear order can be effected.

6. The extremal problem for general H

In this section we describe what is known about the function $c(H)$ for general H. Up until recently nothing was known, but although the situation at the time of writing is still a little fluid, the following description should be fairly accurate. Throughout this section t will stand for the number of vertices of H.

We would like to answer the following questions: (a) how does the function $c(H)$ behave, (b) is there some reasonable structural property that determines its value and (c) do the extremal graphs continue to be pseudo-random?

The answer to these questions appears to be that the function $c(H)$ behaves very similarly to $c(t)$ (indeed, for most graphs H, $c(H)$ is indistinguishable from $c(t)$) and that, at least for graphs with more than $t^{1+\varepsilon}$ edges, the extremal graphs behave in much the same way as before. When asking for a structural property that determines $c(H)$ we have in mind the classical situation of the Erdős-Stone-Simonovits theorem [8], in which the extremal function (for whether H must appear as an ordinary subgraph) is determined by the chromatic number of H.

The fact that the extremal graphs here are pseudo-random, however, makes the situation more complicated than the classical case, for two reasons. First of all, the results must necessarily be of an asymptotic kind (that is, as $|H| \to \infty$, as opposed to the classical case where perhaps $n \to \infty$ but H is allowed to be fixed). Secondly, the extremal function will be insensitive to small changes in the structure of H, such as the addition of an edge, or a handful of edges. This is because such a change in H will have a negligible effect on whether H appears as a minor of a random graph, and random graphs are the extremal graphs. This insensitivity to change is in marked contrast to the classical case, where of course the addition of a single edge can increase the chromatic number and so dramatically affect the extremal function.

As evidenced by Theorem 1.3, $c(H)$ can be described in terms of the parameter $\gamma(H)$ defined in §1.2. The implication of the previous remarks is that some leeway is possible in the definition; if $\gamma'(H)$ were another parameter with $\gamma'(H) = \gamma(H) + o(1)$, where $o(1)$ denotes something tending to zero as $t \to \infty$, then $\gamma'(H)$ could be used just as well as $\gamma(H)$ in all the results. The definition given is chosen because it seems to be the cleanest one that works, and its form is easily related to the appearance of H as a minor in $G(n,p)$, as we noted in §3.

6.1. General H minors

Here are two theorems that generalize Theorems 3.1 and 4.1 to general H. The way we state them, though, is slightly different to before.

Theorem 6.1 ([28]). *Given* $\varepsilon > 0$ *there exists* $T = T(\varepsilon)$ *with the following property.*

Let H *be a graph with* $t > T$ *vertices and with* $\gamma(H) \geq \varepsilon$. *Let* $\varepsilon \leq p \leq 1 - \varepsilon$, *let* $q = 1 - p$ *and let* $n = \left\lfloor \gamma(H)\, t \sqrt{\log_{1/q} t} \right\rfloor$. *Then* H *is a minor of a random graph* $G(n, p - \varepsilon)$ *with probability less than* ε.

The essence of the proof of this theorem has already been given in §3. More work is needed to prove the next theorem, in which the density of G, as usual, means $\left| E(G) \right| / \binom{n}{2}$.

Theorem 6.2 ([28]). *Given* $\varepsilon > 0$ *there exists* $T = T(\varepsilon)$ *with the following property.*

Let H *be a graph with* $t > T$ *vertices and with* $\gamma(H) \geq \varepsilon$. *Let* $\varepsilon \leq p \leq 1 - \varepsilon$, *let* $q = 1 - p$ *and let* $n = \left\lfloor \gamma(H)\, t \sqrt{\log_{1/q} t} \right\rfloor$. *Let* G *be a graph of order* n, *density* $p + \varepsilon$ *and connectivity* $\kappa(G) \geq n(\log \log \log n)/(\log \log n)$. *Then* H *is a minor of* G.

Theorems 3.1 and 4.1 show that the threshold probability p at which an H minor appears in $G(n,p)$ is the threshold density at which H minors appear in every reasonably connected graph of density p. This fact is at the heart of why Theorem 1.3 is true.

The modification to the proof of Theorem 4.1 needed to prove Theorem 6.2 is that the size of the parts W_u varies, being in fact proportional to the optimal weight $w(u)$ that determines $\gamma(H)$. This is the reason behind the change of approach remarked upon in §4.

Arguments similar to those in §4.1, in particular the separate treatment of dense and sparse minimal graphs and the finding of large complete minors in sparse minimal graphs, can be used to derive the extremal function $c(H)$ from Theorem 6.2, so proving Theorem 1.3. The discussion in §5 can also be carried over to general H minors, showing that, apart from a change in

constants, the extremal graphs have the same pseudo-random structure as
they do when H is complete.

6.2. Estimating $\gamma(H)$

It is straightforward to evaluate $\gamma(H)$ when H is complete or complete
bipartite, but otherwise it appears to be difficult. We know, though, that
if H has $t^{1+\tau}$ edges then assigning weight $\sqrt{\tau}$ to every vertex shows that
$\gamma(H) \leq \sqrt{\tau}$. Suppose that w is an optimal weighting of $V(H)$ that realizes
$\gamma(H)$. Then there cannot be a significant proportion of edges uv such that
$w(u)w(v) < \tau$. So, if we group together vertices of roughly equal weight,
there will be almost no edges between the class containing u and the class
containing v if $w(u)w(v) < \tau$. This leads us to approximate H as a subgraph
of a blowup of a small graph, in the following way.

A *shape* is defined to be a pair (F, f), where F is a graph (in which
loops, but not multiple edges, are allowed) and $f : V(F) \to \mathbf{R}^+$ is a function
assigning non-negative numbers to the vertices such that $\sum_{a \in V(F)} f(a) = 1$.
We say that the graph H of order t is an ε-*fit* to shape (F, f) if there is
a partition of $V(H)$ into sets V_a, $a \in V(F)$, such that $\lfloor f(a)t \rfloor \leq |V_a| \leq
\lceil f(a)t \rceil$, and

$$\left| \{ uv \in E(H) : u \in V_a,\ v \in V_b \text{ and } ab \notin E(F) \} \right| \leq t^{-\varepsilon} \left| E(H) \right|.$$

So H is an ε-fit to (F, f) if there is a partition of H into classes indexed
by $V(F)$ and of sizes proportional to f, so that all but a tiny fraction of
the edges of H lie between classes corresponding to edges of F. The fact
that F might have loops allows H to have edges within the corresponding
classes; in particular, every H fits the shape consisting of a single vertex
with a loop.

The parameter of the shape (F, f) that is related to $\gamma(H)$ is the parameter $m(F, f)$, given by

$$m(F, f) = \max_{x \cdot f = 1} \min_{ab \in E(F)} x(a)x(b).$$

Here the maximum is over all functions $x \in [0, \infty)^{V(F)}$ of $V(F)$, and $x \cdot f$
stands for the standard inner product $\sum_{a \in F} x(a)f(a)$. This definition allows
$x(a) > 1$ even though we always have $f(a) \leq 1$. The constant function
$x(a) = 1$ satisfies $x \cdot f = 1$ and so $m(F, f) \geq 1$ always holds. Also, if F has
a single vertex a with a loop then $f(a) = 1$ and $m(F, f) = 1$.

Some calculation then supplies the crucial fact that, if H has $t^{1+\tau}$ edges, then H is an ε-fit to some shape (F, f) with $|F| \leq (1/\varepsilon)$ and $\gamma(H) \geq \sqrt{\tau/m(F,f)} - 4\sqrt{\varepsilon}$. So a lower bound on $\gamma(H)$ can be given by checking that H is not an ε-fit to any small shape (F, f) with $m(F, f)$ large. In so doing it is necessary only to check *critical* shapes: these are shapes (F, f) for which $m(F', f') < m(F, f)$ for any F' resulting from F either by the addition of an edge or by the *merger* of two vertices of F. (The merger of $a, b \in F$ is the replacement of a and b by a single vertex c joined to every vertex previously joined to either a or b, with $f'(c) = f(a) + f(b)$ and $f' = f$ on the other vertices of F'.) This is because if H is an ε-fit to (F, f) then it is also an ε-fit to (F', f').

What makes these observations useful is that the check required is quite short; there are very few critical shapes, and we can describe them explicitly.

Theorem 6.3 ([28]). *A shape (F, f) with $|F| = k+1$ is critical if and only if F is the* half-graph *of order $k + 1$ that is,*

$$V(F) = \{0, 1, \ldots, k\} \quad \text{and} \quad E(F) = \{ij : i + j \geq k\},$$

and moreover f satisfies

$$\frac{f(k)}{f(0)} < \frac{f(k-1)}{f(1)} < \frac{f\big(k - \lfloor (k-1)/2 \rfloor\big)}{f\big(\lfloor (k-1)/2 \rfloor\big)} < 1.$$

For these shapes,

$$m(F, f) = \left\{ \sum_{i=0}^{k} \sqrt{f(i)f(k-i)} \right\}^{-2}.$$

So, if we know the structure of H, it is fairly easy to check whether H is an ε-fit to a small critical shape, and hence to get a lower bound on $\gamma(H)$. The simplest, and commonest, case is where H fails to fit any shape apart from the shape with one vertex and a loop. This case can be reformulated in the statement that H has a *tail*, which is a large subset T whose neighbours lie almost entirely inside a smaller subset S; here is a precise version.

Theorem 6.4 ([28]). *Let $\varepsilon > 0$ and let H be a graph of order $t \geq 1/\varepsilon^2$ with $t^{1+\tau}$ edges such that $\gamma(H) \leq \sqrt{\tau} - 5\sqrt{\varepsilon}$. Then H has an ε-tail — that is, $V(H)$ has a partition $R \cup S \cup T$, with $|T| > |S| + \varepsilon t$, such that $|E(T, T \cup R)| \leq t^{1+\tau-\varepsilon}$.*

Now regular graphs cannot have a tail, nor indeed can graphs that are almost regular, and this includes almost all graphs. We have the following conclusion.

Corollary 6.5. *All regular graphs and almost all graphs H of order t with $t^{1+\tau}$ edges have $\gamma(H) = \sqrt{\tau} + o(1)$.*

As a further corollary we can evaluate $\gamma(H)$ for, for example, complete multipartite graphs; these all have $\gamma(H) \approx 1$ unless the largest part has size βt with $\beta > 1/2$, in which case $\gamma(H) \approx \sqrt{4\beta(1-\beta)}$.

It should be pointed out, however, that this method for approximating $\gamma(H)$ can sometimes give a bound much less than the correct value. This is because the property of being an ε-fit to a shape is insensitive to the introduction of a very sparse subgraph H^*, though this subgraph might be what actually determines $\gamma(H)$. The situation is analogous to that in the classical extremal theory where the chromatic number of H might be determined by $\chi(H^*)$ and not just by the chromatic number of some dense subgraph. An example is when H is the union of $K_{t/8,7t/8}$ with a $t^{1/2}$-regular graph H^* on the same vertex set. We know that $\gamma(K_{t/8,7t/8}) = \sqrt{7}/4 + o(1)$ whereas $\gamma(H^*) = 1/\sqrt{2} + o(1)$. So $\gamma(H) \geq \max(\sqrt{7}/4, 1/\sqrt{2}) + o(1) = 1/\sqrt{2} + o(1)$. But, for every $\varepsilon > 0$, if t is large this graph is an ε-fit to a two vertex shape with $f = (1/8, 7/8)$ and $m(F, f) = 16/7$, so our lower bound method gives only $\gamma(H) \geq \sqrt{7}/4 + o(1)$.

We conclude this section with another lower bound on $\gamma(H)$ based just on the density of the graph. This shows that $\gamma(H)$ can never be close to zero for graphs of positive density.

Theorem 6.6 ([28]). *Let H be a graph of order $t \geq (1/\varepsilon)^{1/\varepsilon}$ and density p. Then $\gamma(H) \geq p - 5\sqrt{\varepsilon}$.*

7. Linking

A graph G is said to be *k-linked* if, for any sequence $s_1, \ldots, s_k, t_1, \ldots, t_k$ of distinct vertices, we can find s_i–t_i paths P_i that are disjoint, $1 \leq i \leq k$. Larman and Mani [20] and Jung [13] noticed that if $\kappa(G) \geq 2k$ and G contains a subdivided complete graph of order $3k$ then G is k-linked. Mader [22] proved that if the average degree of a graph exceeds $2^{\binom{k}{2}}$ then it contains a subdivided K_k, and so, if $\kappa(G)$ is sufficiently large, G is k-linked. (For a survey of subdivisions of graphs, see Mader [24].)

Robertson and Seymour [30], as part of their deep study of graph minors, established a connection between linking and graph minors; they strengthened the above remarks by showing that G is k-linked if $\kappa(G) \geq 2k$ and $G \succ K_{3k}$. It follows from Theorem 1.1 that the connectivity required to force k-linking is only $O(k\sqrt{\log k})$.

Bollobás and Thomason [3] weakened the condition $G \succ K_{3k}$ still further to $G \succ H$ where H is *any* graph such that $2\delta(H) \geq |H| + 4k - 2$. In consequence of Lemma 2.2 they could then show that G is k-linked provided $\kappa(G) \geq 22k$.

The reason we point this out in this survey is to contrast the average degree required to obtain some *specific* H with $2\delta(H) \geq |H| + 4k - 2$, with that needed to achieve just *some* H. By Theorem 1.3 and Theorem 6.6 the former would still require average degree $\Theta(k\sqrt{\log k})$, whereas Lemma 2.2 shows the latter to hold given average degree only $\Theta(k)$.

Added in proof. Thomas and Wollan have recently shown that G is k-linked if $\kappa(G) \geq 10k$.

8. MINORS AND GIRTH

The simple fact underlying the observations in §2 is that contracting an edge of a graph tends to increase the average degree unless the edge lies in many triangles. In particular, if a graph has large girth then many edges can be contracted, each contraction increasing the average degree.

Thomassen [36] made a systematic study of this phenomenon — his aim was to show that many consequences of a graph having large average degree could be derived also for graphs having minimum degree only three but having large girth. His fundamental tool was the following theorem, whose simple and elegant proof we include here. We use $g(G)$ to denote the girth of G.

Theorem 8.1 (Thomassen [36]). *If $\delta(G) \geq 3$ and $g(G) \geq 4k - 5$ then $G \succ H$ where $\delta(H) \geq k$.*

Proof. We may assume $k \geq 4$. Take a partition $A_1, \ldots, A_t$ of $V(G)$ with t maximal such that $G[A_i]$ is connected and $|A_i| \geq 2k - 3$ for $1 \leq i \leq t$. If $G[A_i]$ contains a cycle C, then $|C| \geq 4k - 5$, so by splitting C into two paths we can partition A_i into A_i^1 and A_i^2, with $G[A_i^l]$ is connected and $|A_i^l| \geq 2k - 3$ for $l = 1, 2$; the maximality of t thus implies $G[A_i]$ must in fact be a tree. Suppose now we could find A_i and A_j with $1 \leq i < j \leq t$ for which there were three edges between A_i and A_j. Then we could find vertices $u \in A_i$ and $v \in A_j$ together with three disjoint u–v paths P_1, P_2, P_3 in $G[A_i \cup A_j]$. Any two of these paths have at least $4k - 5$ edges between them and so in particular two of them, say P_1 and P_2, must have length at least $2k - 2$. So we could partition $A_i \cup A_j$ into three sets A^1, A^2, A^3, with A^l containing $2k - 3$ vertices from $P_l - \{u, v\}$, $l = 1, 2$, and A^3 containing the rest of $P_1 \cup P_2 \cup P_3$, such that $G[A^l]$ is connected and $|A^l| \geq 2k - 3$ for $l = 1, 2, 3$. Hence the maximality of t implies that there are at most two edges between A_i and A_j for $1 \leq i < j \leq t$.

Now, of course, we contract each A_i to a single vertex a_i. In the resultant multigraph H^*, every pair of vertices is joined by at most two edges; throw away one edge from each double edge to obtain a graph H. The degree of a vertex a_i in H^* is at least $3|A_i| - 2(|A_i| - 1) \geq 2k - 1$, and so its degree in H is at least $\lceil (2k - 1)/2 \rceil = k$, as desired. ∎

Diestel and Rempel [5] have reduced the girth required here to $6 \log_2 k + 4$. More recently, Kühn and Osthus [18] reduced it to $4 \log_2 k + 27$. They obtained results close to best possible for minors with specified minimum degree and girth; an example is this.

Theorem 8.2 (Kühn and Osthus [18]). *Let $k \geq 1$ and $d \geq 3$ be integers, and let $g = 4k + 3$. If $g(G) \geq g$ and $\delta(G) \geq d$ then $G \succ H$ where $\delta(H) \geq (d - 1)^{(g+1)/4}/48$.*

As a further consequence of their methods they also show that Hadwiger's conjecture holds for graphs of girth at least 19 (Kawarabayashi [14] also found this result).

One natural way of weakening the constraint of large girth is to forbid $K_{s,s}$ as a subgraph, in the hope that this constraint still yields complete minors in graphs of low average degree. (Note that forbidding a non-bipartite subgraph will not help, since the extremal graphs for complete minors contain bipartite subgraphs with at least half as many edges.) Kühn and Osthus [19] have investigated this condition, obtaining the following result, which is again close to best possible provided a standard conjecture about the extremal function for $K_{s,s}$ is true..

Theorem 8.3 (Kühn and Osthus [19]). *Given $s \geq 2$ there exists a constant $c = c(s)$, such that every $K_{s,s}$-free graph of average degree at least r has a K_t minor for $t = \lfloor cr^{1+2/(s-1)}(\log r)^{-3} \rfloor$.*

As might be expected, the proofs of these results are much more substantial than the proof of Theorem 8.1.

9. MINORS AND CONNECTIVITY

Large average degree is the simplest property forcing a graph to have a K_t minor. Robertson and Seymour, in their series of papers on Graph Minors, have investigated more complex structural properties that give rise to minors; one of their fundamental results [29] is that a graph has large tree-width if and only if it contains a large grid minor. Diestel, Jensen, Gorbunov and Thomassen [6] gave a short proof of this result, and introduced the notion of *external connectivity*: a set $X \subset V(G)$ is *externally k-connected* if $|X| \geq k$ and for all subsets $Y, Z \subset X$ with $|Y| = |Z| = k$ there are $|Y|$ disjoint Y-Z paths in G without inner vertices or edges inside X. A large grid that has high external connectivity yields a large complete minor; Kühn [17] has shown that the same conclusion holds even if the large grid is replaced by a large number of large disjoint binary trees, each having an extra vertex joined to its leaves.

There is a simple, and as yet unsolved, problem relating (ordinary) connectivity to complete minors. What connectivity is needed to force a K_t minor? Since $\kappa(G) \leq \delta(G)$ with equality for random graphs, the answer to this question is $(2\alpha + o(1)) t\sqrt{\log t}$, by Theorem 1.1. But the only examples achieving this are pseudo-random graphs of bounded (in t) order; the extremal graphs of larger order for Theorem 1.1 have very low connectivity. It might well be that, for graphs of large order, a lower connectivity will suffice for a K_t minor. We therefore make the following conjecture.

Conjecture 9.1. There is an absolute constant C and a function $n(t)$ such that if $|G| \geq n(t)$ and $\kappa(G) \geq Ct$ then $G \succ K_t$.

Perhaps even $\kappa(G) \geq t + 1$ is enough (though $\kappa(G) = t$ is not, as a 5-connected planar graph joined to K_{t-5} shows). For $t = 6$ Jørgensen [12] (see also [31]) has a related conjecture, that every 6-connected graph with no K_6 minor has a vertex joined to all the others.

Myers [27] has a partial result in this area; if t is odd, a $(t+1)$-connected graph G, with a long sequence of cutsets $S_1, S_2, \ldots$ of size $t + 1$ such that S_j separates $S_1, \ldots, S_{j-1}$ from $S_{j+1}, S_{j+2}, \ldots$, has a K_{t-3} minor if the $G[S_j]$'s are 2-edge-connected.

Added in proof. Böhme, Kawarayabashi, Maharry and Mohar have recently shown that every large $23t$-connected graph has a K_t minor.

REFERENCES

[1]   N. Alon, personal communication.

[2]   B. Bollobás, P. Catlin and P. Erdős, Hadwiger's conjecture is true for almost every graph, *Europ. J. Combin. Theory,* **1** (1980), 195–199.

[3]   B. Bollobás and A. Thomason, Highly linked graphs, *Combinatorica* **16** (1996), 313–320.

[4]   F. R. K. Chung, R. L. Graham and R. M. Wilson, Quasi-random graphs, *Combinatorica,* **9** (1989), 345–362.

[5]   R. Diestel and C. Rempel, *Dense minors in graphs of large girth,* (preprint).

[6]   R. Diestel, T. Jensen, K. Gorbunov and C. Thomassen, Highly connected sets and the excluded grid theorem, *J. Combinatorial Theory Ser. B,* **75** (1999), 61–73.

[7]   G. A. Dirac, Homomorphism theorems for graphs, *Math. Ann.,* **153** (1964), 69–80.

[8]   P. Erdős and M. Simonovits, A limit theorem in graph theory, *Studia Sci. Math. Hungar.,* **1** (1966), 51–57.

[9]   W. Fernandez de la Vega, On the maximum density of graphs which have no subcontraction to K^s, *Discrete Math.,* **46** (1983), 109–110.

[10] H. Hadwiger, Über eine Klassifikation der Streckenkomplexe, *Vierteljahresschr. Naturforsch. Ges. Zürich,* **88** (1943), 133–142.

[11] L. Jørgensen, Contractions to complete graphs, *Annals of Discrete Maths.,* **41** (1989), 307–310. **18** (1994), 431–448.

[12] L. Jørgensen, Contractions to K_8, *J. Graph Theory,* **18** (1994), 431–448.

[13] H. A. Jung, Verallgemeinerung des n-fachen zusammenhangs für Graphen, *Math. Ann.,* **187** (1970), 95–103.

[14] K. Kawarabayashi, *Hadwiger's conjecture on girth condition,* (pre-print).

[15] A. V. Kostochka, The minimum Hadwiger number for graphs with a given mean degree of vertices (in Russian), *Metody Diskret. Analiz.*, **38** (1982), 37–58.

[16] A. V. Kostochka, A lower bound for the Hadwiger number of graphs by their average degree, *Combinatorica*, **4** (1984), 307–316.

[17] D. Kühn, Forcing a K_r minor by high external connectivity, *J. Graph Theory*, **39** (2002), 241–264.

[18] D. Kühn and D. Osthus, *Minors in graphs of large girth* (preprint).

[19] D. Kühn and D. Osthus, *Complete minors in $K_{s,s}$-free graphs* (preprint).

[20] D.G. Larman and P. Mani, On the existence of certain configurations within graphs and the 1-skeletons of polytopes, *Proc. London Math. Soc.*, **20** (1970), 144–160.

[21] M. Las Vergnas and H. Meyniel, Kempe classes and the Hadwiger conjecture, *J. Combin. Theory Ser. B*, **31** (1981), 95–104.

[22] W. Mader, Homomorphieeigenschaften und mittlere Kantendichte von Graphen, *Math. Ann.*, **174** (1967), 265–268.

[23] W. Mader, Homomorphiesätze für Graphen, *Math. Ann.*, **178** (1968), 154–168.

[24] W. Mader, Topological minors in graphs of minimum degree n, Contemporary trends in discrete mathematics (Štiřín Castle, 1997), *DIMACS Ser. Discrete Math. Theoret. Comput. Sci.*, **49**, Amer. Math. Soc., Providence, RI, (1999), 199–211.

[25] W. Mader, personal communication.

[26] J. S. Myers, Graphs without large complete minors are quasi-random, (to appear in *Combinatorics, Probability and Computing*).

[27] J. S. Myers, Extremal theory of graph minors and directed graphs, Ph.D. thesis, University of Cambridge (2002).

[28] J. S. Myers and A. Thomason, *The extremal function for non-complete minors* (in preparation).

[29] N. Robertson and P. D. Seymour, Graph Minors. V. Excluding a planar graph, *J. Combin. Theory Ser. B*, **41** (1986), 92–114.

[30] N. Robertson and P. D. Seymour, Graph Minors. XIII. The disjoint paths problem, *J. Combin. Theory Ser. B*, **63** (1995), 65–110.

[31] N. Robertson, P. Seymour and R. Thomas, Hadwiger's conjecture for K_6-free graphs, *Combinatorica*, **13** (1993), 279–361.

[32] A. Thomason, An extremal function for contractions of graphs, *Math. Proc. Cambridge Phil. Soc.*, **95** (1984), 261–265.

[33] A. Thomason, Pseudo-random graphs, in: *Random Graphs '85* (M. Karonski and Z. Palka, Eds), Annals of Discrete Math., **33** (1987), 307–331.

[34] A. Thomason, The extremal function for complete minors, *J. Combinatorial Theory Ser. B*, **81** (2001), 318–338.

[35] A. Thomason, Complete minors in pseudo-random graphs, *Random Structures and Algorithms*, **17** (2000), 26–28.

[36] C. Thomassen, Girth in graphs, *J. Combinatorial Theory Ser. B*, **35** (1983), 129–141.

[37] B. Toft, A survey of Hadwiger's conjecture, Surveys in graph theory (San Francisco, CA, 1995), *Congr. Numer.*, **115** (1996), 249–283.

[38] K. Wagner, Über eine Eigenschaft der ebenen Komplexe, *Math. Ann.*, **114** (1937), 570–590.

[39] K. Wagner, Beweis einer Abschwächung der Hadwiger-Vermutung, *Math. Ann.*, **153** (1964), 139–141.

Andrew Thomason

Department of Pure Mathematics and
Mathematical Statistics
Centre for Mathematical Science
Wilberforce Road
Cambridge CB3 0WB, U.K.

A.G.Thomason@dpmms.cam.ac.uk

BOLYAI SOCIETY
MATHEMATICAL STUDIES, 15

Conference on Finite
and Infinite Sets
Budapest, pp. 381–405.

PERIODICITY AND ALMOST-PERIODICITY

R. TIJDEMAN

Dedicated to Vera T. Sós

Periodicity and almost-periodicity are phenomena which play an important role in most branches of mathematics and in many other sciences. This is a survey paper[1] on my work in this area and on related work. I restrict myself to periodicity questions in combinatorics on words (the main dish), but I start with a periodicity problem from number theory (the entree) and at the end there is an Appendix by Imre Ruzsa containing a partial answer to one of my problems (the dessert). Sections 1–10 concern one-dimensional results and open problems. Sections 11–16 deal with multi-dimensional analogues. I do not claim completeness in any sense.

Books providing background material and additional references for this paper are Lothaire 1 [30], Lothaire 2 [31], and the Marseille book [5].

1. ENTREE

It is a problem to characterise the periodic functions $f : \mathbb{N} \to \mathbb{Z}$ such that $\sum_{n=1}^{\infty} \frac{f(n)}{n} = 0$. In his memoir [16] Dirichlet stated that every arithmetic progression in which initial term and difference have no common factor, includes infinitely many primes. The proof, which he completed few years later (cf. [15], p. 1), is based on the fact that the Dirichlet L-series is non-zero at 1. The Dirichlet series at $s = 1$ is of the form $\sum_{n=1}^{\infty} \frac{f(n)}{n}$ with f periodic modulo some positive integer q and completely multiplicative and such that

---

[1]This paper is an elaborate version of a talk given in Budapest on 10 June 2002 at a workshop sponsored by the Netherlands Organization for Scientific Research (NWO) and the Hungarian Organization for Scientific Research (OTKA).

$f(n) = 0$ if $\gcd(n, q) > 1$. A function f is called completely multiplicative if $f(mn) = f(m)f(n)$ for all positive integers m, n. The conditions under which $\sum_{n=1}^{\infty} \frac{f(n)}{n} \neq 0$ were studied by Chowla and Siegel. Baker, Birch and Wirsing [2] showed that the sum is non-zero if q is prime. Okada [38] gave a necessary and sufficient condition for the vanishing of the sum. The author [58] showed that it suffices that f is periodic and completely multiplicative so that the condition that $f(n) = 0$ if $\gcd(n, q) > 1$ can be dropped. However, the following problem (cf. [26]) still remains open:

Problem 1 (Erdős, 1965). Does there exist an $f : \mathbb{N} \to \mathbb{Z}$ with period q and $\sum_{n=1}^{\infty} \frac{f(n)}{n} = 0$ such that $f(n) = 0$ if $q|n$ and $\left|f(n)\right| = 1$ otherwise?.

The following recent result of Szabolcs Tengely shows that it is possible that the sum vanishes if $\left|f(n)\right| = 1$ for every n. This makes it more likely that the answer to Problem 1 is yes, opposite to Erdős' expectation.

Theorem 1 (Sz. Tengely). *There exists a function $f : \mathbb{N} \to \{-1, 1\}$ with period 36 such that $\sum_{n=1}^{\infty} \frac{f(n)}{n} = 0$.*

Proof. The choice $f(n) = 1, -1, -1, -1, -1, 1, 1, 1, -1, 1, -1, -1, 1, -1,$ 1, -1, -1, 1, 1, 1, -1, 1, -1, -1, 1, -1, -1, -1, -1, 1, 1, 1, 1, 1, -1, 1 for $n = 1, 2, \ldots, 36$ satisfies the conditions of [38], Theorem 10. ■

Actually Tengely showed by an exhaustive search that 36 is the smallest period for which such a solution exists.

ONE-DIMENSIONAL WORDS

2. TILINGS

Let A be a finite set of integers. The basic problem is to decide whether there exists a set $B \subset \mathbb{Z}$ such that every integer n can be written in precisely one way as $a + b$ with $a \in A, b \in B$. We write $\mathbb{Z} = A \oplus B$ and call A a tile if such a decomposition of $\mathbb{Z}$ is possible.

Suppose A is a tile and $A \oplus B = \mathbb{Z}$. Without loss of generality we assume that $0 \in A \cap B$ and that $\gcd_{a \in A} a = 1$. In [52] I proved that if the cardinality of A is n and h is an integer coprime to n then $hA \oplus B = \mathbb{Z}$.

I used it to prove that if n is prime, then every element of B is divisible by n. Actually it follows from a result of Sands [43] that if n is a prime power p^t, then every element of B is divisible by p. An example of Szabó [50] shows that this property need not hold for general n. Already in 1977 D. J. Newman [37] had given a necessary and sufficient condition for a finite set A to be a tile when the cardinality n of A is a prime power. Recently Coven and Meyerowitz [14] did so in case n has at most two prime factors. The problem for general n is still open.

It follows from the box principle that if A is a tile, then B is periodic. This principle yields that there is an upper bound for the minimal period of B which is exponential in the diameter of A. However, the best example I know has linear dependence on the diameter of A. For example, let m be some positive integer and consider the tile $A = \{0, 1, 2m, 2m + 1\}$. Then every complementary set B has to have period at least $4m$ which is about twice the diameter of A. An example of such a B is: $\{0, 2, 4, \ldots, 2m - 2\} \oplus 4m\mathbb{Z}$. The gap between upper and lower bound is huge.

Problem 2. What is the best upper bound for the period in terms of diam (A)?

During the workshop Imre Ruzsa found the upper bound $\exp\left(c\sqrt{D \log D}\right)$ where $D = \operatorname{diam}(A)$ and c is some constant. The proof of this result is given in the Appendix.

3. THE FINE AND WILF THEOREM

We consider functions $f : I \to X$ where I is $\mathbb{Z}$ or $\mathbb{N}$ or some finite block of integers and X is arbitrary. Suppose f has period q, that is $f(n+q) = f(n)$ whenever $n, n+q \in I$. We call q the (minimal) period if no smaller q has this property. Actually f is now determined by its values at a block of length q.

Now suppose f has two periods, p and q. If the cardinality $|I|$ of I is large, then f has period $\gcd(p, q)$, which implies that f has periods p and q. Fine and Wilf [20] proved in 1965 that the minimal value of $|I|$ for which this holds equals $p + q - \gcd(p, q)$. Hence if p and q are coprime, then there exists a non-constant word w of length $p+q-2$ with periods p and q. By distinguishing different residue classes mod (p, q) it is no restriction to assume that p and q are coprime.

Example 1. Consider $p = 16$, $q = 9$. According to the theorem of Fine and Wilf there exists a non-constant word w with periods 16 and 9 of length 23, but not of length 24. Because of the smaller period 9 it suffices to construct the first 14 symbols:

$$w = e\,d\,c\,b\,a\,i\,h\,g\,f\,e\,d\,c\,b\,a \mid i\,h\,g\,f\,e\,d\,c\,b\,a.$$

Moreover, the first 14 symbols should have periods 9 and 16 - 9 = 7, because $w_k = w_{k+16} = w_{k+7}$ for $k \leq 14$. Hence $h = a, i = b$. Because of the period 7 it suffices to know the first 7 symbols;

$$w = g\,f\,e\,d\,c\,b\,a \mid g\,f\,e\,d\,c\,b\,a \mid b\,a\,g\,f\,e\,d\,c\,b\,a.$$

By a similar reasoning as above we have period 2 there. Hence $g = e = c = a, f = d = b$. Thus the extremal word reads:

$$w = a\,b\,a\,b\,a\,b\,a\,a\,b\,a\,b\,a\,b\,a\,b\,a\,a\,b\,a\,b\,a\,b\,a.$$

This is a non-constant word with periods 9 and 16 indeed. Note that this procedure is closely related to the continued fraction expansion of 16/9:

$$\frac{16}{9} - 1 = \frac{7}{9}, \quad \frac{9}{7} - 1 = \frac{2}{7}, \quad \frac{7}{2} - 3 = \frac{1}{2}.$$

4. BI-SPECIAL WORDS

The extreme Fine and Wilf words have several nice properties which make them occur in various contexts. This explains that they have various names: PER-sequences, Hedlund words, bi-special words. In the sequel we shall use the latter expression which is nowadays the most common name. We state some properties (cf. [32], [54], [24] Ch. 2).

1) If the places of the letters are numbered, then the places where the a's occur form a Beatty sequence ($\lfloor \alpha n + \beta \rfloor$). Of course, the same is true for the places where the b's occur. The value of α can be chosen by using the euclidean algorithm to solve $px - qy = \pm 1$ with $0 < y < p$, $0 < x < q$ and then taking $\frac{q}{x}$ or $\frac{p}{y}$ or any number in between. If the solutions are (x_1, y_1) and (x_2, y_2), then the number of a's in the extreme word equals $x_2 + y_2 - 1$ and the number of b's $x_1 + y_1 - 1$. In Example 1 we have

equation $16x - 9y = 1$ with solution $x_1 = 4, y_1 = 7$. The places of the b's are given by $\lfloor \frac{9i}{4} \rfloor = \lfloor \frac{16i-1}{7} \rfloor$ for $i = 1, \ldots, 10$. Similarly $16x - 9y = -1$ has solution $x_2 = 5, y_2 = 9$ and the places of the a's are given by $\lfloor \frac{9i-1}{5} \rfloor = \lfloor \frac{16i}{9} \rfloor$ for $i = 1, \ldots, 13$. In the extreme word there are $5 + 9 - 1 = 13$ a's and $4 + 7 - 1 = 10$ b's.

2) The extreme words are balanced. A word consisting of the letters a and b is called balanced if the number of occurrences of a in any two subwords of equal lengths differs by at most 1. It is called left-special if the word is still balanced both when the word is extended on the left side by an a and when it is extended on the left side by a b. Right-special is defined analogously. If the word is both left-special and right-special, it is called bi-special. The extreme Fine and Wilf words are bi-special and every bi-special word is an extreme Fine and Wilf word for some coprime integers p and q. The numbers p and q can be computed from the numbers of a's and b's in the bi-special word as follows: Let $m - 1$ be the number of a's and $n - 1$ the number of b's in the bi-special word. Solve the equation $mx - ny = \pm 1$ in integers x, y with $0 < x < n, 0 < y < m$. Let (x_1, y_1) and (x_2, y_2) be the solutions. Then the periods are $x_1 + y_1$ and $x_2 + y_2$. In Example 1 we have $m = 14, n = 11$. The equation $14x - 11y = \pm 1$ admits the solutions $(4, 5)$ and $(7, 9)$. This yields the periods 9 and 16.

3) Bi-special words are palindromes. It does not matter whether you read them from left to right or from right to left.

4) [39]) If you extend a bi-special word on the right by ab and compute the corresponding p and q as in 2), then the first $p - 2$ letters form a palindrome as well as the last $q + 2$ letters. Only bi-special words have this property. In Example 1 we have palindromes $a\ b\ a\ b\ a\ b\ a\ a\ b\ a\ b\ a\ b\ a$ and $b\ a\ a\ b\ a\ b\ a\ b\ a\ a\ b$ of lengths $p - 2 = 14$ and $q + 2 = 11$, respectively. Of course, (a, p) and (b, q) can be interchanged: if we extend by ba on the right then we can split into the palindromes $a\ b\ a\ b\ a\ b\ a$ and $a\ b\ a\ b\ a\ b\ a\ b\ a\ a\ b\ a\ b\ a\ b\ a\ b\ a$ of lengths $q - 2 = 7$ and $p + 2 = 16$, respectively.

5. Balanced words

Consider a word $f : \mathbb{Z} \to \{a, b\}$. As mentioned above f is called balanced if for each two finite subwords of equal lengths the numbers of occurring

a's differ by at most 1. By the work of Morse and Hedlund [34], [35] and of Coven and Hedlund [13] the balanced words are completely classified (cf. [54]). It suffices to classify all balanced $\mathbb{Z}$-words, since every subword of a balanced word is obviously balanced and, conversely, every balanced I-word can be extended to a balanced $\mathbb{Z}$-word (see e.g. [24] Theorem 2.3). The word f is given by the sequence of the places where an a is read. There are three classes of balanced words:

(a) (periodic case)
The places form a Beatty sequence $\left(\lfloor \alpha n + \beta \rfloor \right)_{n \in \mathbb{Z}, \; \alpha n + \beta \in I}$ with $\alpha \in \mathbb{Q}_{>1}$, $\beta \in \mathbb{R}$;

(b) (irrational case)
The places form a sturmian sequence $\left(\lfloor \alpha n + \beta \rfloor \right)_{n \in \mathbb{Z}, \alpha n + \beta \in I}$ or $\left(\lceil \alpha n + \beta \rceil \right)_{n \in \mathbb{Z}, \alpha n + \beta \in I}$ with $\alpha \in \mathbb{R}_{>1} \setminus \mathbb{Q}$, $\beta \in \mathbb{R}$;

(c) (skew case)
The places form a periodic Beatty sequence apart from one irregularity. Skew words correspond two-to-one to bi-special words. The latter words are by definition words of length $p + q - 2$ with two coprime periods p, q. Skew words are obtained by extending a bi-special word on one side with $a\,b$, on the other side with $b\,a$, and subsequently extending into both directions with period $p + q$.

Example 2. Let $p = 4$ and $q = 7$. Then the corresponding bi-special word is isomorphic to

$$a\ a\ b\ a\ a\ a\ b\ a\ a.$$

Therefore the corresponding skew $\mathbb{Z}$-words are isomorphic to

$$\ldots b\ a\ a\ a\ b\ a\ a\ a\ b\ a\ a\ b\ a\ a\ a\ b\ a\ a\ a\ b\ a\ a\ a\ b\ a\ a\ b\ a\ a\ a\ b\ a\ a\ a\ b\ldots$$

and

$$\ldots a\ b\ a\ a\ b\ a\ a\ a\ b\ a\ a\ a\ b\ a\ a\ b\ a\ a\ a\ b\ a\ a\ b\ a\ a\ a\ b\ a\ a\ a\ b\ a\ a\ b\ a\ldots .$$

In the former word there is a triple $a\ a\ a$ too many, in the latter word such a triple is missing, when comparing the words with the periodic word with periodic part $a\ a\ a\ b\ a\ a\ a\ b\ a\ a\ b$.

A classical way to construct balanced sequences is by approximating a line in the plane as well as possible by a discrete line. The principle of the so-called cutting line is as follows (cf. [45]). Let the line $y = \alpha x + \beta$ in the x-y-plane be given where $\alpha \in \mathbb{R}_{>0}$ and $\beta \in \mathbb{R}$. Start at some integer point (x_0, y_0) under the line, but at distance less than 1 from the line. If $y_0 + 1 \leq \alpha x_0 + \beta$ then move to $(x_0, y_0 + 1)$ and write an a, otherwise move to the point $(x_0 + 1, y_0)$ and write a b. Iterate the procedure. Of course, the word can also be extended into the negative direction so forming a $\mathbb{Z}$-word. If α is rational, then this yields a word from class (a). If α is irrational, then we obtain a word from class (b) with the $\lfloor \ \rfloor$ brackets. If we require $y_0 + 1 < \alpha x_0 + \beta$ instead, then we get a word from class (b) with the $\lceil \ \rceil$ brackets. The skew case is the case where α is rational and the line passes through an integer point where on the one hand of that point the strict inequality criterion is used and on the other hand the $\leq$-criterion.

6. FRAENKEL WORDS

It is obvious that if a word of a's and b's is balanced with respect to a, then it is also balanced with respect to b. We call a general word $f : I \to A$ balanced if the word is balanced with respect to each letter from A. All balanced words on two letters a, b have been classified in the previous section. We have seen that the letter a always has some density and that every density in $[0, 1]$ can occur. Hence there are uncountably many balanced words on two letters.

What are the balanced words $f : \mathbb{Z} \to A$ when $|f(\mathbb{Z})| > 2$? Obviously each letter has again a density. If the densities of two letters are equal, then they can first be identified as one letter with double density, and then the latter letter can be replaced alternately by the first and second letter. It is therefore a crucial question to determine the balanced words the letters of which have distinct densities, so-called Fraenkel words. The following conjecture of Fraenkel suggests that Fraenkel words are very rare.

Problem 3 ([21], [18] p. 19). Prove that for $n = 3, 4, \ldots$ the only balanced word on n letters having distinct densities is isomorphic to the periodic word with periodic part F_n inductively defined by

$$F_1 = 1, \qquad F_n = F_{n-1} \, n \, F_{n-1} \quad \text{for} \quad n = 2, 3, \ldots .$$

So for $n = 3$ we find the periodic part 1213121 and for $n = 4$ the periodic part 121312141213121. Graham [23] showed that the densities of a Fraenkel word have to be rational. The Fraenkel conjecture has been proved in case $n = 3$ by Morikawa [33], cf. [53], in case $n = 4$ by Altman, Gaujal, and

Hordijk [1], and in case $n = 5, 6$ by the author [56]. Related results have been obtained by a.o. Fraenkel, Morikawa, and Simpson, cf. [57].

Recently Fagnot and Vuillon [19] have studied balancedness with respect to subwords of given length instead of letters.

7. STIFF WORDS

Another way to measure the regularity of a word is to compute its complexity. For $n = 1, 2, \ldots$ the complexity $P(n)$ of a word is defined as the number of distinct subwords of length n. In our notation we suppress the dependence of P on the word.

Example 3. We apply the substitution $a \to ab$, $b \to a$ starting with an a:

a

$a\ b$

$a\ b\ a$

$a\ b\ a\ a\ b$

$a\ b\ a\ a\ b\ a\ b\ a$

This leads to the limit word

$a\ b\ a\ a\ b\ a\ b\ a\ a\ b\ a\ a\ b\ a\ b\ a\ a\ b\ a\ b\ a\ a\ b\ a\ a\ b\ a\ b\ a\ a\ b\ a\ a\ b\ \ldots$.

This is the famous Fibonacci word. The density of the a's is successively 1, $1/2$, $2/3$, $3/5$, $5/8$ with the limit value $(\sqrt{5} - 1)/2$. Hence the word is not periodic. In fact it is a sturmian word (class (b)). For the complexity we find: $P(2) = 3, P(3) = 4, \ldots, P(n) = n + 1, \ldots$. Coven and Hedlund [13] proved:

if $P(n) \le n$ for some n, then the word is ultimately periodic.

So the Fibonacci word is a non-periodic word with minimal complexity.

We call a word stiff if $P(n) \le n + 1$ for all n. Thus the Fibonacci word is stiff. In fact all balanced words are stiff. Again we only have to study stiff $\mathbb{Z}$-words, since on the one hand it is obvious that a subword of a stiff word is stiff, and on the other hand it is true that every stiff I-word can be extended to a stiff $\mathbb{Z}$-word ([24], Theorem 2.4). Apart from the balanced and therefore stiff words, which have been classified in Section 5, there is one class of unbalanced stiff $\mathbb{Z}$-words (cf. [54], [24] Theorem 2.6):

(d) (Hedlund case)

Extreme Fine and Wilf words which extend into one direction with the one period and into the other direction with the other. So again we start with a bi-special word f of length $p + q - 2$ where p and q are coprime and f has periods p and q. Then we extend into the positive and the negative direction with the different periods.

Example 4. Let again $p = 4$ and $q = 7$. Then the corresponding bi-special word is isomorphic to

$$a\ a\ b\ a\ a\ a\ b\ a\ a.$$

Therefore the corresponding Hedlund words are isomorphic with

$$\ldots a\ a\ a\ b\ a\ a\ a\ b\ a\ a\ a\ b\ a\ a\ a\ b\ a\ a\ a\ b\ a\ a\ b\ a\ a\ a\ b\ a\ a\ b\ a\ a\ a\ b\ a\ \ldots$$

and its reversed word. A Hedlund word has a left density of a's which is different from the right density of the a's. Therefore the word cannot be balanced.

Following a suggestion of Jean-Paul Allouche I shall call a word f repetitive (instead of recurrent) if every subword w occurs infinitely often and uniformly repetitive (instead of uniformly recurrent or almost periodic) if for every subword there exists a number C such that every subword of length C contains w as a subword, i.e. the "distance" between occurrences of any subword w is bounded. It is easy to check that in the given classification the classes (a) and (b) contain only uniformly repetitive words, but that the words from classes (c) and (d) are not even repetitive, since they contain only one copy of the bi-special word we started with.

Many stiff words are ultimately periodic. By the mentioned theorem of Coven and Hedlund non-periodic stiff words have to satisfy $P(n) = n + 1$ for all n. The class (b) is the only class of stiff words which is not ultimately periodic. The classes (c) and (d) are ultimately periodic, but nevertheless they satisfy $P(n) = n + 1$ for all n. If we restrict our attention to $\mathbb{N}$-words, then class (b) is the only class with $P(n) = n + 1$ for all n.

8. THREE DISTANCES THEOREMS

The following construction method for sturmian words is very useful for the study of the structure of such words. Let $\alpha > 1$ and β be real numbers.

Denote by $\{x\}$ the fractional part $x - \lfloor x \rfloor$ of x. For $n \in \mathbb{Z}$ write an a at place n if $\{n\alpha + \beta\} < \alpha$ and otherwise b. This yields a $\mathbb{Z}$-word f which is sturmian if α is irrational and a periodic balanced word if α is rational. From the definition it is clear that the letter at place n equals a or b according to whether $\lfloor n\alpha + \beta \rfloor - \lfloor (n-1)\alpha + \beta \rfloor$ equals 1 or 0.

Suppose α is irrational. An easy argument shows that the complexity of f equals $P(n) = n + 1$ for every n: Consider $\{0\}, \{\alpha\}, \ldots, \{n\alpha\}$. These $n+1$ points split the torus $[0, 1]$, where the points 0 and 1 are identified, into $n + 1$ half-open intervals. Observe that the sequence $f_m, f_{m+1}, \ldots, f_{m+n-1}$ of letters of f at places $m, m + 1, \ldots, m + n - 1$ is completely determined by to which half-open interval $m\alpha + \beta$ belongs. Hence there are at most $n + 1$ distinct subwords of length n. It follows from the theorem of Coven and Hedlund that $P(n) > n$, since otherwise f would be ultimately periodic which is impossible in view of the irrational density.

A remarkable fact occurs when one studies the lengths of the half-open intervals. As Sós [47] and also others observed at most three distinct lengths occur, one being the sum of the two others.

The relevance of this fact for the structure of Sturmian words becomes clear from the following result. Let n be a positive integer and $f : \mathbb{Z} \to \{a, b\}$ a sturmian word. Consider the $n + 1$ distinct subwords of length n of f. Compute the densities of a in these words. Then there are at most three distinct densities, one being the sum of the others. Berthé [4] has given explicit expressions for the occurring frequencies and the cardinality of each frequency.

9. LINEAR COMPLEXITY WORDS

In fact the three densities result in the last paragraph of the previous section is a special case of the following result of Boshernitzan [9]: Let $f : \mathbb{Z} \to \{a, b\}$ be a repetitive word of complexity P. Then the densities of the subwords of length n attain at most $3\big(P(n + 1) - P(n)\big)$ values.

Many papers have been written on non-periodic words having linear complexity. Words having complexity function $P(n) = n + k$ for some constant k were already studied by Coven [12] in 1975. Such words are said to be of minimal block growth and the minimal k is called the stiffness of the word. Coven characterised the structure of non-repetitive $\mathbb{Z}$-words of minimal block growth. Heinis [24], Ch. 3, has given explicit formulas for the stiffness of words of minimal block growth.

Let t be any positive integer. It is easy to construct a word $f : \mathbb{Z} \to \{a, b\}$ such that f has complexity $P(n) = tn + 1$ for all n. Hence for every positive integer t there exists a word f the complexity of which satisfies $\lim_{n \to \infty} \frac{P(n)}{n} = t$. It was a surprise when Heinis [24] showed that there are no words f such that its complexity P satisfies $\lim_{n \to \infty} \frac{P(n)}{n} \in (1, 2)$. He also showed that if $\liminf_{n \to \infty} \frac{P(n)}{n} = \frac{3}{2}$ then $\limsup_{n \to \infty} \frac{P(n)}{n} \geq \frac{5}{3}$ and that the value $\frac{5}{3}$ is optimal. It is an open question whether there are words for which the limit exists and attains a non-integral value:

Problem 4. Is the limit value $\lim_{n \to \infty} \frac{P(n)}{n}$, if it exists, necessarily an integer?

The definition of complexity can be extended to words $f : \mathbb{Z} \to A$ where $A = \{1, 2, \ldots, q\}$ is an alphabet on $q > 2$ letters. For every positive integer $q > 1$ there are words on q letters with complexity function $P(n) = (q - 1)n + 1$ for every n. For $q = 2$ this is the case for sturmian sequences, for $q = 3$ for example for the so-called Arnoux–Rauzy sequences. More precisely, for every pair of positive integers q, t with $t \leq q$ there exist words on q letters such that the irrationality degree of the densities of the q letters is t and the complexity of the word equals $P(n) = (n - 1)(t - 1) + q$ for every n. Such a word can be constructed in a similar way as sketched at the beginning of the previous section. The following result shows that the mentioned complexity is minimal [55]. If $f : \mathbb{Z} \to A$ is a word such that it contains q distinct letters and the irrationality degree of the densities of the letters equals t, then

$$P(n) \geq (n - 1)(t - 1) + q$$

for every n. For $q = t = 2$ this reduces to the theorem of Coven and Hedlund.

Other notions to measure complexity have also been proposed. Let $P_\infty(n)$ denote the number of distinct subwords of length n which occur infinitely often in the given word f. Both Nakashima, Tamura and Yasutomi [36] and Heinis and Tijdeman [25] have studied this asymptotic complexity. The former authors gave characterizations of both $\mathbb{N}$-words and $\mathbb{Z}$-words having small P_∞-complexity. They showed that in the case of $\mathbb{N}$-words there is no difference between "asymptotically balanced" and "asymptotically stiff". The latter authors characterized all $\mathbb{N}$-words which are asymptotically stiff, that is, satisfy $P_\infty \leq n + 1$ for every n:

(i) *The* $\mathbb{N}$-*word* $f = f_1 f_2 \ldots$ *with rational density* α *is asymptotically stiff if and only if there exists an ultimately monotonic sequence* $\{g_n\}_{n=1}^{\infty}$ *with* $g_{n+1} - g_n \to 0$ *as* $n \to \infty$ *such that* $f_n = \lfloor n\alpha^{-1} + g_n \rfloor$ *for* $n \in \mathbb{N}$.

(ii) *The* $\mathbb{N}$-*word* f *with irrational density* α *is asymptotically stiff if and only if there exists a sequence* $\{g_n\}_{n=1}^{\infty}$ *with* $g_{n+1} - g_n \to 0$ *as* $n \to \infty$ *such that* $f_n = \lfloor n\alpha^{-1} + g_n \rfloor$ *for* $n \in \mathbb{N}$.

Kamae and Zamboni [28] studied complexity not referring to a block of places, but to a fixed pattern of places. Let $f : \mathbb{N} \to \{0, 1\}$ be a word. Let $P^*(k_1, k_2, \ldots, k_n)$ denote the number of distinct vectors

$$\big(f(m + k_1), f(m + k_2), \ldots, f(m + k_n) \big)$$

for $m \in \mathbb{N}$ and define the pattern complexity as

$$P^*(n) = \sup\nolimits_{k_1, \ldots, k_n} P^*(k_1, \ldots k_n).$$

Kamae and Zamboni showed that if $P^*(n) < 2n$ for some n, then the word is ultimately periodic, that every sturmian word satisfies $P^*(n) = 2n$ for every n, but that there exist non-sturmian words with $P^*(n) = 2n$ for every n. Again sturmian words are minimal non-periodic words, but the complete set of non-periodic words which are minimal in this sense is not yet known.

Problem 5 (Kamae).   Characterize all words having $P^*(n) = 2n$ for every n.

10. Fine and Wilf words for several periods

Let $p_1, \ldots, p_r$ be positive integers. Let $w = w_1 \ldots w_n$ be a word with periods $p_1, \ldots, p_r$. This means that $w_{i+p} = w_i$ for $i = 1, \ldots, n - p$ and $p \in \{p_1, \ldots, p_r\}$. Suppose that w does not have period $\gcd(p_1, \ldots, p_r)$. The case $r = 2$ has been treated in Section 3. In 1999 Castelli, Mignosi and Restivo [11] studied the case $r = 3$. They defined some function $h(x, y, z)$ such that if a word f has periods p_1, p_2, p_3 and length $\geq h(p_1, p_2, p_3)$, then f has period $\gcd(p_1, p_2, p_3)$. They further showed that under suitable conditions their bound $h(x, y, z)$ is the best possible. They also showed that the set of subwords of these maximal words coincides with the set of factors of the Arnoux–Rauzy sequences. Justin [27] generalized the results of Castelli, Mignosi and Restivo to words with more than three periods.

Notice that the periods $p_1, \ldots, p_r$ can only induce relations between letters at places i and j when i and j are in the same residue class modulo $\gcd(p_1, \ldots, p_r)$. It is therefore no restriction to consider such a residue class. Hence we may assume without loss of generality that $\gcd(p_1, \ldots, p_r) = 1$. We shall do so in the sequel. If the maximal length of w under the gcd-condition is n, then the maximal length in the general case equals $(n+1)\gcd(p_1, \ldots, p_r) - 1$.

Tijdeman and Zamboni [59] have developed an algorithm to compute the extreme n and w for any given periods $p_1, \ldots, p_r$ subject to $\gcd(p_1, \ldots, p_r) = 1$. Here we illustrate the algorithm by an example. Starting from the six periods 127, 189, 222, 235, 243, 248 the method reveals that the non-constant word of maximal length having these periods has length $m = 254$, that an extreme word can have at most three distinct letters and in that case has to be isomorphic with the constructed word. In Table 1 each time the smallest positive period is underlined and subtracted from the others. Its index is written in the column $g[k]$ and the subtracted number is added in the column $m[k]$. The procedure is continued until in all columns with an underlined number the value is at most 1. The last found value of $m[k]$ is the sought maximal length m. Subsequently the column for $n[k]$ is filled by computing $n[k] := m - m[k]$ for every k. In Example 5 we have $m = m[9] = n[0] = 254$.

Example 5. We construct the extremal word for periods

$$p_1 = 127, \quad p_2 = 189, \quad p_3 = 222, \quad p_4 = 235, \quad p_5 = 243, \quad p_6 = 248.$$

$p_1[k]$	$p_2[k]$	$p_3[k]$	$p_4[k]$	$p_5[k]$	$p_6[k]$	k	$g[k]$	$m[k]$	$n[k]$
127	189	222	235	243	248	0		0	254
127	62	95	108	116	121	1	1	127	127
65	62	33	46	54	59	2	2	189	65
32	29	33	13	21	26	3	3	222	32
19	16	20	13	8	13	4	4	235	19
11	8	12	5	8	5	5	5	243	11
6	3	7	5	3	0	6	4	248	6
3	3	4	2	0	-3	7	2	251	3
1	1	2	2	-2	-5	8	4	253	1
1	0	1	1	-3	-6	9	1	254	0

Table 1: Computation of the maximal length

The extreme word is found as follows. Write down the column number of the lowest underlined number. In each next step compare the underlined

number l one row higher with the number of letters in the already constructed word $f[k]$. If the latter number is smaller, then write the column
number of the underlined number l as next letter of the word and repeat
$f[k]$. Otherwise repeat the last l letters of $f[k]$. By | we indicate the stage
reached after each step; the symbols | are not part of the word. The number
of added letters at level k equals $n[k] - n[k+1]$. Therefore the number of
letters of $f[k]$ equals $n[k]$ for $k = K - 1, K - 2, \ldots, 0$, respectively. Thus
the resulting extremal word has length $m = n[0]$.

$$1|41|141|41141|14141141|4114114141141|3$$
$$1\ 41\ 141\ 41141\ 14141141\ 4114114141141|$$
$$141\ 41141\ 14141141\ 4114114141141\ 3$$
$$1\ 41\ 141\ 41141\ 14141141\ 4114114141141|$$
$$1\ 41\ 141\ 41141\ 14141141\ 4114114141141\ 3$$
$$1\ 41\ 141\ 41141\ 14141141\ 4114114141141$$
$$141\ 41141\ 14141141\ 4114114141141\ 3$$
$$1\ 41\ 141\ 41141\ 14141141\ 4114114141141|$$

Extreme word expressed in letters 1, 3, 4

In [59] it is shown that the word found by the algorithm is indeed
the non-constant word of maximal length and among such words with a
maximum number of distinct letters. Furthermore, it is proved that the
extreme word is a palindrome and unique apart from isomorphy. The
case that the number of letters in the extreme word equals the number
of periods is precisely the case in which Castelli, Mignosi and Restivo in
case of three periods and Justin in case of more than three periods proved
that their bounds are the best possible ones. It is interesting that the
multi-dimensional continued fraction corresponding with the algorithm also
occurs in the study of ergodic properties of a dynamical system arising from
percolation theory.

MULTIDIMENSIONAL WORDS

From now on we consider multi-dimensional words $f : \mathbb{Z}^k \to Q$ where we
usually will have $k = 2$, $Q = \{0, 1\}$.

11. TILINGS

Let A be a finite set in $\mathbb{Z}^2$. The problem is to decide whether for a given set A there exists a set B such that $A \oplus B = \mathbb{Z}^k$. There are many results about this problem most of which require that A has some connectedness or convexity property. A rather general result is due to Beauquier and Nivat [3] who characterized all tilings in case A is a polyomino and proved that in that case B is periodic. We call a set B in $\mathbb{Z}^k$ periodic if there exists a $\vec{v} \neq 0$ such that $B + \vec{v} = B$. I state two open problems.

Problem 6 (cf. Lagarias and Wang [29]). Is it true that for every finite set $A \subset \mathbb{Z}^k$ it is possible to determine whether there is a set B such that $A \oplus B = \mathbb{Z}^k$ in time bounded in terms of diam (A)?

Problem 7 (Periodic Tiling Conjecture [29]). Is it true that for every tile $A \subset \mathbb{Z}^k$ there exists a periodic set B such that $A \oplus B = \mathbb{Z}^k$?

The answer to both problems is yes if $k = 1$, as mentioned in Section 2. Szegedy [51] has provided algorithms for Problem 6 if $|A|$ equals 4 or is a prime number. In these cases the Periodic Tiling Conjecture holds true.

When more than one distinct tile may be used, then non-perodic sets B are possible. This leads to the theory of quasi-crystals.

12. BALANCEDNESS

We call a word $f : \mathbb{Z}^k \to \{a, b\}$ balanced if for any two k-dimensional finite blocks with hyperfaces parallel to the axes and of the same shape, the number of symbols a occurring in them differs by at most 1. Berthé and Tijdeman [6] have given a complete characterisation of balanced words. They prove that for each $k > 1$ there are only finitely many isomorphy classes of balanced words and that all of them with density α of a not equal to 0 or 1 are fully periodic, that is, have k linearly independent period vectors. In fact, α has to be rational with denominator in $\{1, 2, 3, 5\}$ where the 5 can only occur when $k = 2$.

From this result it follows that f has to be unbalanced if the letter a has irrational density in f. It is not clear how unbalanced such a word has to be.

Problem 8. Consider $f : \mathbb{Z}^2 \to \{a, b\}$ such that the density of a's tends to some irrational number θ when considering blocks $[-m_1, m_2) \times [-n_1, n_2)$ with $m_1, m_2, n_1, n_2 \to \infty$ in any way. What is the minimal measure of unbalancedness as a function of $m_1 + m_2$ and $n_1 + n_2$?

13. Complexity

Let $f : \mathbb{Z}^2 \to \{a, b\}$. Define the complexity $P(m, n)$ of f to be the number of distinct patterns of $f(x, y)$ of size $m \times n$, that is, the number of distinct arrays $\big(f(x, y) \big)_{k \leq x < k+m, \, l \leq y < l+n}$ for $k, l \in \mathbb{Z}$. The following problem is still open:

Problem 9 (Nivat–Vuillon). Suppose $P(m, n) \leq mn$ for some $m, n \in \mathbb{N}$. Does it follow that f is periodic?

The answer in case $m = 1$ is yes because of the theorem of Coven and Hedlund. Sander and Tijdeman [42] showed that the answer is also yes in case $m = 2$. For $m > 2$ the answer is open. Epifanio, Koskas and Mignosi [17] proved the slightly weaker result that if $P(m, n) \leq \frac{mn}{100}$ for some m, n, then f is periodic. Sander (cf. [41] Example 5) gave a simple example that the corresponding question has a negative answer when $k > 2$:

Example 6. Let $k > 2$ and let $m_i > 1$ for $1 \leq i \leq k$ be given integers. Let

$$\vec{a} = \big\{ (a_1, \ldots, a_k) \in \mathbb{Z}^k : 0 \leq a_i < m_i \quad (1 \leq i \leq k) \big\}.$$

Hence A is a k-dimensional block with volume $|A| = M := \prod_{i=1}^{k} m_i$. Define $f : \mathbb{Z}^k \to \{a, b\}$ by setting for $\vec{x} = (x_1, \ldots, x_k)$,

$$f(\vec{x}) := 1 \iff x_1 = x_2 \cdots = x_{k-1} = 0 \quad \text{or} \quad x_2 = m_2, \, x_3 = \cdots = x_k = 0.$$

It is easy to see that

$$\big| P(A) \big| = \frac{M}{m_k} + \frac{M}{m_1} + 1.$$

Consequently we have $\big| P(A) \big| \leq |A|$ for $m_k > 2$. But apparently f is not periodic.

Sander and Tijdeman [41] also studied the multi-dimensional analogue of pattern complexity. Cassaigne [10] characterised all two-dimensional words with complexity $P(m, n) = mn + 1$. This can be compared with the classification of stiff words in the one-dimensional case.

14. Fine and Wilf words

In this section some results are mentioned which generalise those mentioned in Section 3. Here we consider functions $f : V \to \{a, b\}$ where $V \subset \mathbb{Z}^k$ has period vectors $\vec{v}_0, \ldots, \vec{v}_k$. Whereas in Section 3 we assumed without loss of generality that the periods were coprime, we assume in this section that the period vectors generate the full lattice $\mathbb{Z}^k$.

Initially only periodicity lemmas were obtained. A periodicity lemma is a statement that a function f defined on the integer points in some region and having certain period vectors has to be constant, without indicating how far the region can be reduced without affecting the conclusion. Papers by Amir and Benson, Galil and Park and recently Mignosi, Restivo and Silva provided periodicity lemmas for parallelograms and similar domains in $\mathbb{R}^2$. Regnier and Rostami provided a framework for the study of periodicity lemmas in case of multi-dimensional patterns. Simpson and Tijdeman [46] obtained the following periodicity lemma for arbitrary dimension:

Suppose $\vec{v}_1, \ldots, \vec{v}_k \in \mathbb{Z}^k$ generate $\mathbb{Z}^k$ and $\vec{v}_0 \in \mathbb{Z}^k$ given by

$$\vec{v}_0 = \mu_1 \vec{v}_1 + \cdots + \mu_k \vec{v}_k \quad with \quad \mu_i > 0 \quad for \quad i = 1, \ldots, k$$

is an integer point. Put

$$V^* = \{\lambda_1 \vec{v}_1 + \cdots + \lambda_k \vec{v}_k \in \mathbb{Z}^k : 0 \leq \lambda_i < l_i\}$$

where $l_i \geq 1 + \mu_i$ for $i = 1, \ldots, k$. Let f be periodic modulo $\vec{v}_0, \ldots, \vec{v}_k$ on V^*. Then f is constant on V^*.

It is an obvious question to ask how much V^* can be reduced without affecting the conclusion, in other words, what the k-dimensional generalisation of the Fine and Wilf theorem is. Giancarlo and Mignosi [22] gave a multi-dimensional generalisation of this theorem for connected subsets of Cayley graphs. Simpson and Tijdeman [46] gave a generalisation of the Fine and Wilf theorem of the following type. We use the notation as above. Define

$$W = \{\lambda_0 \vec{v}_0 + \lambda_1 \vec{v}_1 + \cdots + \lambda_k \vec{v}_k : 0 \leq \lambda_i \leq 1, \ \lambda_i \in \mathbb{R} \text{ for } i = 0, \ldots, k\}.$$

Then some explicitly given set $V \subset W \cap \mathbb{Z}^k$ with cardinality equal to the Lebesgue measure of W has the property that if $f : V \to \{a, b\}$ has period vectors $\vec{v}_0, \ldots, \vec{v}_k$ generating $\mathbb{Z}^k$, then f is constant. The assertion remains true if one point of V is removed, but it is no longer true if two points of V are removed which do not differ by some $\vec{v}_i$. For $k = 1$ the result coincides with the Fine and Wilf theorem.

Problem 10. Let $\vec{v}_1, \ldots, \vec{v}_k \in \mathbb{Z}^k$ be given. How small in size can $V \subset \mathbb{Z}^k$ be if only constant functions $f : V \to \{a, b\}$ admit periods $\vec{v}_1, \ldots, \vec{v}_k$ on V?

15. FROBENIUS' LINEAR DIOPHANTINE PROBLEM

In Frobenius' classical Linear Diophantine Problem, also known as the Postage Stamp Problem and as the Coin-changing Problem, we are given positive integers $a_0, \ldots, a_k$ with greatest common divisor 1, and asked to find the least integer n such that every integer greater than n can be written as a sum of non-negative multiples of $a_0, \ldots, a_k$. In the case $k = 1$ the answer $n = n_0 := a_0 a_1 - a_0 - a_1$ is due to Sylvester [49]. The case $k = 2$ has been settled by Selmer and Beyer [44], see also Rödseth [40]. For $k > 2$ the answer is only known in special cases and various estimates exist for the general case.

Suppose $\vec{v}_0, \ldots, \vec{v}_k$ defined as above generate $\mathbb{Z}^k$ and have the property that $\vec{0}$ cannot be written as a non-trivial non-negative linear combination of $\vec{v}_0, \ldots, \vec{v}_k$. In other words, the period vectors $\vec{v}_0, \ldots, \vec{v}_k$ are on the same side of some hyperplane. Let d_0 be the smallest positive integer for which positive integers $d_1, \ldots, d_k$ exist with

$$d_0 \vec{v}_0 = d_1 \vec{v}_1 + \cdots + d_k \vec{v}_k.$$

As an application of the results mentioned in the previous section Simpson and Tijdeman [46] derived the following generalisation of Sylvester's formula in case of $k + 1$ vectors in $\mathbb{Z}^k$:

Put $\vec{w} = d_0 \vec{v}_0 - (\vec{v}_0 + \cdots + \vec{v}_k)$. Every point in

$$X := \{s_1 \vec{v}_1 + \cdots + s_k \vec{v}_k + \vec{w} : s_1 > 0, \ldots, s_k > 0\} \cap \mathbb{Z}^k$$

can be written as $\lambda_0 \vec{v}_0 + \cdots + \lambda_k \vec{v}_k$ where $\lambda_0, \ldots, \lambda_k$ are non-negative integers, but an integer point of the form $s_1 \vec{v}_1 + \cdots + s_k \vec{v}_k + \vec{w}$ with $s_1 \geq 0, \ldots, s_k \geq 0$, $s_1 s_2 \cdots s_k = 0$ can be written in this way unless and only unless $s_1, \ldots, s_k \in \mathbb{Z}$.

For $k = 1$ the obtained value is the one due to Sylvester. In general X is an open sector with corner point $\vec{w}$ the shape of which is determined by the outer vectors $\vec{v_1}, \ldots, \vec{v_k}$.

16. BV-WORDS

By approximating a plane in $\mathbb{Z}^3$ by a discrete plane as well as possible in a similar way as explained at the end of Section 5 for a discrete line, one obtains a two-dimensional analogue of Sturmian sequences. This kind of words have been studied by Berthé and Vuillon [60], [7], [8], and are often called BV-words. BV-words are words $f : \mathbb{Z}^2 \to \{a, b, c\}$ such that the densities of the letters a, b, c are linearly independent over $\mathbb{Q}$. Vuillon [60] proved that the complexity of the BV-words satisfy $P(m, n) = mn + m + n$ for every m, n. He also derived a formula for the complexity on a triangle in place of a rectangle. Berthé and Vuillon [7] characterised the doubly uniformly repetitive words $f : \mathbb{Z}^2 \to \{a, b\}$ with complexity $P(m, n) = mn + n$ for all large m. These words are obtained by identifying two letters in a BV-word. They further showed that these words f are the only non-periodic words from $\mathbb{Z}^2$ to $\{a, b\}$ such that the density of a is irrational and the complexity satisfies $P(m, 2) \leq 2m + 2$. They also classified all words having complexity $P(m, n) = m + n$. One of the results of [8] deals with the distribution of the frequencies of the various subwords of size $m \times n$ and is in the same vein as the three frequencies theorem in Section 8.

Problem 11. Let $f : \mathbb{Z}^k \to \{1, 2, \ldots, k\}$ such that the densities of the letters exist and are linearly independent over $\mathbb{Z}$. Compute lower bounds for the complexity of f.

From the beginning on, in 1772 by Bernoulli, the study of sturmian sequences has been closely related with the theory of continued fractions. In fact, the continued fraction expansion of a number α provides a recipe to construct sturmian words in which the letter a has density α, as a limit of finite words. See e.g. the survey paper of Stolarsky [48]. In a similar way BV-words can be constructed as a limit of finite two-dimensional words where each finite word is defined on the fundamental domain of a lattice and where the sequence of lattices has to do with multi-dimensional continued fractions. The construction, which can be given for any dimension, is subject of a paper by Berthé and Tijdeman. (See added in Proof.)

APPENDIX BY I. Z. RUZSA

Proof of the result stated at the end of Section 2.
Let $A \subset \mathbb{Z}$ be a tile. Then there is a $B \subset \mathbb{Z}$ such that $A \oplus B = \mathbb{Z}$, that is, every integer is represented exactly once as a sum of an element of A and a element of B. Assume that the smallest element of A is 0. Let

$$B^+ = B \cap [0, \infty),$$

and let

$$C = \{n \in \mathbb{Z} : n \geq 0, \ n = a + b, \ a \in A, \ b \in B, \ b < 0\}.$$

Put

$$f(x) = \sum_{a \in A} x^a, \quad g(x) = \sum_{b \in B} x^b, \quad h(x) = \sum_{c \in C} x^c.$$

We have

$$(1) \qquad\qquad f(x)g(x) + h(x) = \frac{1}{1 - x}.$$

Note that f and h are polynomials. An integer k is a period of B if $g(x)(1 - x^k)$ is a polynomial. We know that such a k exists. We want to show that there is a small one. Let n be the largest element of A. Then $\deg f = n$. By (1) we have

$$(2) \qquad\qquad g(x) = \frac{\frac{1}{1-x} - h(x)}{f(x)} = \frac{1 - (1 - x)h(x)}{(1 - x)f(x)} = \frac{p(x)}{q(x)},$$

where p, q are coprime polynomials. Hence

$$g(x)(1 - x^k) = \frac{p(x)(1 - x^k)}{q(x)}.$$

So

$$(3) \qquad\qquad q(x) \mid 1 - x^k = \prod_{d \mid k} \psi_d(x)$$

where ψ_d is the d-th cyclotomic polynomial. Therefore

$$q(x) = \prod_{d \in D} \psi_d(x),$$

where D is a set of divisors of k. Conversely, if (3) holds and we define $k = \operatorname{lcm}[d \ : \ d \in D]$, then $q(k) \mid 1 - x^k$, so

$$g(x)(1 - x^k) = p(x)\frac{1 - x^k}{q(x)}$$

is a polynomial. By (2) and (3) we have

$$n + 1 \geq \deg q = \sum_{d \in D} \deg \psi_d = \sum_{d \in D} \phi(d).$$

Let $p_1^{a_1}, \ldots, p_r^{a_r}$ be the prime powers $> L$ dividing k; we will specify L later. Each $p_i^{a_i}$ divides some $d \in D$. By a repeated application of the inequality $xy > x + y$, valid for $x, y > 2$, we find that

$$n + 1 \geq \sum \phi(p_i^{a_i}) \geq \frac{1}{2}\sum p_i^{a_i}$$

(we assume $L > 2$, which yields $\phi(p_i^{a_i}) > 2$). Hence $r \leq 2(n + 1)/L$ and

$$\prod p_i^{a_i} \leq (n + 1)^{2(n+1)/L} < n^{c_1 n/L}.$$

Consequently

$$k \leq \prod p_i^{a_i} \operatorname{lcm}[l \ : \ l \leq L - 1] < n^{c_1 n/L} e^{c_2 L} < e^{c_3 \sqrt{n \log n}}$$

if $L = \sqrt{n \log n}$.

REFERENCES

[1] E. Altman, B. Gaujal and A. Hordijk, Balanced sequences and optimal routing, *J. ACM,* **47** (2000), 752–775.

[2] A. Baker, B. J. Birch and E. A. Wirsing, On a problem of Chowla, *J. Number Th.,* **5** (1973), 224–236.

[3] Beauquier and M. Nivat, On translating one polyomino to tile the plane, *Discrete Comput. Geom.,* **6** (1991), 575–592.

[4] V. Berthé, Complexité des facteurs des suites sturmiennes, *Theor. Comput. Sci.,* **165** (1996), 295–309.

[5] *Substitutions in Dynamics, Arithmetics and Combinatorics,* N. Pytheas Fogg, ed. by V. Berthé, S. Ferenczi, C. Mauduit, A. Siegel, Lect. Notes Math. 1794, Springer-Verlag (2002).

[6] V. Berthé and R. Tijdeman, Balance properties of multi-dimensional words, *Theor. Comput. Sci.,* **273** (2002), 197–224.

[7] V. Berthé and L. Vuillon, Suites doubles de basse complexité, *J. Th. Nombr. Bordeaux,* **12** (2000), 179–208.

[8] V. Berthé and L. Vuillon, Tilings and rotations on the torus: a two-dimensional generalization of sturmian sequences, *Discrete Math.,* **223** (2000), 27–53.

[9] M. Boshernitzan, A condition for minimal interval exchange maps to be uniquely ergodic, *Duke Math. J.,* **52** (1985), 723–752.

[10] J. Cassaigne, Double sequences with complexity $mn + 1$, *J. Autom. Lang. Comb.,* **4** (1999), 153–170.

[11] M. G. Castelli, F. Mignosi and A. Restivo, Fine and Wilf's theorem for three periods and a generalization of Sturmian words, *Theor. Comput. Sci.,* **218** (1999), 83–94.

[12] E. M. Coven, Sequences with minimal block growth II, *Math. Systems Th.,* **8** (1975), 376–382.

[13] E. M. Coven and G. A. Hedlund, Sequences with minimal block growth, *Math. Systems Th.,* **7** (1973), 138–153.

[14] E. M. Coven and A. Meyerowitz, Tiling the integers with translates of one finite set, *J. Algebra,* **212** (1999), 161–174.

[15] H. Davenport, *Multiplicative Number Theory,* Springer-Verlag, New York etc., 2nd edition, 1980.

[16] Dirichlet P. G. Lejeune, *Werke I,* pp. 315–342, 411–496.

[17] C. Epifanio, F. Mignosi and M. Koskas, On a conjecture on bi-dimensional words, *Theor. Comput. Sc.,* **299** (2003), 123–150.

[18] P. Erdős and R. L. Graham, *Old and New Problems and Results in Combinatorial Number Theory,* Monogr. **28**, L'Enseignement Math., Genève, 1980.

[19] I. Fagnot and L. Vuillon, Generalized balances in sturmian words, *Discr. Appl. Math.*, **121** (2002), 83–101.

[20] N. J. Fine and H. S. Wilf, Uniqueness theorems for periodic functions, *Proc. Amer. Math. Soc.*, **16** (1965), 109–114.

[21] A. S. Fraenkel, Complementing and exactly covering sequences, *J. Combin. Th. A*, **14** (1973), 8–20.

[22] R. Giancarlo and F. Mignosi, Generalizations of the periodicity theorem of Fine and Wilf, *CAAP '94*, Lect. Notes Comput. Sc. **787**, 130–141.

[23] R. L. Graham, Covering the positive integers by disjoint sets of the form $\{[n\alpha+\beta]$: $n = 1, 2, \dots\}$, *J. Combin. Th. A*, **15** (1973), 354–358.

[24] A. Heinis, *Arithmetics and combinatorics of words of low complexity*, Ph.D. Thesis, Leiden University, 2001, 91 pp., `www.math.leidenuniv.nl/~tijdeman/`

[25] A. Heinis and R. Tijdeman, Characterisation of asymptotically sturmian sequences, *Publ. Math. Debrecen*, **56** (2000), 415–430.

[26] A. E. Livingston, The series $\sum_1^\infty f(n)/n$ for periodic f, *Canad. Math. Bull.*, **8** (1965), 413–432.

[27] J. Justin, On a paper by Castelli, Mignosi, Restivo, *Rairo: Theor. Informatics Appl.*, **34** (2000), 373–377.

[28] T. Kamae and L. Zamboni, Sequence entropy and the maximal pattern complexity of infinite words, *Ergodic Th. Dynamical Systems*, **22** (2002), 1191–1199.

[29] J. C. Lagarias and Y. Wang, Tiling the line with translates of one tile, *Invent. Math.*, **124** (1996), 341–365.

[30] *Combinatorics on Words*, ed. by M. Lothaire, Addison Wesley, Reading MA, 1983.

[31] *Algebraic Combinatorics on Words*, ed. by M. Lothaire, Encyclopedia of Mathematics and its Applications 90, Cambridge University Press, 2002.

[32] A. de Luca and F. Mignosi, Some combinatorial properties of Sturmian words, *Theor. Comput. Sci.*, **136** (1994), 361–385.

[33] R. Morikawa, On eventually covering families generated by the bracket function, *Bull. Fac. Liberal Arts*, Natural Science, Nagasaki Univ., **23** (1982), 17–22.

[34] M. Morse and G. A. Hedlund, Symbolic dynamics, *Amer. J. Math.*, **60** (1938), 815–866.

[35] M. Morse and G. A. Hedlund, Symbolic dynamics II – Sturmian trajectories, *Amer. J. Math.*, **62** (1940), 1–42.

[36] I. Nakashima, J. Tamura and S. Yasutomi, Modified complexity and *-Sturmian word, *Proc. Japan Acad.*, **75** Ser. A (1999), 26–28.

[37] D. J. Newman, Tesselation of integers, *J. Number Th.*, **9** (1977), 107–111.

[38] T. Okada, On a certain infinite series for a periodic arithmetical function, *Acta Arith.*, **40** (1982), 143–153.

[39] R. M. Robinson, Problem E 3156, *Amer. Math. Monthly*, **93** (1986), 482. Solution: *ibid.*, **95** (1988), 954-955.

[40] Ö. Rödseth, On a linear diophantine problem of Frobenius, *J. reine angew. Math.*, **301** (1978), 171–178.

[41] J. W. Sander and R. Tijdeman, The complexity of functions on lattices, *Theor. Comput. Sci.*, **246** (2000), 195–225.

[42] J. W. Sander and R. Tijdeman, The rectangle complexity of functions on two-dimensional lattices, *Theor. Comput. Sci.*, **270** (2002), 857–863.

[43] A. D. Sands, On the factorisation of finite abelian groups, *Acta Math. Acad. Sci. Hungar.*, **8** (1957), 65–86.

[44] E. S. Selmer and Ö.Beyer, On the linear diophantine problem of Frobenius in three variables, *J. reine angew. Math.*, **301** (1978), 161–170.

[45] C. Series, The geometry of Markoff-numbers, *Math. Intel.*, **7** (3) (1985), 20–29.

[46] R. J. Simpson and R. Tijdeman, Multi-dimensional versions of a theorem of Fine and Wilf, *Proc. Amer. Math. Soc.*, **22** (2002), 1191–1199.

[47] V. T. Sós, On the distribution mod 1 of the sequence $n\alpha$, *Ann. Univ. Sci. Budapest. Eötvös Sect. Math.*, **1** (1958), 127–134.

[48] K. B. Stolarsky, Beatty sequences, continued fractions, and certain shift operators, *Canad. Math. Bull.*, **19** (1976), 473–482.

[49] J. J. Sylvester, Mathematical questions, with their solutions, *Educational Times*, **41** (1884), 21.

[50] S. Szabó, A type of factorization of finite abelian groups, *Discrete Math.*, **45** (1985), 121–124.

[51] M. Szegedy, Algorithms to tile the infinite grid with finite clusters, preprint, `www.cs.rutgers.edu/˜˜szegedy/`

[52] R. Tijdeman, Decomposition of the integers as a direct sum of two subsets, *Number Theory*, ed. by S. David, Cambridge University Press, 1995, pp. 261–276.

[53] R. Tijdeman, On complementary triples of sturmian bisequences, *Indag. Math. N.S.*, **7** (1996), 419–424.

[54] R. Tijdeman, Intertwinings of periodic sequences, *Indag. Math. N.S.*, **9** (1998), 113–122.

[55] R. Tijdeman, On the minimal complexity of infinite words, *Indag. Math. N.S.*, **10** (1999), 123–129.

[56] R. Tijdeman, Fraenkel's conjecture for six sequences, *Discrete Math.*, **222** (2000), 223–234.

[57] R. Tijdeman, Exact covers of balanced sequences and Fraenkel's conjecture, in: *Algebraic Number Theory and Diophantine Analysis*, Proc. Intern. Conf. Graz Austria, ed. by F. Halter-Koch and R. F. Tichy, Walter de Gruyter, Berlin etc., 2000, pp. 467–483.

[58] R. Tijdeman, Some applications of diophantine approximation, *Number Theory for the Millenium,* III, A. K. Peters, Natick MA (2002), pp. 261–284.

[59] R. Tijdeman and L. Zamboni, The Fine and Wilf theorem for any number of periods, *Indag. Math. N.S.*, **14** (2003), 135–147.

[60] L. Vuillon, Combinatoire des motifs d'une suite sturmienne bidimensionelle, *Theor. Comput. Sci.,* **209** (1998), 261–285.

Added in proof (May 2005)

Section 2 and Appendix: A further improvement of Ruzsa's result has been obtained by András Biró in a paper entitled: Divisibility of integer polynomials and tilings of the integers. In this paper the bound for $\log k$ is improved to $n^{\frac{1}{3}+\varepsilon}$ for any positive ε and $n > n_0(\varepsilon)$.

Section 6: Fraenkel's conjecture for $n = 7$ has been proved in J. Barát and P. P. Varjú, Partitioning the positive integers to seven Beatty sequences, *Indag. Math.* N.S., **14** (2003), 149–161.

Section 9: A. Heinis has shown that in Problem 4 the limit value cannot be in the open interval $(2, 3)$ (personal communication).

Section 13: The result by Epifanio, Koskas and Mignosi has been improved by A. Quas and L. Zamboni in the paper Periodicity and local complexity, *Theor. Comput. Sc.,* **319** (2004), 169–174.

Section 16: The construction mentioned in the last sentence of the section can be found in V. Berthé and R. Tijdeman, Lattices and multidimensional words, *Theor. Comput. Sc.,* **319** (2004), 177–204.

Rob Tijdeman
Mathematisch Instituut
Universiteit Leiden
Niels Bohrweg 1
Postbus 9512
2300 RA Leiden
the Netherlands

tijdeman@math.leidenuniv.nl

Készült: Regiszter Kiadó és Nyomda Kft.